THE ICI POLYURETHANES BOOK

SECOND EDITION

GEORGE WOODS

WILEY

THE ICI POLYURETHANES BOOK

SECOND EDITION

GEORGE WOODS

Published jointly by

 Polyurethanes

and

 JOHN WILEY & SONS
Chichester - New York - Brisbane - Toronto - Singapore

Library of Congress Cataloguing-in-Publication Data:

Woods, George
    The ICI Polyurethanes book   /   George Woods. – 2nd ed.
       p.   cm.
    Includes bibliographical references and index.
    ISBN 0-471-92658-2 (cloth)
    1. Polyurethanes.   I. ICI Polyurethanes (Firm) II. Title.
TP1180.P8W66 1990
668.4'239–dc20   90-43732
                                    CIP

British Library Cataloguing-in-Publication Data

   Woods, George

   The ICI Polyurethanes Book. – 2nd ed.

   1. Bioengineering. Use of materials. Polyurethane
   I. Title
   668.4239

   ISBN 0 471 92658 2

Printed and bound in The Netherlands

Introduction

Polyurethanes make a varied and increasing contribution to our daily lives – from foam insulation to shoe soles, car seats to abrasion-resistant coatings. I hope that this book will reflect the versatility of these materials, and convey to the reader an understanding of their production, properties and the considerable potential they still have to fulfil.

In preparing this book we have held a number of different readers in mind. Foremost are those who wish to obtain a better understanding of polyurethanes – without necessarily having to study the subtleties of the chemistry and physics involved. Designers may learn of the opportunities offered by polyurethanes for the manufacture of complex composites, students and teachers should find this a helpful introduction to an important family of materials, and users of polyurethanes will have an easily accessible source of important information on processing, safety and the properties of polyurethanes. We believe a significant section of the readership of the book will be found in the Third World, where polyurethanes are set to make a genuine contribution to development and the enhancement of the quality of life.

In the two years following the publication of the first edition of this book, in 1987, it became clear that the entire print run would sell by 1990. Therefore a second edition was put in hand. This was timely as in this period much progress and many changes have taken place. The most obvious are the change in use of CFCs, and the advent of polyurea elastomers.

A new chapter has been written – within ICI Polyurethanes – cross referenced to the original text, and bringing these important aspects fully up to date. It is printed at the end of the book.

The *ICI Polyurethanes Book* should prove of great value to all who are engaged in product design, development and manufacture. By promoting a better understanding and broader knowledge of polyurethanes, I hope it will make a genuinely worthwhile contribution to the exciting future of polyurethanes in the decade to come.

David Sparrow
Research Manager ICI Polyurethanes
Everberg, Belgium
July 1990

Acknowledgements

In the production of a book such as this, a great deal of behind-the-scenes work goes on and a number of people have been involved. Thanks are due in particular to Paul Chapon of ICI France who sowed the seed for the enterprise, and Roger Natan and Judith Dobbs in Belgium for suggestions and advice in the early stages.
We would like to thank Dr. Dennis Allport who, as technical editor, brought his wide experience of polyurethanes to assist George Woods in the major task of planning and writing this book. Dr. Allport also guided the late John Barrett who, as editorial consultant, assisted in the presentation of George's work.
We are indebted to Jim Pattison and the staff of Dunholm Publicity in England for nurturing the manuscript over many months.
Many members of the ICI Polyurethanes staff have helped with contributions and particularly with thorough checking of the drafts. Dr. David Sparrow, Dr. Arun Watts, Mr. Bill Green and Dr. Mike Jeffs and their teams assisted greatly particularly David Thorpe and Graham Carroll. Much of the data on physical testing has been revised and completely updated by John Partington and Benoit Fraeys de Veubeke.
Final editorial guidance has been provided by Clare Hunt and Graham Look of Communication Systems in Brussels.
The design and layout of the book, typesetting and print production supervision has been carried out by Aad Schram in Holland.

Second edition
In this second edition George Woods and the technical editor David Sparrow have completely reviewed the book and prepared a detailed update which is printed as Chapter 13. We are extremely grateful to them for the considerable effort they have made.

Again we are indebted to Jim Pattison at Dunholm Publicity for progressing the whole project and shepherding it through its many stages, also to Graham Look of Communication Systems for editorial assistance.

Robert Genge
Managing Editor

Contents

Second edition.
Additional information prepared for this edition is
given in Chapter 13, the contents of which are listed
at the end of this section.

1   An introduction to polyurethanes

Polyurethanes are all around us, playing a vital role in many industries – from shipbuilding to footwear; construction to cars. They appear in an astonishing variety of forms, a variety that is continuously increasing.

Rigid polyurethane foam is one of the most effective practical thermal insulation materials, used in applications ranging from buildings to the modest domestic refrigerator. Comfortable and durable mattresses and car and domestic seating are manufactured from flexible foam. Items such as shoe soles, sports equipment, car bumpers and 'soft front ends' are produced from different forms of polyurethane elastomers. And many of us rely on polyurethanes – elastic threads of the material are found in underwear and other clothing.

All polyurethanes are based on the exothermic reaction of polyisocyanates with polyol molecules, containing hydroxyl groups. Relatively few basic isocyanates and a range of polyols of different molecular weights and functionalities are used to produce the whole spectrum of polyurethane materials. Additionally several other chemical reactions of isocyanates are used to modify or extend the range of isocyanate-based plastic materials. The chemically efficient polymer reaction may be catalysed, allowing extremely fast cycle times and quantity production. No unwanted by-products are given off and, because the raw materials react completely, no 'after cure' treatment is necessary.

Cost and processing advantages

Although the unique advantage of using polyurethanes lies in the wide variety of high performance plastics that can be produced, polyurethanes may often compete with low cost polymers. This is because raw material costs are not the only considerations in the total costs of producing an article.

Factors of at least equal importance are cycle times, the cost of tooling and finishing as well as reject rates and opportunities for recycling. As polyurethane reaction moulding requires only low pressures, moulds can be made of less expensive materials. This allows the simple production of inexpensive prototypes for the development of new products or the refinement of established ones.

Figures 1-1 to 1-8
Polyurethanes are used for a variety of applications

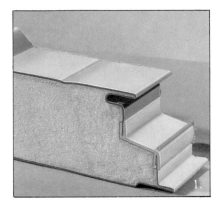

1. Metal faced building panels
2. Shoe and boot soles

3. Car exterior panels
4. Car seating

5. Housings for electronic equipment
6. Buoyancy in boats

7. Refrigerator insulation
8. Structural foam furniture

Polyurethanes also differ from most other plastic materials because they allow the processor to control the nature and the properties of the final product. This is possible because most polyurethanes are made using reactive processing machines. These mix together the polyurethane chemicals which then react to make the polymer required. The polymer is usually formed into the final article during this polymerisation reaction. This accounts for much of the versatility of polyurethanes: they can be tailored with remarkable accuracy to meet the needs of a particular application.

Summary of the properties of polyurethanes

Polyurethanes can be manufactured in an extremely wide range of grades, in densities from 6 kg/m^3 to 1220 kg/m^3 and polymer stiffnesses from very flexible elastomers to rigid, hard plastics. Although an oversimplification, the following chart illustrates the broad range of polyurethanes, with reference to density and polymer stiffness.

Figure 1-9   Property matrix of polyurethanes

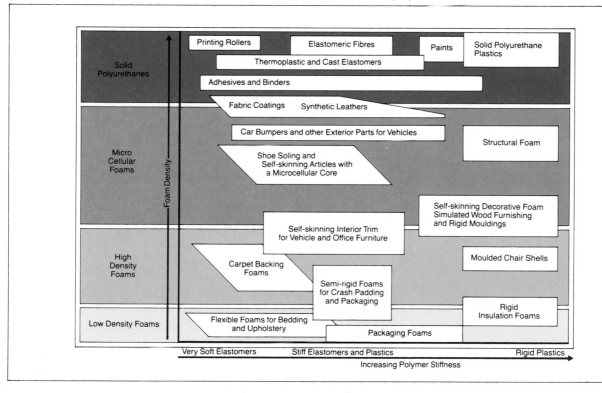

Polyurethane reaction mixtures have another important property – they are powerful adhesives. This enables simple manufacture of strong composites such as building panels and laminates, complete housings for refrigerators and freezers, crash padding for vehicles and reinforced structures in boats and aircraft.

Types of polyurethanes

A consideration of particular properties of certain grades of polyurethanes and the way in which these are used will serve to demonstrate their versatility.

Foams

By itself the polymerisation reaction produces a solid polyurethane. Foams are made by forming gas bubbles in the polymerising mixture. This is called 'blowing'.

Foam manufacture can be carried out continuously, to produce continuous laminates or slabstock, or discontinuously, to produce moulded items or free-rise blocks.

Flexible foams can be produced easily in a variety of shapes by cutting or moulding. They are used in most upholstered furniture and mattresses. Flexible foam moulding processes are used to make comfortable, durable, seating cushions for many types of seats and chairs. The economy and cleanliness of flexible polyurethane foams are important in all upholstery and bedding applications. Strong, low density rigid foams can be made and, when 'blown' using fluorocarbons, closed-cell structures are produced that have low thermal conductivity. Their superb thermal insulation properties have led to their widespread use in buildings, refrigerated transport vehicles, refrigerators and freezers.

Rigid and flexible foam articles having an integral-skin that is both decorative and wear-resistant, are produced by a fast, simple, moulding process. Fine surface detail can be reproduced by the integral skin of the foam, allowing the simple manufacture of instrument housings, articles with a simulated wood finish and padded steering wheels.

There are three foam types that, in quantity terms, are particularly significant: low density flexible foams, rigid foams, self-skinning foams and microcellular elastomers (high density flexible foams).

- *Low density flexible foams* are materials of densities 10-80 kg/m^3, composed of lightly cross-linked, open-cells. In other words, air may flow through the structure very easily. Essentially flexible and resilient padding materials, flexible foams are produced as slabstock or individually moulded cushions and pads. Semi-rigid variants also have an open-cell structure but different chemical formulations.
- *Low density rigid foams* are highly cross-linked polymers with a closed-cell structure – each bubble within the material has unbroken walls so that gas movement is impossible. These materials offer good structural strength in relation to their weight, combined with outstanding thermal insulation properties. A chlorofluoromethane gas is usually contained within the cells, and

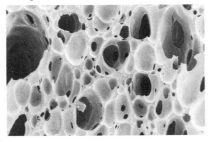

Figure 1-10 Photomicrograph showing the open cells of a flexible foam

Figure 1-11 Photomicrograph showing the closed cells of a rigid foam

Table 1-1 **Relative importance of three foams types.**

Foam type	Density kg/m^3)	Approx. usage (millions of tonnes/ year)
Low density flexible	10–80	2
Low density rigid	30–80	0.9
Self-skinning foams		
Microcellular elastomers	400–800	0.6

as these substances have a much lower thermal conductivity than air, such closed-cell foams have a significantly lower thermal conductivity than any open-celled foam. However, if this low thermal conductivity is to be retained, the chlorofluoromethane gas must not leak away. Consequently, rigid polyurethane foam insulation must have at least 90 percent of closed cells and a density above about 30 kg/m^3.

– *High density flexible foams* are defined as those having densities above 100 kg/m^3. The range includes moulded self-skinning foams and microcellular elastomers. Self-skinning foam systems are used to make moulded parts having a cellular core and a relatively dense, decorative skin. There are two types, those with an open-cell core and an overall density in the range up to about 450 kg/m^3 and those with a largely closed cell or microcellular core and an overall density above about 500 kg/m^3. Microcellular elastomers have a substantially uniform density in the range from about 400-800 kg/m^3 and mostly closed cells which are so small that they are difficult to see with the naked eye. The biggest applications of self-skinning foams and microcellular elastomers are in moulded parts for upholstery and vehicle trim and for shoe-soling. Microporous elastomers are microcellular foam foils or sheets having a proportion of open, communicating cells. Chapters five and six describe self-skinning and microcellular foams, the former concentrating on high density flexible foams while the latter describes the RIM (Reaction Injection Moulding) process – a commonly used and growing method of moulding microcellular elastomers and self-skinning foams. RIM is used for other types of polyurethanes such as structural foam and substantially solid elastomers.

Solid polyurethanes

Although foamed polyurethanes form some 90% by weight of the total market for polyurethanes, there is a wide range of solid polyurethanes used in many, diverse applications.

Solid polyurethane elastomers. Most polyurethane elastomers have excellent abrasion resistance with good resistance to attack by oil, petrol and many common non-polar solvents. They may be tailored to meet the needs of of specific applications, as they may be soft or hard, of high or low resilience, solid or cellular. A wide range of polyurethane elastomers is described in chapter eight.

Adhesives, binders, coatings and paints. Polyurethanes are also used in flexible coatings for textiles and adhesives for film and fabric laminates. Polyurethane paints and coatings give the highest wear resistance to surfaces such as floors and the outer skins of aircraft. They are also becoming widely used for high quality finishes on automobiles.

Applications of polyurethanes

Figure 1-12 World polyurethane markets.

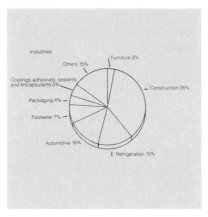

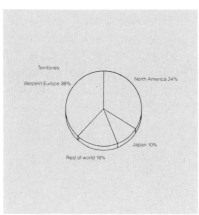

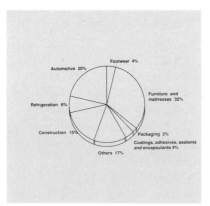

Top: World MDI polyurethane market by industry.
Centre: World polyurethane market by territory.
Bottom: Total world polyurethane market by industry.

A detailed breakdown of the polyurethanes industry by application is shown in figure 1-12. The versatility of polyurethanes is demonstrated by a summary of their applications in five important areas.

Automotive
In recent years, polyurethanes have found increasing use in this area, to the benefit of both the manufacturer and customer. Applications include seating, interior padding, exterior body panels, complete soft front ends, components mounted in the engine space and accessories such as mirror surrounds and spoilers. This is a particularly exacting market as materials are carefully selected to meet tough specifications.

Furniture
The market for cushioning materials is dominated by polyurethane flexible foams. And where strong – tough but decorative – integral-skinned flexible or rigid foam structures are needed, polyurethanes are also ideal. Polyurethanes compete with rubber latex foam, cotton, horsehair, polyester fibre, metal springs, wood, expanded polystyrene, polypropylene and PVC.

Construction
When sandwiched between metal, paper, plastics or wood, polyurethane rigid foam plays an important role in the construction industry. Such composites can replace conventional structures of brick, cement, wood or metal, particularly when these latter materials are used in combination with other insulating materials such as polystyrene foam, glass fibre or mineral wool.

Thermal insulation
Rigid polyurethane foams offer unrivalled technical advantages in the thermal insulation of buildings, refrigerators and other domestic appliances, and refrigerated transport. Competitive materials include cork, wood, glass fibre, mineral wool, foamed (expanded) polystyrene, urea formaldehyde and phenol formaldehyde.

Footwear
Soles and some synthetic uppers for many types of footwear are produced from polyurethanes. These compete with traditional leather and rubber, polyvinyl chloride and poly(ethylene-vinyl acetate). Polyurethane adhesives are widely used in shoe and slipper manufacture. Polyurethane coatings are used to improve the appearance and wear resistance of shoe uppers made from both real leather and from PVC leathercloth.

2   Making Polyurethanes

In this chapter the basic principles and methods of manufacture of the wide variety of polyurethane materials are described. Those readers requiring a detailed discussion of polyurethane chemistry should also read Chapter 3. Other readers can pass directly from this chapter to Chapter 4.

Figure 2-1   A polyurethane is formed when two liquid chemicals, an isocyanate and a polyol, react on being mixed together in the presence of suitable catalysts and additives.

There is a fundamental difference between the manufacture of polyurethanes and many other plastics materials. The majority of thermoplastics, unlike most polyurethanes, are polymerised in large chemical factories and then sold to the user as polymer granules or powders. These are then converted to useful articles by one of the thermoplastic processing techniques which involve heating, shaping the melted polymer under pressure, and cooling. The properties of articles made in this way are almost entirely dependent upon the properties of the thermoplastic polymer purchased from the manufacturer. The polymerisation of most polyurethanes, however, is carried out by the converter and not by the chemical manufacturer. The latter sells polyurethane chemical components, usually in the form of liquids, and provides advice and information to enable the customer to make the polyurethane he requires by mixing the chemical components in specified proportions.

Specially blended polyurethane chemical components are also sold by chemical manufacturers to make polyurethanes for particular applications. The purchase of such pre-blended chemical components or 'systems' enables the customer to make polyurethanes by a simple mixing process, but nevertheless it is the customer who makes the polyurethane polymer in addition to shaping it into a useful article. The exceptions are some polyurethane elastomers which are made by the chemical manufacturers. These include elastomeric polyurethane fibres (spandex fibres), which are sold directly to textile manufacturers, and thermoplastic polyurethane elastomers, which are supplied in granular or sheet form for conversion into articles by injection moulding, extrusion, blow-moulding, calendering, lamination or some other established thermoplastic conversion process.

Table 2-1   **Some highlights in the development of polyurethanes**

1937-40	Otto Bayer and co-workers made polymers by polyaddition processes from various diisocyanates with glycols and/or diamines.	**1960**	ICI introduces the first polymeric MDI-based semi-rigid energy absorbing foam for vehicles.
1940-5	Development of millable polyurethane elastomers and adhesives in Germany (I.G. Farben), U.K. (ICI) and the U.S.A. (du Pont). Polyurethane coatings for barrage balloons (ICI), synthetic polyurethane bristles (I.G. Farben).	**1960-5**	Rigid foam blowing by chlorofluoromethanes.
		1962	First production line moulded 'deep seat' flexible polyurethane car cushions at Austin-Morris (U.K.).
		1963	ICI demonstrates production line manufacture of refrigerators using MDI-based polyurethane foam.
1945-7	Manufacture of millable polyurethane elastomers, coatings and adhesives.	**1963**	First cold-store built entirely from metal-faced polyurethane rigid foam laminate made continuously (Australia).
1950	Cast elastomers from polyester diols, diisocyanate prepolymers and chain extenders.	**1964**	ICI inverse- and floating-platen systems for the continuous manufacture of rigid polyurethane foam-cored building boards in production.
1953	First flexible polyurethane foam manufacture with a Bayer system using a high pressure machine, a polyester polyol and TDI.	**1965**	First commercial production of self-skinning flexible foam (Soc. Quillery, France).
1956	First manufacture of polyether-based flexible polyurethane foam in the U.S.A. using a two stage or 'pre-polymer' process.	**1968**	ICI introduces isocyanurate rigid foams.
		1968	General Motors make the first polyurethane microcellular bumper for the Pontiac G.T.O.
1957	ICI introduces the first commercially available polymeric MDI composition for rigid polyurethane foam manufacture.	**1973**	MDI-based 'soft-face' bumpers made by RIM system for Chevrolet taxis.
1959	ICI introduces the first rigid foam system based on polymeric MDI and a polyether polyol.	**1979**	ICI introduces wholly-MDI-based systems for flexible foam moulding.
1959	'One-shot' system for flexible polyether-based foam introduced in the U.S.A.	**1983**	ICI introduces system to make dual-hardness, moulded seating from MDI-based, flexible foam.

The chemicals

A urethane group is formed by the chemical reaction between an alcohol and an isocyanate. Polyurethanes result from the reaction between alcohols with two or more reactive hydroxyl groups per molecule (diols or polyols) and isocyanates that have more than one reactive isocyanate group per molecule (a diisocyanate or polyisocyanate). This type of polymerisation is called addition polymerisation.

The urethane-producing reaction was well known in the nineteenth century, but only as a laboratory curiosity. It was not until the late 1930s that the commercial potential of polyurethanes as fibres, adhesives, coatings and foams began to be recognised by Otto Bayer, his co-workers, and others. In fact, most polyurethanes' applications have been developed during the past 30 years. Table 2-1 illustrates that progress.

Most polyols and polyisocyanates used in the manufacture of polyurethanes are liquid at ambient temperatures (ca. 18°C) and are easily handled. The reactions which produce a solid polymer are rapid and substantially complete within two minutes. The solid product can generally be handled within five minutes from the start of mixing, although rates can be varied enormously by the choice of catalyst and its concentration. The reactions are exothermic – the heat generated may be used to vaporise a liquid 'blowing agent' such as a fluorocarbon, when the reacting chemicals will polymerise and expand to produce a polyurethane foam. Alternatively, some water may be incorporated in the polyol so that it reacts with the polyisocyanate to release carbon dioxide gas.

The reactions which produce polyurethane foam may be summarised as figure 2-2 below.

Figure 2-2   The polyurethane foam reaction

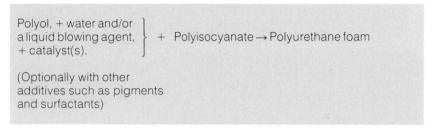

Polyols

Some 90 percent of the polyols used in making polyurethanes are polyethers with terminal hydroxyl groups. Hydroxyl-terminated polyesters are also used to obtain polyurethanes with special properties. Polyesters are usually more expensive than polyethers, but competitively priced polyesters (made from recovered diacids), have been developed, particularly for use in making rigid foams. The polyols that are used to make polyurethanes have been developed to have the required reactivity with commercially available isocyanates and to produce polyurethanes with specific properties. The choice of polyol, especially the size and flexibility of its molecular structure and its functionality (the number of isocyanate-reactive hydroxyl groups per molecule of polyol) controls, to a large extent, the degree of cross-linking achieved in the polymer that is formed in the reaction with the polyisocyanate. That degree of cross-linking has a dominant effect on the stiffness of the polymer: to obtain a rigid foam there must be a stiff polymer network and, hence, a high degree of cross-linking; for flexible foam a proportionally lesser degree of cross-linking is needed.

The characteristics of the polyols to make the two principal classes of polyurethanes are:

Table 2-2   Characteristics of polyols

Characteristic	Flexible foams and elastomers	Rigid foams, rigid solids, and stiff coatings
Molecular weight range	1,000 to 6,500	400 to 1,200
Functionality range	2.0 to 3.0	3.0 to 8.0

Isocyanates

Another major method of varying the properties of the final polyurethane is by varying the type of isocyanate used. Isocyanates may be modified in many ways to give products with differing physical and chemical properties.

Several aromatic and aliphatic isocyanates are available, but about 95 percent of all polyurethanes are based on two of them. These are toluene diisocyanate (TDI), and diisocyanato-diphenylmethane (MDI) and its derivatives. Both materials are derived from readily-available petrochemical intermediates and are manufactured by well understood and closely defined chemical processes.

Most of the TDI used is a mixture of two molecular forms called isomers – the 2,4 and 2,6 isomers in an 80:20 mix (known as 80:20 TDI). A 65:35 mix is also available. TDI is used mainly in the production of low density foams for cushioning, generally as slabstock. TDI is also used for moulded cushions, sometimes mixed with various MDI products. The pure 2,4 isomer is used in some elastomer manufacture.

Development of MDI was necessary as the volatility of TDI caused severe problems when TDI-based rigid foams were sprayed in enclosed spaces, such as the holds of ships. The production chemistry of MDI is considerable more complex than that of TDI. However, this complexity confers a significant degree of freedom on the chemical manufacturer to modify and optimise grades of MDI to meet specific user needs.

MDI is produced from aniline and formaldehyde, reacted together using hydrochloric acid as a catalyst. This condensation

Figure 2-3   Manufacturing route to MDI

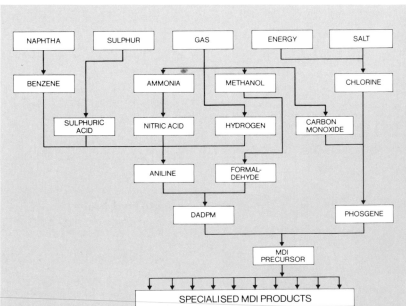

reaction produces a complex mixture of polyamines which are phosgenated to obtain a polyisocyanate mixture. The product, known as polymeric MDI, is principally used for manufacture of rigid foams.

Such polyisocyanate mixtures can be processed further to give a variety of compositions, known as MDI variants, which have a wide range of applications.

Pure MDI is a solid of melting point 38°C which is produced by separation from a polymeric MDI precursor. It can be used in its basic form or modified chemically to give a composition that is liquid at ambient temperatures.

To make polyurethanes as cost-effective as possible, producers of isocyanates have developed processes for tailoring the MDI composition to the requirements of the end-product. The polyisocyanate mixtures are formulated to offer a range of differing functionalities. Functionality is defined as the average number of chemically-reactive groups on each individual molecule present.

Table 2-3 **Range of MDI variants**

Average functionality	Product description	Polyurethane type	Main applications
2.0	Pure MDI	High performance elastomers	Shoe-soling. Spandex fibres. Flexible coatings. Thermoplastics.
2.01 - 2.1	Modified, liquid pure MDI	High performance elastomers. Microcellular elastomers.	Shoe-soling. Flexible coatings. RIM and RRIM. Cast elastomers.
2.1 - 2.3	Liquid, low functionality polyisocyanates.	Flexible, semi-rigid and rigid (structural) foams. One-component froth.	Automotive parts. Cabinets for electronic equipment. Insulating sealants. Cast elastomers.
2.5	Low viscosity liquid polyisocyanates.	High density flexible foams. Structural foams.	Foam-backs for carpets/vinyls. Computer cabinets and other moulded articles.
2.7	Low viscosity polymeric MDI.	Low density rigid foams. Semi-rigid foams. Isocyanurate foams. Particle binders.	Insulating foams. Energy absorbing foams. Isocyanurate foam building panels. Mine-face consolidation. Chipboard and foundry sand binders.
2.8 - 3.1	High functionality polymeric MDI.	Rigid polyurethane and isocyanurate foams.	Continuous lamination of rigid foam and rigid foam slabstock.

Table 2-3 shows that the functionalities of the products available range from 2.0 to about 3.0.

Diisocyanate products are required for making elastomers, whilst high functionality MDI – rich in polyisocyanates – is desirable for the manufacture of rigid foams and binding materials. Within the latter group, wide variation in molecular weight distribution and

11

functionality is possible, giving considerable control over reactivity, viscosity, and application properties.

Between these two extremes of functionality lie several MDI variants specially selected to meet the needs of a variety of polyurethane products – structural foam, integral-skin foam and semi-rigid foam. It is this tailoring of the isocyanate component that enables the polyurethanes producer to choose optimum properties of handling, processing and polymer characteristics for a particular application. In addition to TDI and MDI, speciality isocyanates are produced, but in much smaller quantities. Other aromatic diisocyanates include 1,5-diisocyanato-naphthalene (naphthalene diisocyanate or NDI), and 1,4-diisocyanato-benzene (*p*-phenylene diisocyanate or PPDI) which are used to make polyurethane elastomers. Polyurethanes based on aromatic diisocyanates tend to yellow gradually on exposure to daylight, a process that can be slowed down – but not prevented completely – by the use of additives.

Aliphatic diisocyanates, which are much less reactive than the aromatic diisocyanates, are therefore used for applications requiring high resistance to yellowing. Those most widely available are 1-isocyanato-3-isocyanatomethyl-3,5,5,-trimethylcyclohexane (isophorone diisocyanate or IPDI), 1,6-diisocyanato-hexane (hexamethylene diisocyanate or HDI) and 4,4-diisocyanato-dicyclohexylmethane (hydrogenated MDI or HMDI). The isomeric diisocyanates, 1,4- and 1,3-di(isocyanato-dimethyl-methyl)-benzene (*m*- and *p*-tetramethyl xylene diisocyanate or *m*- and *p*-TMXDI), which behave like aliphatic diisocyanates, are also available.

Figure 2-4   Annual sale of MDI and TDI

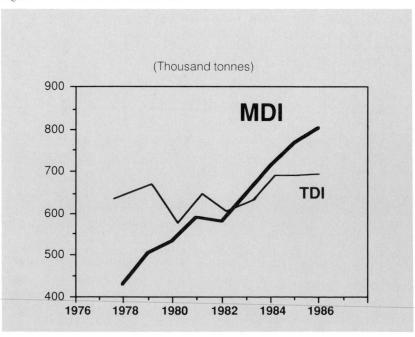

Additives

Apart from polyisocyanates and polyols – the basic materials for polyurethane production – a wide variety of auxiliary chemicals may be added in order to control and modify both the polyurethane reaction itself and the properties of the final polymer. These additives include catalysts, cross-linking agents, chain extending agents, blowing agents, surfactants, colouring materials, fillers, smoke suppressants and flame retardants. All practical polyurethane systems include at least some of these auxiliary chemicals, table 2-4.

Table 2-4 **Reasons for using additives**

Additive	Type of material	Purpose
Catalysts	Tertiary amines Organometallic compounds	To speed up the reaction of isocyanate and polyol
Cross-linking chain-extending agents	Polyols Polyamines	To give polymer cross-linking or to introduce specialised polymer segments
Blowing agents	Water (reacts with isocyanate giving carbon dioxide gas). Chlorofluoromethanes	To produce foamed structures
Surfactants	Silicone fluids	To aid and help foam-forming processes
Colours	Various pigments Carbon black	To identify different foam grades and for aesthetic reasons
Fillers	Particulate inorganic materials Fibres (chopped, milled or as continuous fibres, nets or scrims)	To modify properties (stiffness, fire performance etc.)
Flame retardants	Phosphorus or halogen-containing molecules	To reduce flammability
Smoke suppressants	Particulate inorganic and/or organic materials. (polycarboxylates, hydrated oxides, borates, etc.)	To reduce the amount of smoke or to slowdown the rate of smoke production on burning.

The development of systems

The possible number of variables in the formation of polyurethanes is very large. Clearly, the development or selection of a chemical system that gives satisfactory and reproducible processing, with a high quality product at the minimum cost, is not simple. It requires not only a knowledge of polyurethane chemistry, but also skill in evaluating both process and product. Fortunately there are other ways of ensuring satisfactory polyurethane manufacture.

Chemicals together with polyurethane production advice, are available from chemical manufacturers and chemical suppliers. Complete, automated blending equipment, for batch or in-line formulation blending, is offered by specialist polyurethane machinery makers.

Fully blended systems are also available to meet the process and product requirements of particular customers. Such systems include those for rigid and flexible foams, self-skinning foams and microcellular materials, elastomers for casting, potting, spraying and coating, adhesives and sealants. Compounded systems may be designed specifically for use with a particular type of processing machinery but many have wider applications, or may be modified easily to suit differing scales and methods of manufacture.

The polyurethane manufacturer may thus choose between buying individual polyurethane chemicals and compounding them in-house, or buying fully or partially compounded chemical systems. The choice will depend upon the volume of his production, on his technical capability and experience, and on the cost of factory space and formulating equipment.

The manufacture of polyurethanes

The chemical reaction between a polyol and a diisocyanate or polyisocyanate, starts spontaneously when they are blended together in the presence of suitable catalysts. Many polyurethanes may be made on a small scale by simply mixing the required ratios of the chemicals in a cup – using a suitable stirrer – and hand pouring into a mould. The polymerisation, once started, is exothermic to completion. Most polyurethanes, however, are made using machines that have been specially designed to measure the required ratio of chemicals, mix them together and dispense the reaction mixture in pre-determined amounts. Polyurethane dispensing machines for either continuous or discontinuous operation may be obtained in a wide range of capacities from a few grams/minute to over 500 kilogrammes/minute. Whichever method of mixing and dispensing is used, the same basic requirements for satisfactory polyurethane manufacture apply. These may be summarised as:

> – *Temperature control and conditioning*
> – *Accurate metering or weighing*
> – *Uniform and reproducible mixing*
> – *Suitable production line methods*
> – *Curing*
> – *Freedom from contaminants*
> – *Safety*

Temperature control and conditioning

Effect of temperature. The viscosity, density and chemical reactivity of the polyols and the isocyanates vary with the temperature. Controlling the temperature of the chemical components is therefore essential to the reproducible processing of polyurethanes. Although some dispensing machines meter the components by weight, most meter the chemicals by volume, thus any change in the temperature – and therefore in the density – of the chemicals may also give rise to an error in the materials ratio. Large variations in temperature, that cause proportionally larger changes in the viscosity of the chemical components, may affect both the performance of the metering pump and the efficiency of the final mixer.

The effects of temperature variation on the polyurethane reaction may be complex. In most polyurethane reaction mixtures, simultaneous and sequential chemical reactions take place together. The various chemical reactions may be affected, to a different extent, by a change in the temperature of the components. Consequently, changes in temperature may not only change the rate at which the polyurethane polymerisation takes place, but they may also affect the quality of the final product.

Conditioning. The conditioning of the chemicals before use is done in various ways depending on the scale of manufacture and the temperature requirement of the particular process. Some processes, such as elastomer manufacture from isocyanates and polyols that are solid at room temperature, may be operated at elevated temperatures, but most polyurethanes are made from isocyanates and polyols that are liquid at room temperature. Many of these polyurethane processes may be initiated, by adjusting the catalyst level, at any convenient temperature ranging from about 15°C to about 30°C, although there is often an optimum range of temperatures that gives the best product or the lowest costs. For most polyurethane processes using liquid isocyanates, batch-to-batch reproducibility requires that the chemical components are within ± 2°C of the standard process temperature.

A secondary, but important, aspect of chemical conditioning before use is the release of excess entrained bubbles of air or nitrogen gas. Liquids containing such gas bubbles cannot be metered accurately by volumetric pumps and the presence of excessive gas bubbles may give rise to defects and variations in the final product. Drums of isocyanate and polyol, for use on machinery that draws the materials directly from the drums, are best conditioned by storage – preferably on pallets – in a temperature-controlled room with fan-assisted air circulation. Drums of isocyanate should be stored in a separate, vented room or an isolated outbuilding (see Chapter 10). Drums containing solidified isocyanate or isocyanate prepolymers, that require to be melted at elevated temperatures, must be handled according to the manufacturer's instructions. The best method of conditioning the chemical components is in enclosed vessels such as bulk storage tanks, intermediate conditioning tanks or, on many machines, in the smaller machine tanks. Temperature conditioning may be done using double-walled tanks having a thermostatically controlled heat-transfer fluid circulated in the jacket. Most systems use water as the heat-transfer fluid but some elastomer production using solid polyesters may require temperatures around 130°C and use electrically-heated oil. Electric immersion heaters are sometimes used because of their low initial cost but care must be exercised in their design and high recirculation rates adopted if local overheating of the chemicals is to be avoided.

Figure 2-5   Chemical materials flow on a continuous rigid foam plant.

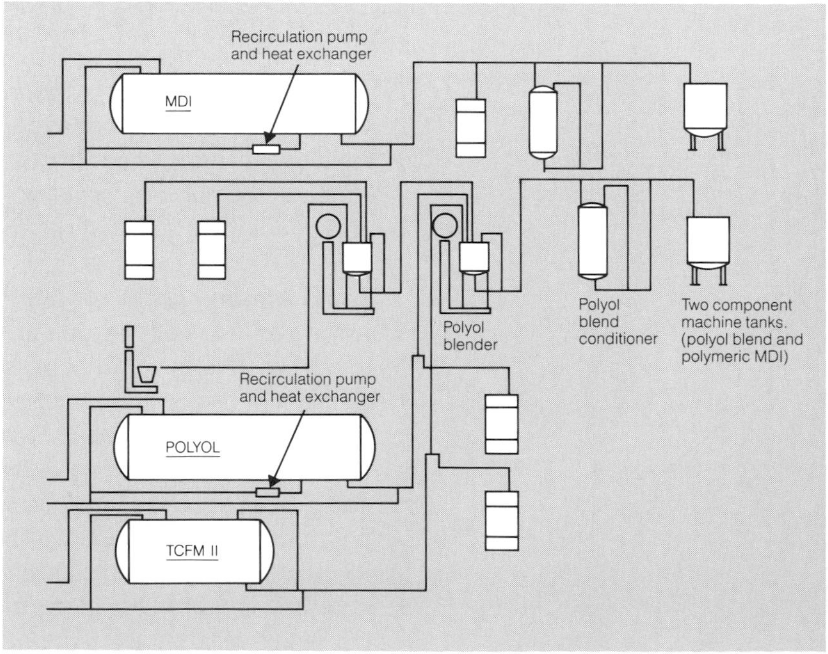

All these methods of temperature conditioning require mechanical recirculation of the contents of the tank in order to obtain a sufficiently uniform temperature. For low viscosity systems the simplest and most easily controlled temperature adjustment is obtained by the use of insulated tanks and recirculation of the

contents through an external multi-tube heat exchanger.
Polyurethane machinery suppliers usually offer a range of heat
exchangers with capacities to match the various sizes of dispensing
machine. High capacity in-line tubular heat exchangers may be used
on high output plants for continuous foam-making in order to
permit manufacture directly from bulk storage tanks without the
intermediate conditioning and blending tanks often used on smaller
capacity units. A typical scheme is shown, diagrammatically, in
figure 2-5.

Chemical conditioning may also necessitate in-line removal of gas
bubbles and sometimes – for polyols to be used for solid elastomer
production for instance – the removal of traces of water. Combined
de-gassing and dehydration units are available for this purpose. A
well-tried design uses a system incorporating a spinning disc and
sieve to pour a thin film of polyol down the heated walls of a vacuum
tank. A typical flow diagram, for a polyester polyol which is solid at
room temperature, is illustrated in figure 2-6.

Figure 2-6   Conditioning a solid polyester polyol for elastomer manufacture.

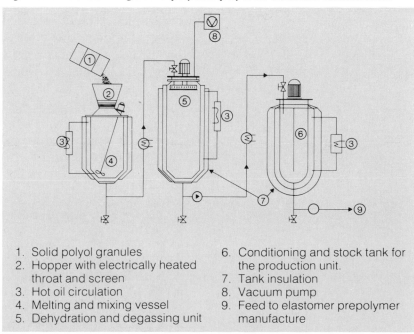

1. Solid polyol granules
2. Hopper with electrically heated
   throat and screen
3. Hot oil circulation
4. Melting and mixing vessel
5. Dehydration and degassing unit
6. Conditioning and stock tank for
   the production unit.
7. Tank insulation
8. Vacuum pump
9. Feed to elastomer prepolymer
   manufacture

Accurate metering or weighing

Once the component chemicals have been conditioned to the
required temperature, they must be metered to the mixer in the ratio
required by the formulation, mixed together and dispensed into the
mould or cavity before the isocyanate has reacted significantly with
the other components of the reaction mixture.

Variations in mixing, whether attributable to mechanical factors
or to changes in the viscosity of the chemical components, may give

rise to variations in the uniformity and the reactivity of the mixture and to variations in the final product.

To make a polyurethane, the conditioned chemicals must be mixed together in the proportions required by the formulation. Weighing the chemical components is potentially the most accurate method of obtaining the correct proportions; it is used for making small laboratory-scale preparations and also industrially in the batch manufacture of some elastomers and of rigid and flexible foams. The usual methods of polyurethane manufacture, both continuous and discontinuous, require the use of metering pumps to feed the chemical components to a mixer in the correct proportions and at the required rate. Many types of positive feed pump are used, the choice depending on the required output, the properties of the chemicals, the number of components, the degree of rate adjustment required, the feed and delivery pressures and other considerations such as any servicing and durability restrictions. The separate metering of minor chemical additives, such as water, catalysts and surfactants, is usually practicable only on machines with high outputs (100 kg/min to 500 kg/min) for the continuous manufacture of flexible and rigid foams. Most machines with lower outputs, and those with intermittent dispensing units use one or more pre-blended component streams in order to reduce the number of chemical components metered on the machine, and to avoid the need to meter relatively small amounts of additives. For intermittent, single-shot metering at low outputs, piston and plunger pumps are widely used for dispensing the chemicals used for making solid and microcellular elastomers, self-skinning, semi-rigid and rigid foams.

Figure 2-7   Simple cylinder metering arrangements.

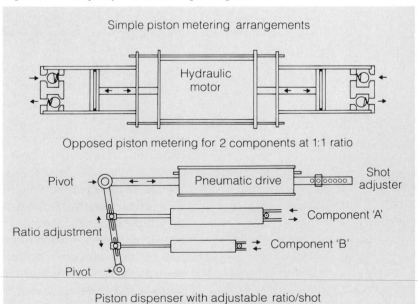

18

A common arrangement employs two separate cylinders, each of which handles one of the two components that are usual in these fields. Several methods of ratio control are used e.g. simple purpose-designed machines use two cylinders having their diameters proportional to the required component ratio, the pistons being operated by a single hydraulic or pneumatic cylinder. No adjustment of the component ratio is possible but total shot weight may be controlled by limit stops on the operating cylinder. Many other arrangements of simple metering cylinder pumps are used. Some of these are illustrated in figure 2-7.

High-pressure machines use electrically driven piston pumps to meter the chemical components at pressures over the range 100 bar to 200 bar. Mixing is done, wholly or partially, by the impingement of high velocity jets of the components in a small volume mixing chamber. **Low-pressure machines** use pumps and pipelines operating at 2 bar or 3 bar (except with very viscous materials, such as some polyester polyols, when pressures may range from 20 bar to 30 bar). Low-pressure machines usually use mechanical mixers, driven by an electric or hydraulic motor.

Low- and high-pressure dispensing machinery. Polyurethane metering and mixing machinery is usually classified as *low pressure* or *high pressure* machinery depending on the metering pressure and the type of mixing.

Most machines of either type use electrically-driven metering pumps. The latter are usually gear pumps, diaphragm pumps or piston pumps. Other types of pump, such as turbine pumps and screw pumps, may be used to feed the metering pump or to recirculate the components. Vertical piston pumps, adapted from in-line, multi-cylinder fuel injection pumps, have been used for metering low viscosity isocyanates for over 30 years, but are now being replaced for many applications by axial piston pumps. The *metering rate* on simple machines is controlled by varying the pump speed. The output of the pump is measured over a range of speeds and under specific operating conditions of temperature, feed pressure and output pressure.

Figure 2-8   A 'Rotameter' type flowmeter.

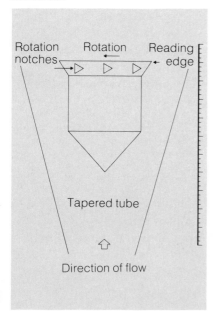

Flowmeters. Most large continuous machines and many modern foam dispensing machines use flowmeters to monitor the actual *flow rate* of the liquid components. The flow rate signal can be fed back electronically to adjust the pump drive speed in order to maintain a pre-set flow rate. Many modern machines may be programmed to maintain and adjust both the overall output and the ratios of the chemical components. Flowmeters, down-stream from the metering pumps, monitor the flow rates and transmit electrical pulses to a microprocessor where the ratios of the signals are compared with the pre-set figure. Differences are used to adjust the pump drive speed until the required ratios and output are obtained.

Several types of flowmeter may be used, the main ones being rotameter, turbine meters, positive displacement gear meters and differential-pressure flow measuring devices. In any flowmeter used in polyurethanes production processes, the main criterion is reproducibility rather than fundamental accuracy. Instruments of the *rotameter type* are variable area, constant head, devices. A spinning float in a tapered vertical tube rises with increasing flow-rate as illustrated in figure 2-8.

Rotameters are widely used in continuous metering systems.

19

Figure 2-9 Flow measurement by
determining the pressure differential
across a flow restrictor.

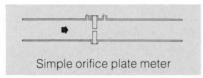

Simple orifice plate meter

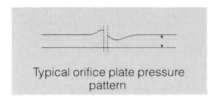

Typical orifice plate pressure
pattern

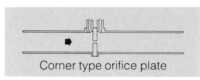

Corner type orifice plate

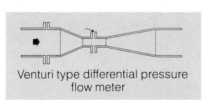

Venturi type differential pressure
flow meter

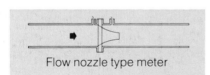

Flow nozzle type meter

They are simple, the float position is linearly proportional to the flow-rate and the reading is largely independent of changes in the density of the liquid. The working range is up to 10:1 with an accuracy better than ± 0.5 per cent. The main applications in polyurethanes production are for monitoring the flow of TDI, chloro-fluoromethane-11, water and air or nitrogen.

In-line *turbine flowmeters* have been used to measure TDI flow-rates on continuous machines. The rotational speed of the turbine depends on its design, the quality of its bearings, on the properties of the liquid and its flow velocity.

The rotating turbine gives a pulse-train output that can be adapted easily to give a remote indication of flow-rate or fed to a control mechanism. Used for TDI flow, turbine meters can be accurate over a wide range to within ± 0.25 to ± 0.5 per cent but, when calibrated for typical production flow-rates, an accuracy of ± 0.1 per cent is attainable.

Positive displacement, *gear-type flow-meters* have larger friction losses and a higher pressure drop than turbine meters. They are used on continuously operating machines to measure the flow-rate of polyols and other materials of relatively high viscosity being the most accurate type of flowmeter for this purpose. An output signal from a typical oval-gear flowmeter is an electrical pulse train that is used to indicate or control flow.

Differential-pressure flow-measurement – the pressure change across a restriction in the liquid flow is proportional to the flow-rate. The method is versatile. It is used for both continuous and intermittent metering and is the method most easily adapted for high pressure systems. Various types of flow restrictor are used (see Figure 2-9). The change in pressure caused by the restrictor is measured by transducers. The flow-rate range for accurate measurement is restricted to about 4:1.

Calibration at about the required flow-rates should give flow-rates to an accuracy of about ± 0.5 per cent. A major advantage of differential pressure transducer systems, on compact or portable machines, is their small size compared with mechanical systems.

Uniform and reproducible mixing

Mixers used for polyurethanes are of two types, static and mechanical:

Static mixers. Such mixers include high-pressure-jet mixers. The streams of metered components are mixed together by the turbulence resulting from the directly-opposed impingement of the high-velocity jets that result from forcing the components through small orifices in a mixing chamber of very low volume (about 1 cm^3). Many self-cleaning designs are available (see Chapters 4 and 6). The method is most satisfactory with low-viscosity chemical components and with systems having component ratios close to 1:1. High viscosity materials and high component ratios much different from 1:1 are better handled by mechanical mixers.

Another type of static mixer operates by repeatedly dividing and re-combining the stream of materials. Such mixers are most satisfactory for mixing when the ratio of the components is nearly 1:1, especially those for making solid polyurethanes. They are less effective for high-ratio component mixtures and are unsatisfactory for foam-making reaction mixtures that require uniform bubble nucleation.

Mechanical mixers. Mechanical mixers consisting of a rotor in a mixing chamber are available in a very wide range of designs and capacities. The rotor is driven at speeds from about 1,500 to about 10,000 rev/min depending on the design and capacity of the mixer. The volume of the free space in the mixing chamber may vary from a few cubic centimetres up to about one litre, depending on the application and throughput required.

Simple mechanical mixers, designed for a specific polyurethane system, are operated at a fixed speed. Most other mixers have a variable speed drive, so enabling the efficiency of the mixing to be varied with changes in the chemical system and the rate of throughput. The work done during mixing is converted to heat and this will affect the reaction rate of the polyurethane. Varying the stirring rate of a mechanical mixer, or the pressure drop in an impingement mixer, will thus affect the reaction rate of the dispensed mixture.

Suitable production line methods

Flexible and rigid foams are made by a variety of continuous processes (see Chapters 4, 5 and 7).

The polyurethane reaction mixture is dispensed:
– Intermittently into moulds
– Continuously on to conveyors or into troughs carried on conveyors
– Sprayed or spread
depending on the system and its application

Figure 2-10   Controlling the pour pattern from a simple dispensing machine.

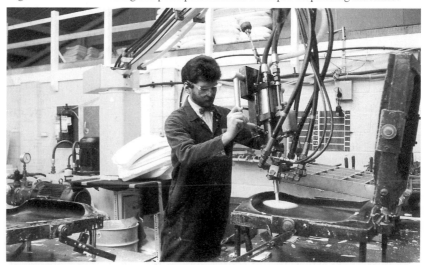

facing materials to form a strong, rigid composite. Some laminates

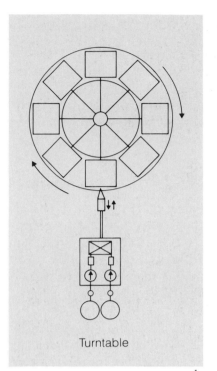

Turntable

Figure 2-11   Moulding line
arrangements, moving moulds – 1 and
2, stationary moulds – 3, 4 and 5.

1.

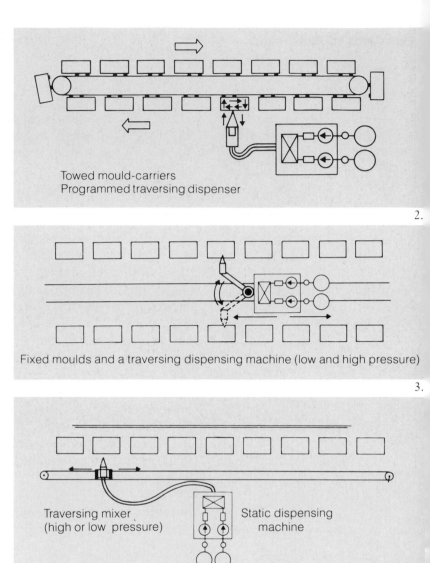

Towed mould-carriers
Programmed traversing dispenser

2.

Fixed moulds and a traversing dispensing machine (low and high pressure)

3.

Traversing mixer
(high or low pressure)

Static dispensing
machine

Static moulds and static dispensing machine with a traversing mixer

4.

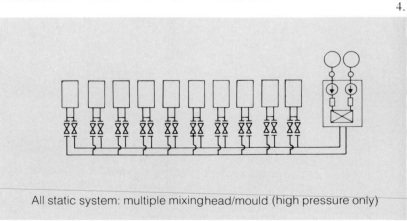

All static system: multiple mixinghead/mould (high pressure only)

5.

Figure 2-12   Carousel arrangement: static dispenser and moving moulds

Many polyurethanes, from low density foams to solid elastomers and plastics, are made by dispensing into moulds. These may be made from glass-fibre-reinforced polyester or epoxy-resins, metal (steel, aluminium, alloys, sprayed metal or electro-deposited metal shells) or elastomers such as silicone or polyurethane rubber. Wooden and glass moulds are sometimes used.

The polyurethane reaction mixture may be *poured into an open mould* or *injected into a closed mould* depending on the polyurethane system, the component design, the design and construction of the mould, and the method of mould manipulation. Mould conditioning, before filling, includes the adjustment of the mould temperature and the application of a suitable release agent to its internal surfaces. The time from dispensing the polyurethane reaction mixture into the mould to the demoulding of the moulded article varies from less than one minute to over an hour, but most production line systems are demoulded in minutes.

The method of handling the moulds depends largely upon the rate of manufacture. Low production rates may require nothing more elaborate than trolley-mounted moulds that are wheeled to the vented dispensing-machine area, where they stand after filling until the reaction is complete. A wide range of mould-handling equipment is available depending on the number of moulds and the temperature conditioning requirements.

Curing

In general, polyurethane mouldings do not develop their optimum properties until some hours after demoulding. Many polyurethanes cure satisfactorily and develop optimum properties simply through storage at room temperature. Some systems benefit from a post cure heating cycle. The processing and curing conditions required for the different types of polyurethane manufacture are described in relevant later chapters.

Freedom from contaminants

Water. An important requirement in handling, storing and conditioning polyurethane chemicals is to avoid contamination with water or atmospheric moisture. Isocyanates react with water to form a gas (carbon dioxide) and a solid, insoluble polyurea derivative. (In the production of low density flexible foams and some types of low density rigid foams, the main blowing agent is the carbon dioxide produced by the reaction of water with isocyanate). Most of the polyols used in polyurethane manufacture are hygroscopic i.e. they absorb moisture from the air. Storage tanks and conditioning tanks must be so designed that any contact of the contents with moist air is prevented. This is usually done by venting the tanks to atmosphere through a suitable chemical or freeze dryer, or by blanketing the contents with a small over-pressure of dry nitrogen or dry air. The precise control of the water content of polyols and other polyurethane chemicals is important because quite small variations in the water content of these materials may have a major effect on the properties of the final product. This results from the low equivalent weight of water, i.e. 1 part of water reacts with nearly 10 parts of TDI or nearly 14 parts of MDI.

Other contaminants. Defects in polyurethane manufacture may be caused by contaminants such as some oils and greases, especially silicone greases and antifoams. Contamination by heavy metals, especially from the copper in brass fittings in contact with blends of polyols and tertiary amine catalysts or tertiary amines and water, can alter the rate of the polymerisation reaction significantly. Contamination of polyols or polyol blends by any material that reacts with isocyanates may cause faults, especially in foam manufacture. Particulate reactive matter is likely to cause major defects in the structure of polyurethane foams. When a uniform foam structure is important, the conditioning unit should include an efficient filtration system.

Safety
The importance of safety equipment and proper safety procedures in the handling of reactive chemicals is obvious, but the handling of isocyanates requires special care because it is important that the operatives avoid breathing isocyanate vapours. The hazard is lowest when handling such isocyanates as polymeric MDI that have a low vapour pressure, that is, they do not produce much vapour at room temperature. In many countries there are statutory requirements for handling isocyanates. Methods of handling isocyanates and monitoring vapour levels are given in Chapter 10.

The main principles of safety are as follows:

- The people involved should be medically screened and thoroughly trained
- Good working practices with high levels of ventilation are necessary to control exposure to isocyanate vapour
- Spillage and leaks of isocyanate must be neutralised at once.

3   The chemistry and materials of polyurethane manufacture

Reactions of isocyanates

With polyols: polyurethanes are addition polymers formed by the reaction of di- or poly-isocyanates with polyols:

Figure 3-1   Polyurethane addition reaction

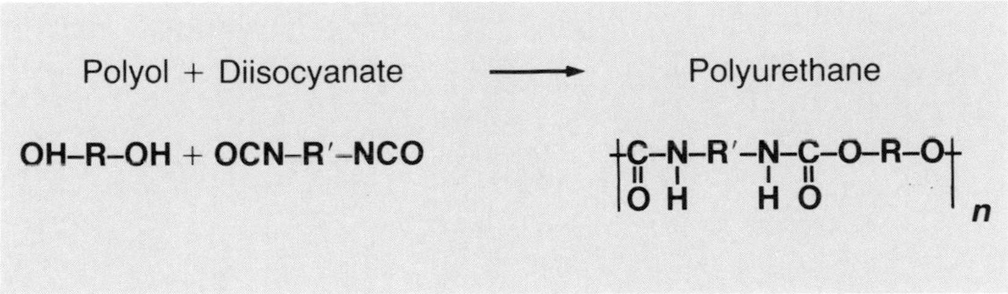

$$\text{Polyol} + \text{Diisocyanate} \longrightarrow \text{Polyurethane}$$

$$\text{OH–R–OH} + \text{OCN–R'–NCO} \longrightarrow \left[\!\!\begin{array}{c} \underset{\overset{\|}{O}}{C}\text{–N–R'–N–}\underset{\overset{\|}{O}}{C}\text{–O–R–O} \\ \;\;HH \end{array}\!\!\right]_{n}$$

The reaction is exothermic. The rate of the polymerisation reaction depends upon the structure of both the isocyanate and the polyol. Aliphatic polyols with primary hydroxyl end-groups are the most reactive. They react with isocyanates about 10 times faster than similar polyols with secondary hydroxyl groups. Phenols react with isocyanates more slowly and the resulting urethane groups are easily broken on heating to yield the original isocyanate and the phenol. This reversible reaction is used in the manufacture of 'blocked' isocyanates which are activated by heating. The term polyurethane or even simply 'urethane' is often used to describe any polymer that has been chain extended by reaction with di- or poly-isocyanate. The isocyanate group (–NCO) can react with any compound containing 'active' hydrogen atoms (i.e. those that react with methyl magnesium iodide to form methane), and diisocyanates may therefore be used to modify many other products.

With water: after the reaction with polyols, the next most important reaction of isocyanates is that with water. The reaction of isocyanates and water yields a substituted urea and carbon dioxide. This reaction provides the principal source of gas for blowing in the manufacture of low density flexible foams. The initial product of the reaction with water is a substituted carbamic acid which breaks down into an amine and carbon dioxide. The amine then reacts with further isocyanate to yield the substituted urea: see figure 3-2.

27

Figure 3-2

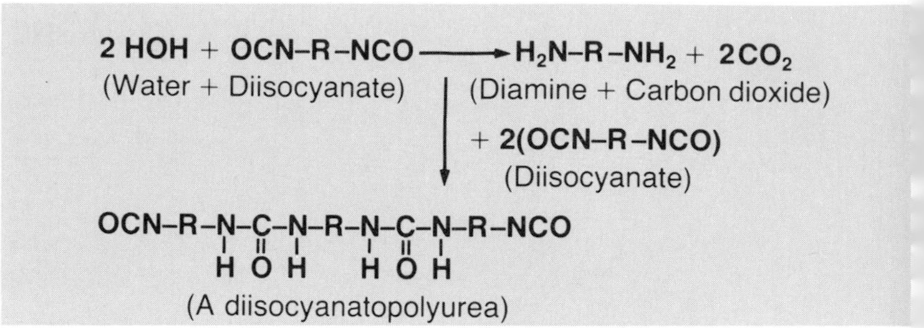

$$2\ HOH + OCN\text{--}R\text{--}NCO \longrightarrow H_2N\text{--}R\text{--}NH_2 + 2CO_2$$
(Water + Diisocyanate)   (Diamine + Carbon dioxide)

$$+ 2(OCN\text{--}R\text{--}NCO)$$
(Diisocyanate)

OCN–R–N–C–N–R–N–C–N–R–NCO
     | || | | || |
     H O H   H O H

(A diisocyanatopolyurea)

Diisocyanates having isocyanate groups of similar reactivity, such as MDI, p-phenylene diisocyanate and naphthalene-1,5-diisocyanate, tend to give complex crystalline polymeric ureas. On the other hand, 2,4-TDI has a isocyanate group in the shielded 2-position which is less reactive than that in the 4-position and the simple di-substituted urea (figure 3-3) is the main product of the reaction with water at temperatures below 50°C.

Figure 3-3

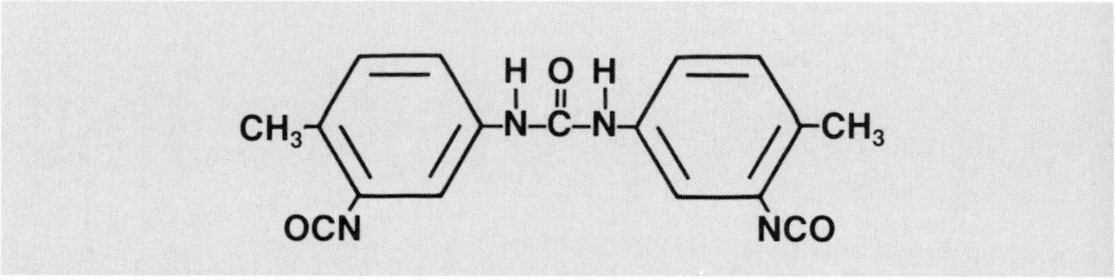

In the absence of catalysts, both TDI and MDI – because of their low solubility – react with water quite slowly. Aqueous emulsions of polymeric MDI have thus a sufficiently long life for them to be used as a binder in the manufacture of particle board.

With amines: the reaction of diisocyanates with primary and secondary amine compounds, especially diamines, is the third most important reaction in practical polyurethane chemistry. Diamines are used as chain-extending and curing agents in polyurethane manufacture. The effect of selected diamine addition is to increase the reactivity of the reaction mixture and the resulting polyurea segments in the polymer increase the potential for both primary (covalent) and secondary, or hydrogen-bonded, cross-linking. The reaction of unhindered isocyanates with primary amines, at room temperature in the absence of catalysts, is about 100 to 1000 times faster than the reaction with primary alcohols. The reactivity of amines increases with the basicity of the amine, and aliphatic amines react much faster than aromatic amines. Substantial reductions in the reactivity of aromatic amines, useful in elastomer formulation, result from steric hindrance by substituents, such as ethyl groups, ortho to the amine group or from the presence of electron withdrawing groups, such as chlorine, in the molecule. Tertiary amines, because they contain no active hydrogen atoms, do not react with isocyanates, but they are powerful catalysts for many other isocyanate reactions.

Secondary reactions of isocyanates

Isocyanates may react, under suitable conditions, with the active hydrogen atoms of the urethane and urea linkages to form biuret and allophanate linkages, respectively. Both reactions are cross-linking reactions. The reaction of isocyanates with urea groups is significantly faster and occurs at a lower temperature than that with urethane groups.

Figure 3-4

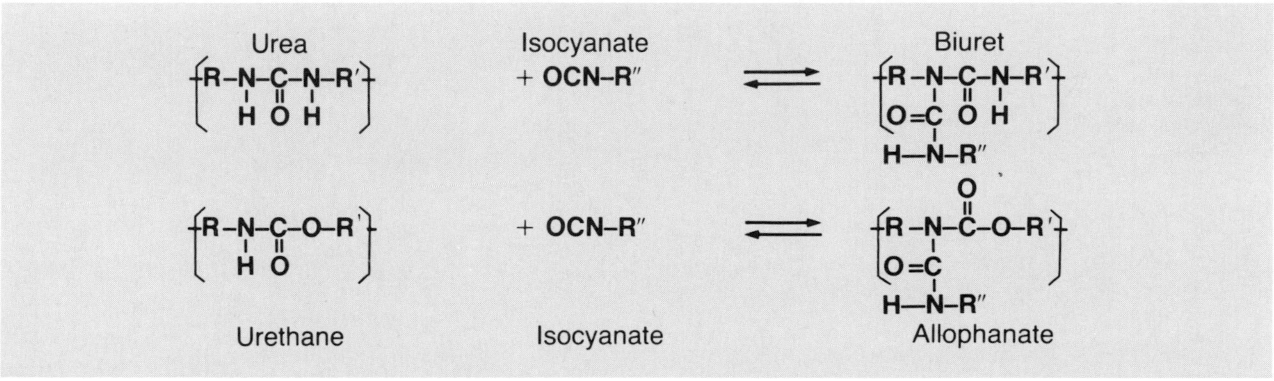

Isocyanate polymerisation reactions

Isocyanates form oligomers, especially in the presence of basic catalysts, giving uretidinediones (commonly called dimers), and isocyanurates (commonly called trimers). Dimer formation arises only from aromatic isocyanates and it is inhibited by ortho substituents. Thus 2,4- and 2,6-TDI do not form dimers at normal temperatures but 4,4'-diphenylmethane diisocyanate (MDI) dimerises slowly when left standing at room temperature. At higher temperatures insoluble polymeric materials are formed.

Figure 3-5

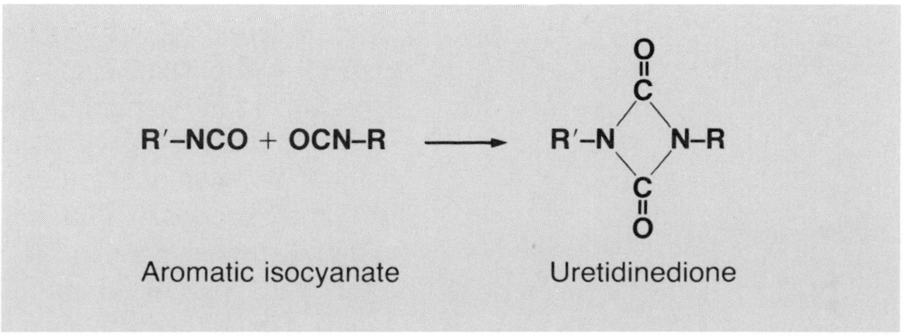

Isocyanurates are formed on heating both aliphatic or aromatic isocyanates. The reaction is accelerated by basic catalysts. Isocyanurate formation in polyurethane manufacture gives very stable branch points as, unlike the uretidinedione, biuret, allophanate and urethane linkages, the reaction is not easily reversed. Isocyanurate foams from both MDI and TDI show little degradation below 270°C.

Figure 3-6   Isocyanurate formation

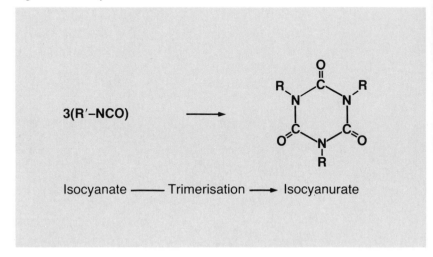

$3(R'-NCO)$

Isocyanate ——— Trimerisation ——→ Isocyanurate

In the presence of special catalysts isocyanates can condense, with the elimination of carbon dioxide, to form carbodiimides which then react reversibly with further isocyanate to give a uretonimine.

Figure 3-7   Uretonimine formation

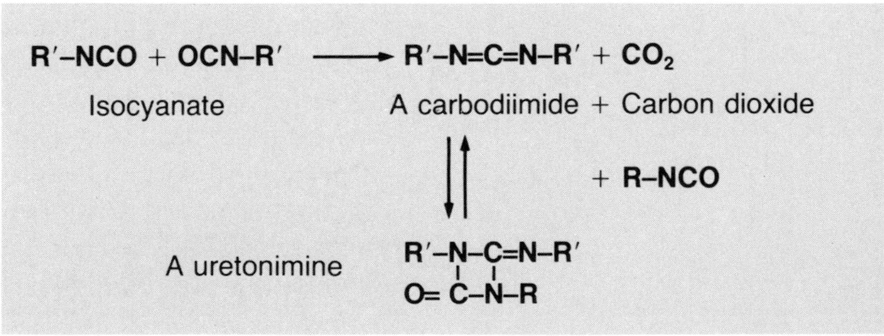

$R'-NCO + OCN-R' \longrightarrow R'-N=C=N-R' + CO_2$

Isocyanate   A carbodiimide + Carbon dioxide

                                                            $+ R-NCO$

A uretonimine $R'-N-C=N-R'$
                               $O= C-N-R$

These reactions of isocyanates are all exothermic. The heat evolved is important in most polyurethane manufacture, but it is especially so in the manufacture of low density flexible and rigid foams. In making the former, the increasing temperature during the foaming reaction facilitates the reaction of the more hindered isocyanate groups in TDI and helps to balance the polymerisation and blowing reactions in making flexible foams. In production of rigid foams, part of the exothermic heat of reaction is used to vaporise the blowing agent.

Cross-linking in polyurethanes

The molecular structures of polyurethane polymers vary from rigid cross-linked polymers to linear, highly-extensible elastomers as illustrated diagrammatically in figure 3-8.

Figure 3-8   Polyurethane polymer structures. (Diagrammatic)

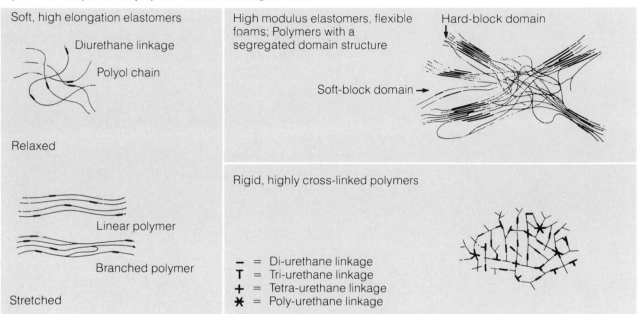

Soft, high elongation elastomers

Diurethane linkage
Polyol chain

Relaxed

Linear polymer

Branched polymer

Stretched

High modulus elastomers, flexible foams; Polymers with a segregated domain structure

Hard-block domain

Soft-block domain →

Rigid, highly cross-linked polymers

− = Di-urethane linkage
T = Tri-urethane linkage
+ = Tetra-urethane linkage
✳ = Poly-urethane linkage

Flexible polyurethanes, flexible foams and many elastomers have segmented structures consisting of long flexible chains (e.g. of polyether or polyester oligomers) joined by the relatively rigid aromatic polyurethane-polyurea segments. Their characteristic properties depend largely upon secondary or hydrogen bonding of polar groups in the polymer chains. Hydrogen bonding occurs readily between the NH-groups (proton donor) and the carbonyl groups (electron donor) of the urethane and urea linkages. Hydrogen bonds are also formed between the NH-groups of the urethane and urea linkages and the carbonyl oxygen atoms of polyester chains. The ether oxygens of polyether chains also tend to align by hydrogen-bonding with the NH-groups, but these bonds are much weaker and more labile than those formed with carbonyl oxygen atoms. The hard segments and especially the stiff polyurea hard segments, display strong secondary bonding, and tend to agglomerate into hard segment domains in structures having long flexible chains.

Rigid polyurethane polymers, in contrast, have a high density of covalent cross-linking. This results from the use of branched starting materials such as polyfunctional alcohols, amines and isocyanates. An alternative method of primary cross-linking, resulting in high thermal stability, requires the use of excess

31

polyisocyanate in conjunction with trimerisation catalysts. There are, of course, many urethane polymers whose properties arise from a combination of both covalent and secondary bonding. These include many semi-rigid foams and polymers used in adhesives, binders and paints.

Many of the basic raw materials used to make polyurethanes are derived from oil. The most important exceptions are the polyols based on natural products such as sucrose which are used to make many rigid polyurethane foams. Most polyurethanes are made from aromatic isocyanates reacted with aliphatic polyols.

Isocyanates

Several aromatic and aliphatic diisocyanates are available but some 95% of all polyurethanes are based upon the two aromatic diisocyanates, toluene diisocyanate (TDI) and diphenylmethane diisocyanate (MDI) and its derivatives. The consumption of MDI overtook that of TDI in 1984.

Most of the TDI used is a mixture of the 2,4- and 2,6-isomers:

Figure 3-9   TDI isomers.

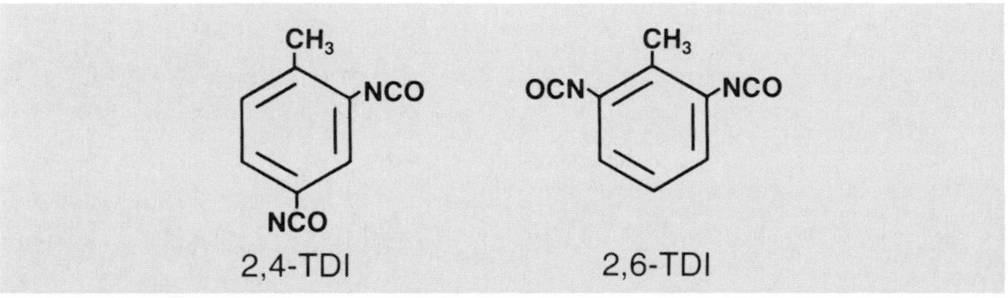

2,4-TDI   2,6-TDI

The 80:20 mixture of 2,4-TDI and 2,6-TDI (80:20-TDI) is today the most important commercial product, but a 65:35 ratio mixture of the 2,4- and 2,6-isomers (65:35-TDI) is also available from some suppliers.

MDI is available in several forms based on two types of product, purified monomeric MDI and polymeric MDI. Pure MDI is substantially 4,4'-diisocyanato-diphenylmethane (4,4'-diphenylmethanediisocyanate or 4,4'-MDI). It usually contains a small amount of the 2,4'-isomer.

Figure 3-10   Pure MDI.

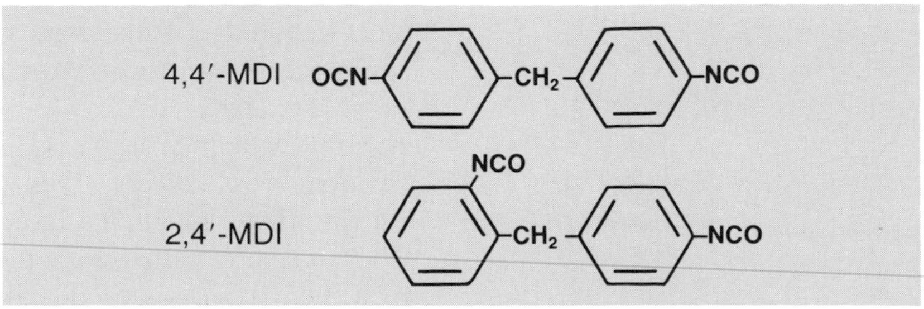

32

Pure MDI is a white to pale yellow solid of melting point about 38°C. It tends to form insoluble dimer when stored. The difficulty of handling solid pure MDI and its increased tendency to form dimer when stored as a liquid at over 40°C, have led to the development of modified pure MDIs which are liquid at ordinary temperatures and have a reduced tendency to dimerise. Two main methods of depressing the melting point of pure MDI are used. Both methods involve reacting part of the pure 4,4'-MDI to form a derivative that is soluble in 4,4'-MDI. One method is to react some of the isocyanate groups with an aliphatic diol having a low molecular weight or with a mixture of such diols, to yield a solution of diurethanes having isocyanate end-groups (e.g. figure 3-11) in 4,4'-MDI.

Figure 3-11   Modified pure MDI.

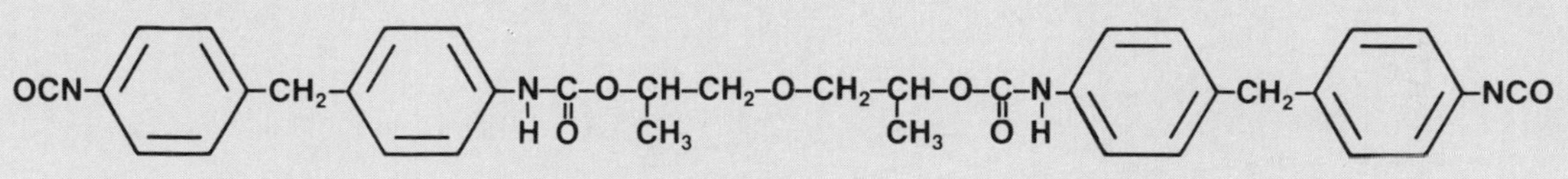

This type of liquid diisocyanate mixture has an effective functionality of 2.0 and is useful in the production of polyurethane elastomers of high quality. The second common method of modifying pure 4,4'-MDI is by converting part of the diisocyanate to uretonimine-linked trifunctional and higher isocyanates, (e.g. figure 3-12).

Figure 3-12   Modified pure MDI.

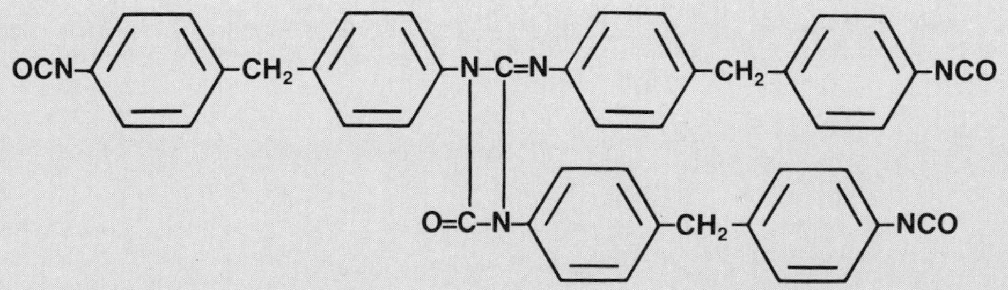

Uretonimine-modified pure MDI is a stable, low viscosity, liquid with a mean functionality usually approaching 2.2. Mixtures of uretonimine-modified and urethane-modified pure MDI are available with lower functionality. The melting point of pure 4,4'-MDI is also reduced by increasing the 2,4'-MDI content. Compositions containing high levels of 2,4'-MDI are not suitable for the production of polyurethane elastomers having a very high modulus but, for many applications, adjustment of the 2,4'-MDI

33

Figure 3-13

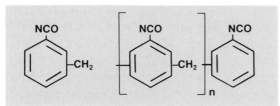

Figure 3-14   The functionality distribution of a typical polymeric MDI.

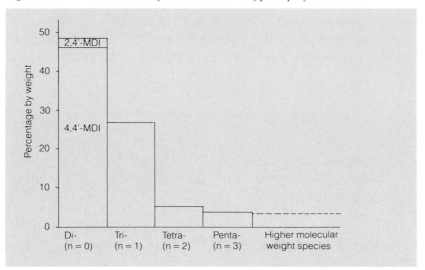

Table 3-1   **MDI-based isocyanates for polyurethane manufacture**

Product name	Isocyanate value (% NCO by wt).	Viscosity (mPa s at 25°C)	Average functionality	Main application area
'Suprasec' MPR	33.4	Solid at 25°C 4.7 at 50°C	2.00	High grade elastomers.
'Suprasec' ML	33.2	5.0 at 50°C	2.00	Binder for rubber granules.
'Suprasec' VM021	23.0	1000	2.01	Polyester-based shoe soling.
'Suprasec' VM051	18.0	190 at 50°C	2.01	Polyether-based shoe soling.
'Suprasec' VM10	26.0	150	2.07	RIM elastomers. Carpet backing foams.
'Suprasec' VM15	27.4	170	2.12	Integral-skin foams.
'Suprasec' VM20	29.0	50	2.13	RIM elastomers.
'Suprasec' VM25	24.3	170	2.21	Cold-cure auto seating cushions.
'Suprasec' VM28	25.0	220	2.28	Cold-cure furniture cushions.
'Suprasec' VM30	28.6	200	2.30	Integral-skin and RIM parts. Structural polyurethane.
'Suprasec' VM50	30.6	130	2.49	Semi-rigid foams. Carpet backing.
'Suprasec' 1042	29.5	275	2.7	Water emulsifiable MDIs for
'Suprasec' 1249	22.0	1300	2.7	chipboard binders.
'Suprasec' DND	30.7	230	2.7	Rigid foams.
'Suprasec' DNR	30.7	230	2.7	Structural foams. Self-skinning foams.
'Hexacal' F	30.7	230	2.7	Rigid isocyanurate foams.
'Hexacal' SN	26.0	2700	2.7	Rigid isocyanurate foams.
'Hexacal' LN	27.0	1300	2.7	Rigid isocyanurate foams.
'Suprasec' VM85 HF	30.4	550	2.9	Rigid foams for construction applications.
'Suprasec' VM90 HF	30.2	900	3.0	Rigid foams for construction applications.

'Suprasec' and 'Hexacal' are trade marks of Imperial Chemical Industries PLC.

level is a useful method of increasing both the liquidity and the storage stability of pure MDI. 2,4'-MDI does not form dimer easily. Polymeric MDIs are undistilled MDI compositions made by the phosgenation of polyamine mixtures. Polymeric MDI compositions are available with effective mean functionalities from about 2.5 to over 3.0. A typical functionality distribution is illustrated diagrammatically in figure 3-14. Polymeric MDI compositions are characterised by their viscosity and their content of reactive isocyanate groups. The viscosity increases with increasing mean molecular weight and polymeric isocyanate content.

The wide range of MDI-based isocyanates available is illustrated in table 3-1 which lists some products offered by ICI Polyurethanes Group.

Polyols

A wide range of polyols is used in polyurethane manufacture. Most of the polyols used, however, fall into two classes: hydroxyl-terminated polyethers, or hydroxyl-terminated polyesters. The structure of the polyol plays a large part in determining the properties of the final urethane polymer. The molecular weight and functionality of the polyol are the main factors, but the structure of the polyol chains is also important. The characteristics of the polyols used to make the two main classes of flexible and rigid polyurethanes are shown in table 3-2.

Table 3-2 **Polyols for polyurethane manufacture**

Characteristic	Flexible foams and elastomers	Rigid foams, rigid plastics and stiff coatings
Molecular weight range	1,000 to 6,500	150 to 1,600
Functionality range	2.0 to 3.0	3 to 8
Hydroxyl value range (mg KOH/g)	28 to 160	250 to 1,000

The 'hydroxyl value' is used as a measure of the concentration of isocyanate-reactive hydroxyl groups per unit weight of the polyol and is expressed as mg KOH/g. Polyols sold for use in polyurethanes are invariably characterised by hydroxyl value as this is convenient for calculation of the stoichiometric formulation. The measured hydroxyl value of a polyol is related to its molecular weight and functionality:

Figure 3-15

$$\text{Hydroxyl value, (mg KOH/g)} = \frac{56.1 \times \text{functionality}}{\text{molecular weight}} \times 1000$$

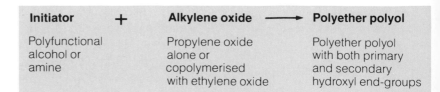

Initiator	+	Alkylene oxide	⟶	Polyether polyol
Polyfunctional alcohol or amine		Propylene oxide alone or copolymerised with ethylene oxide		Polyether polyol with both primary and secondary hydroxyl end-groups

Figure 3-16   The manufacture of polyether polyols.

The polymerisation of Propylene oxide and Ethylene oxide.

$$R-CH_2-O^- + CH_3-\overset{1}{C}H-\overset{2}{C}H_2 \quad \text{base catalyst}$$

$$R-CH_2-O-\overset{2}{C}H_2-\overset{1}{C}H\begin{smallmatrix}CH_3\\OH\end{smallmatrix} \quad \text{Secondary hydroxyl (95\%)}$$

$$R-CH_2-O-\overset{1}{C}H-\overset{2}{C}H_2-OH \quad \text{Primary hydroxyl (5\%)}$$

$$R-CH_2-O-CH_2-CH\begin{smallmatrix}CH_3\\OH\end{smallmatrix} \quad + \; CH_2-CH_2 \rightarrow R-CH_2-O-CH_2-\overset{CH_3}{C}H-O-CH_2-CH_2-OH$$

Secondary hydroxyl   Ethylene oxide   Primary hydroxyl

Polyether polyols

About 90% of the polyols used in polyurethane manufacture are hydroxyl-terminated polyethers. These are made by the addition of alkylene oxides, usually propylene oxide, onto alcohols or amines which are usually called starters or 'initiators'. The addition polymerisation of propylene oxide occurs with either anionic (basic) and cationic (acidic) catalysis although commercial polyol production is usually by base catalysis. The epoxide ring of propylene oxide may theoretically open at either of two positions on reaction but, in practice, the ring opens preferentially at the less sterically-hindered position with base catalysis. Polyethers based on propylene oxide thus contain predominantly secondary hydroxyl end-groups. Secondary hydroxyl end-groups are several times less reactive with isocyanates than primary hydroxyl groups and for some applications polyether polyols based only on propylene oxide may have inconveniently low reactivity.

The primary hydroxyl content may be increased by the separate reaction of the polyoxypropylene polyols with ethylene oxide to form a block copolymer with an oxyethylene 'tip'. By this means the primary hydroxyl end-group content may be varied from about 5% to over 80% of the total hydroxyl end-groups.

Polyether polyols for flexible polyurethanes. Most of the polyether polyols used to make flexible polyurethane foams and elastomers are triols based on trifunctional initiators, mainly glycerol or trimethylolpropane. Polyether diols, made using glycol initiators, are also frequently used, often together with triols, in making high-elongation foams and elastomers. These high molecular weight polyether polyols are made mainly from propylene oxide but are usually modified by the co-polymerisation of 5% to 20% of ethylene oxide.

The many polyether polyols used to make flexible polyurethanes include speciality products for the production of flat-top slabstock foam, polyether polyols for moulded flexible and semi-rigid foams, and specially developed polyethers for elastomers, RIM products, adhesives and coatings.

The majority of flexible foam is made from 80:20-TDI and polyether triols with molecular weights in the range from 3,000 to 4,000. A typical triol with a mean molecular weight of 3500 is made by the addition polymerisation of 50 to 55 moles of propylene oxide and 10 to 15 moles of ethylene oxide onto 1 mole of glycerol or trimethylolpropane. Both the quantity and the position of the oxyethylene groups in the polyoxypropylene chains are important when tailoring the polyether polyol system for use in a particular process for making slabstock or moulded foam, and in obtaining a foam rise that is sufficiently robust to be unaffected by minor variations in processing conditions.

A side reaction in the base catalysed polymerisation of propylene oxide produces unsaturated end-groups that do not react with isocyanates. Polyol manufacturers control this reaction and limit the unsaturation of their products. One effect of unsaturated end-group formation is to reduce the effective functionality of the polyol, but, in practice, this is significant only for polyols with equivalent weights above about 1300.

The epoxide monomers and the polyether polyols are easily oxidised. Air is excluded from the manufacturing process and, when polymerisation is complete, antioxidants are added to prevent oxidation of the polyether. Much work has been done to develop efficient antioxidant systems. Many polyether polyols contain a synergistic mixture of antioxidants that not only protect the polyol from oxidation during storage, but also protect low density foam from oxidative scorch at the high curing temperature immediately after manufacture.

Modified polyether polyols. There are three main types:
1) Polyvinyl-modified polyethers or 'polymer polyols'.
2) Polyols containing polyurea dispersions (PHD polyols (Poly Harnstoff Dispersion)).
3) Polyols containing polyurethane dispersions (PIPA polyols (PolyIsocyanate Poly Addition)).

37

Each of these modified polyether polyols is used to make flexible foams and microcellular elastomers of higher hardness than can be obtained by using unmodified polyether polyols.

Polymer polyols are tipped, polyether triols, containing stable dispersions of polyvinyl fillers made by the in-situ polymerisation of vinyl monomers. Polymer polyols contain three types of polymer: unmodified polyether polyol, the vinyl polymer and some vinyl polymer grafted onto the polyether chains. The latter is important in stabilising the suspension. Stable, low viscosity products are obtained by the in-situ polymerisation of a mixture of styrene and acrylonitrile. Stabilisers, (Polyether polyols, modified to facilitate graft polymerisation) are used to stabilise high concentrations of polyvinyl fillers and also to make polymer polyols containing mostly polystyrene. The latter yield whiter foams with improved resistance to smouldering ignition.

Polyurea-modified polyols are conventional polyether polyols containing dispersed particles of polyurea formed from the reaction of TDI and a diamine. A secondary reaction between excess isocyanate and the polyol may form a polyurea/urethane/polyether reaction product and help to stabilise the polyurea dispersion. These polyols are produced containing up to 20% of dispersed polyureas. The dispersed polyureas may react with isocyanates during polyurethane manufacture to give increased cross-linking of the final polymer.

Pipa polyols are basically similar in concept to PHD polyols but they contain dispersed particles of polyurethane(s) formed by the reaction, in-situ, of an isocyanate and an alkanolamine.

Amine-terminated polyethers. There are several methods of converting the hydroxyl end-groups of polyether polyols to amino end-groups, but the preferred process is by the reductive amination of the secondary hydroxyl groups of polyether polyols having terminal polyoxypropylene blocks. Such amine-terminated polyethers have been available for several years but their use with isocyanates was restricted by their extremely fast reaction. During the last few years however, the use of amine-terminated polyethers has increased rapidly. They are sometimes used with conventional polyether polyols but the main application is their use, together with hindered aromatic diamines as chain-extenders, in the polyurea RIM process.

Polyether polyols for making rigid polyurethanes. The polyether polyols used to make rigid polyurethane foams, rigid polymers and hard surface coatings, have a lower equivalent weight than those used to make flexible polyurethanes. The properties of

38

these polyols vary with the choice of polyol initiator, either an alcohol or an amine and with the composition and the length of the oxyalkylene chain. Polyethers based on aminic initiators have a significantly higher reactivity with isocyanates than similar polyethers based on alcohols. Some typical polymerisation initiators are listed in table 3-3.

Table 3-3 **Polymerisation initiators: Polyether polyols for rigid polyurethanes**

Alcohols	Amines	Functionality
Water Ethylene glycol Diethylene glycol		2 2 2
Glycerol Trimethylolpropane	Triethanolamine	3 3
Pentaerythritol	Toluene diamine Ethylene diamine 4,4'-diaminodiphenyl- methane	4 4 4
	Diethylene triamine	5
Sorbitol		6
Sucrose		8

The weight of the starter may represent over one third of the weight of the polyol. The choice of the starter is thus a very important factor in controlling the cost as well as the performance of the polyol. For these reasons a common starter is sucrose which is available in a pure form at low cost and yields polyols with a satisfactory performance. Blends of polyols are often used to obtain the best combination of processability, cost and the properties of the final product, and for general applications a mixture having an average functionality of 4 to 5 is often used. Mixture of amines and alcohols as starters provide a useful way of adjusting the performance of the polyol in the urethane reaction. Since amine-based polyols are self-catalytic in their reaction with isocyanates compared with nitrogen-free polyols, they tend to be less affected by tertiary amine catalysts and by changes in temperature. Polyols based upon aromatic diamine starters yield polyurethane systems that cure quickly to form strong polymers.

Polyester polyols
Saturated polyesters with terminal hydroxyl-groups are used to make both flexible and rigid polyurethane polymers. Polyester polyols tend to be more expensive than polyether polyols and they are usually more viscous and therefore more difficult to handle. They also yield polyurethane polymers which are more easily

hydrolysed. Consequently they are only used to make polyurethanes for demanding applications where the particular physical properties obtainable from polyesters are important. Polyester-based polyurethane elastomers combine high levels of tensile properties with resistance to flexing and abrasion. They also have good resistance to many types of oil. Flexible foams made with polyester polyols and TDI combine high elongation with resistance to dry cleaning solvents. Polyesters are also less easily oxidised and resist higher temperatures than polyethers. The improved heat resistance, especially of polyesters based on aromatic diacids, is important in many rigid foam systems.

Polyesters are typically made by the condensation reaction between glycols and di-carboxylic acids. Branching can be introduced by the addition of a small amount of a triol to the reaction mixture. As the esterification proceeds the water produced is removed from the reaction. Water commonly represents up to 15% of the reaction product in making oligomeric polyesters so that the maximum yield of useful product is about 85% by weight of the raw materials used. This contrasts with the production of polyether polyols where yields may approach 100%. This lower yield, combined with the fact that the raw materials tend to be more expensive, explains the higher cost of many polyesters compared with polyethers.

Relatively low cost polyester polyols, based on recovered materials, are also available. Mixed adipic, glutaric and succinic acid polyesters are made using purified nylon waste acids (AGS acids). AGS acids are also hydrogenated to make a mixture of 1,4-butanediol, 1,5-pentanediol and 1,6-hexanediol which is used to make polyadipates having a low melting point. Aromatic polyester polyols made from dimethyl terephthalate (DMT), process waste, have been available for several years but, more recently, a number of low-cost polyester polyols have been produced using a proportion of material obtained by depolymerising waste poly(ethylene terephthalate) (PET), recovered from scrap bottles and film.

Polyester polyols for making flexible polyurethanes include:

- Poly (ethylene adipates), which are wax-like solids at room temperature and are mostly used in the manufacture of polyurethane elastomers and adhesives.
- Lightly branched poly (diethyleneglycol adipates), which are used mainly to make flexible foams, and
- A wide range of adipates made with more than one aliphatic diol. These are used to make solid and microcellular elastomers, flexible coatings and adhesives.
- Mixed polyadipates from hydrogenated AGS acids are used to make microcellular elastomers with good hydrolytic stability.
- Aromatic polyester polyols are used in rigid polyurethane and

polyisocyanurate foam. Polyesters based partly on scrap PET and/or DMT process waste, are also used in rigid foam, particularly in North America where some 40,000 tonnes of polyester polyols were used in 1985.

- Polycaprolactone diols have lower viscosities but are more expensive than polyadipates of similar molecular weight. Copolymer diols of polycaprolactones and polyadipates are easy to handle because they are often liquids of relatively low viscosity at room temperature. Polycaprolactone diols give elastomers with relatively good resistance to hydrolysis but similar stability can be obtained with polyadipates by increasing the molecular weight of the glycols.
- Polycarbonate diols, which are waxy solids at room temperature, are also used to make polyurethanes – mainly for coatings and adhesives. They have significantly improved resistance to hydrolysis under humid conditions.
- Halogen containing polyesters, made from halogenated diols and diacids, may be used to reduce the flammability of heat-resistant polyurethanes.

Generally, the hydrolysis resistance of polyesters increases with increasing length of the chains between the polyester linkages and also with reducing residual acidity. The resistance to hydrolysis and to swelling by solvents and oils is also increased by increased branching of the polyester chains.

Table 3-4 **Typical properties of polyester polyols**

Application	Type	Hydroxyl no. (mg KOH/g)	Viscosity (mPa s at 25°C)	Acid value (mg KOH/g)
Flexible foams	Linear	45 to 60	10,000 to 20,000	< 2.0
Flexible foams	Branched	60 to 70	15,000 to 20,000	< 5.0
Low density semi-rigid foams	Highly branched	200	ca. 20,000	< 10
Elastomers	Linear	50 to 60	3,000 to 10,000	< 1.0

Additives

In addition to isocyanates and polyols – the basic materials for making polyurethanes – a wide range of auxiliary chemicals may be added to control and modify both the polyurethane reaction and the properties of the final polymer.

These additives include catalysts, chain extenders, cross-linking agents, surface-active materials, flame-retardants, colouring materials and fillers. All practical polyurethane systems include some of these auxiliary chemicals.

Catalysts. A number of catalysts can be used for the reaction of isocyanates with water and with polyols and these include aliphatic

41

and aromatic tertiary amines, and organo-metallic compounds, especially tin compounds, although compounds of mercury or lead are also used. Alkali metal salts of carboxylic acids and phenols, and symmetrical triazine derivatives are used to promote the polymerisation of isocyanates.

Tertiary amines are the catalysts most widely used in making polyurethane foams. The mechanism of catalysis by a tertiary amine involves the donation of electrons by the tertiary nitrogen to the carbonyl carbon of the isocyanate group thus forming a complex intermediate. The catalytic activity of a tertiary amine depends on its structure and its basicity. The catalytic effect increases with increasing basicity but is reduced by steric hindrance of the aminic nitrogen. A great deal of work has been done to study the effect of catalysts on the reaction of isocyanates with compounds containing active hydrogen, but much of the published results concerns the reaction mechanism of model compounds in dilute solution at a fixed temperature. Polar solvents may increase the reactivity of isocyanates by stabilising the polarisation of the isocyanate group. Water-blown flexible polyurethane foams are usually catalysed with a synergistic mixture of one or more tertiary amines and an organo-tin catalyst. The catalyst mixture is required to maintain a balance between the reaction of isocyanate with the secondary hydroxyl end-group of the polyether polyol, and the reaction of the isocyanate with water. The choice of tertiary amine may also affect the foam properties. Some tertiary amines catalyse chain branching by biuret linkages. The effect of such chain branching within the polyurea segments is to inhibit the formation of well-ordered secondary bonded domains, producing softer foams having higher resilience. Similar effects may be obtained by the addition of isocyanate polymerisation catalysts.
The choice of catalysts for making rigid, closed-cell, polyurethane foams from polymeric MDI is usually concerned with obtaining the gelation/foam rise profile and cure rate most suitable for the process application. Some commonly used tertiary amine catalysts are listed in table 3-5.

Table 3-5 **Some tertiary-amine catalysts**

Catalyst	Application
1 N,N-Dimethylaminoethanol $(CH_3)_2NCH_2CH_2OH$	Inexpensive, low-odour, isocyanate reactive, mobile liquid catalyst used in polyether-based flexible foams.
2 N,N-Dimethylcyclohexylamine, (Catalyst SFC) $N(CH_3)_2$	Liquid with an intense odour. Rigid foams, polyester-based flexible foams and some semi-rigid foams.
3 Bis-(2-dimethylaminoethyl)ether $(CH_3)_2NCH_2CH_2OCH_2CH_2N(CH_3)_2$	Low-odour, mobile liquid used in high resilience and cold-cure flexible foams.
4 N,N,N′,N′,N″-Pentamethyl-diethylene-triamine $(CH_3)_2NCH_2CH_2NCH_2CH_2N(CH_3)_2$ $\qquad\qquad\quad CH_3$	Flexible foams and semi-rigid foams.
5 N,N-Dimethylbenzylamine, (Catalyst SFB) $CH_2N(CH_3)_2$	Liquid with characteristic smell used in polyester-based flexible foams, semi-rigid foams and for prepolymer making.
6 N,N-Dimethylcetylamine $CH_3(CH_2)_{14}CH_2N(CH_3)_2$	Viscous liquid with a low odour used in polyester-based flexible foams and some potting compounds.
7 Diaminobicyclooctane (DABCO) 	Solid, soluble in water, glycols and polyethers. May be used in most types of polyurethanes.
8 N-Ethylmorpholine 	Volatile, low viscosity liquid with characteristic odour. Used as synergistic catalyst in flexible foams and in prepolymer preparation.

Table 3-5 **Continued**

9 Methylene-bis-dimethylcyclo-hexylamine $(CH_3)_2N$—⬡—CH_2—⬡—$N(CH_3)_2$	Low-volatility liquid with much lower odour than 2. Used in polyester-based flexible foams and potting compounds.

10 N,N,N′,N″,N″-pentamethyl-dipropylene-triamine
 	
$(CH_3)_2NC_3H_6NC_3H_6N(CH_3)_2$
           $|$
           CH_3
	Liquid with strong ammoniacal odour used in polyether-based slabstock foams and in semi-rigid foam moulding.

11 N,N′-Diethylpiperazine CH_3CH_2N⬡NCH_2CH_3	Liquid used in flexible foams and RIM elastomers. The more volatile N,N′-Dimethylpiperazine is used in some semi-rigid foam systems.

12 N,N,N,′-Trimethylaminoethyl-ethanolamine
 	
         CH_3
         $|$
$CH_3NCH_2CH_2NCH_2CH_2OH$
         $|$
         CH_3
	Mobile liquid used in flexible, high resilience and microcellular foams. Isocyanate reactive.

13 1-(2-hydroxypropyl) imidazole OH $N{=}CH$ $|$ $|$ NCH_2CHCH_3 $CH{=}CH$	Isocyanate reactive catalyst for polyether-based foams and low density rigid foams.

14 1,4-bis (2-hydroxypropyl)-2-methylpiperazine
 	
  CH_3 CH_2CH_2 CH_3
  $|$ $|$
$HO{-}CHCH_2N$ $NCH_2CH{-}OH$
         $CHCH_2$
         $|$
         CH_3
	Low odour, isocyanate reactive chain extending catalyst for flexible foams and RIM.

Organo-metallic catalysts see table 3-6 are used to accelerate the urethane reaction. The most popular are stannous octoate and dibutyltin dilaurate. Stannous octoate is used in most flexible foam systems, except pre-blended two-pack systems where its low hydrolytic stability is unacceptable. Microcellular elastomers, RIM systems and cast elastomers are often catalysed by low levels of dibutyltin dilaurate or dibutyltin mercaptide. Organo-metallic catalysts form an intermediate complex with an isocyanate group and a hydroxyl group of the polyol. This complex formation is inhibited by steric hindrance of the Sn atom. This steric effect is used in one type of 'delayed action' catalyst, i.e. a catalyst that is not very active at room temperature but becomes effective when the reaction temperature rises.

Table 3-6 **Some commercially available organo-metallic catalysts**

Catalyst	Principal applications
Stannous octoate	Slabstock polyether-based flexible foams, moulded flexible foams.
Dibutyltin dilaurate	Microcellular foams, RIM, two-pot moulding systems, elastomers.
Dibutyltin mercaptide	Hydrolysis resistant catalyst for storage stable two-pot systems.
Dibutyltin thiocarboxylates Dioctyltin thiocarboxylates	Delayed action (hindered) catalysts for RIM and high resilience foams.
Phenylmercuric propionate	In glycol solution for potting compounds, as a powder for delayed action catalysis.
Lead octoate	Urethane chain extension catalyst
Alkali-metal salts, e.g. CH_3COOK, K_2CO_3 $NaHCO_3$ and Na_2CO_3	General catalysts for the urethane reaction and for isocyanate polymerisation.
Calcium carbonate	A common filler with a catalytic effect on the urethane reaction and on the cure rate of polyurethanes.
Ferric acetylacetonate	Catalyst for cast elastomer systems, especially those based on TDI.

Cross-linking agents and chain-extenders. These are low molecular weight polyols or polyamines. They are also sometimes known as curing agents. Chain-extenders are difunctional substances, glycols, diamines or hydroxy amines; cross-linking agents have a functionality of three or more. Chain-extenders are used in flexible polyurethanes such as flexible foams, microcellular elastomers, cast elastomers and RIM systems.

The chain-extender reacts with diisocyanate to form a polyurethane or polyurea segment in the urethane polymer. It is usually added in sufficient amount to permit hard-segment segregation that results in an increase in the modulus and the hard-segment glass transition temperature (Tg) of the polymer. The Tg provides a measure of the polymer softening point and some indication of the upper limit of its working temperature. When diamines are used as additives, instead of glycols of similar molecular weight, they give faster reaction with isocyanates. The resulting polyurea hard-segment has a higher density of secondary bonding so that the Tg and the thermal stability of the polymer are both increased.

Cross-linkers are used to increase the level of covalent bonding in rigid polyurethanes such as some rigid foams and also as additives in many semi-rigid foam systems.

Chain-extenders and curing agents are used in both the single shot and prepolymer processes for making polyurethanes. Aromatic and aliphatic diamines are each used as curing agents but the former are more favoured because of their lower reactivity with isocyanates. A high level of polyurea hard-segment separation is required in order to obtain elastomers having a high modulus and this necessitates at least 3 to 4 molecular equivalents of chain extender per oligomer chain. Simple diamines are, in general, too reactive to permit such a high level of addition in microcellular elastomers even for the very fast RIM systems, and special derivatives have been developed. Aromatic diamines with bulky substituents ortho to the amino groups are available where the reaction with isocyanates is sufficiently hindered to allow the use of high levels of chain-extending or curing agent.

Table 3-7 lists some common chain-extending and cross-linking agents together with the stoichiometric weight of TDI and MDI with which they react. Calculated hydroxyl values in mg KOH/g equivalent are also given in order to assist calculation of the formulation by the standard methods. (see Appendix B).

See also Chapter 13, part 1, page 321

Blowing agents. Cellular polyurethanes are manufactured by using blowing agents to form gas bubbles in the polymerising reaction mixture. Flexible polyurethane foams are usually made using the carbon dioxide formed in the reaction of water and diisocyanate (water blowing) either as the sole blowing agent or as the principal blowing agent in association with trichloromonofluoro-methane (CFM-11) or methylene chloride, or a mixture of both. It is impracticable to blow low density flexible polyurethane foams without some water blowing because, in addition to generating carbon dioxide, the water/isocyanate reaction produces not only polyureas, an essential part of the polymer hard-segment, but also the exothermic heat required to complete the polymerisation and to vaporise any non-reactive blowing agent used. Self-skinning foams, formulated with chain-extenders in addition to polyols of high molecular weight, are substantially water-free systems blown with CFM-11 mixed sometimes with methylene chloride. In these systems the heat of vaporisation of the blowing agent is provided by the exothermic reaction of the chain-extender with the diisocyanate. Rigid foams, because they are made with polyols having higher

Table 3-7 Chain-extending agents, cross-linking agents and curing agents and their diisocyanate equivalents

Additive	Function-ality	Mole weight	OH value (mg KOH/g)	Weight of diisocyanate (g per 100 g of required additive)	
				TDI	MDI
Ethylene glycol $C_2H_4(OH)_2$	2	62.07	1801	280	401
Diethylene glycol $O(C_2H_4OH)_2$	2	106.12	1057	164	235
Propylene glycol $C_3H_6(OH)_2$	2	76.11	1474	229	329
Dipropylene glycol $O(C_3H_6OH)_2$	2	134.18	836	130	186
1,4-Butane diol $C_4H_8(OH)_2$	2	90.12	1245	193	278
Polypropylene glycol 400	2	400	280	43.5	62
m-Phenylene diamine $C_6H_4(NH_2)_2$	2	108.15	1037	161	231
Diethyl toluene diamine $C_6HCH_3(C_2H_5)_2(NH_2)_2$	2	178.27	629	97.7	140
Dimethylthio toluene diamine $C_6HCH_3(SCH_3)_2(NH_2)_2$	2	214.34	523	81.2	116
Water HOH	2	18.01	6230	968	1389
Diethanolamine $HN(CH_2CH_2OH)_2$	3	105.14	1601	248	357
Triethanolamine $N(CH_2CH_2OH)_3$	3	149.19	1128	175	252
Glycerol $CH_2OHCHOHCH_2OH$	3	92.11	1827	284	407
'Daltolac' C4	3	168	1000	155	223
'Daltolac' C5	3	150	1125	175	251
'Daltolac' 50	4	468	480	75	107
'Uropol' G 790	4	280	800	124	178

functionality and a higher order of hydroxyl group content than those used in flexible polyurethanes, yield sufficient exothermic heat from the urethane reaction alone to allow foam expansion simply by vaporising an inert blowing agent. The usual blowing agent for rigid foams is CFM-11, sometimes mixed with dichlorodifluoromethane (CFM-12) which has a lower boiling point. The low thermal conductivity of the CFM-11 vapour which is retained in the closed-cell rigid foam is important in giving a cellular product with better insulating properties than those of most competitive materials.

In addition to its low conductivity, CFM-11 has many advantages as a blowing agent for polyurethanes. It has been made on a large scale for use as a refrigerant and as a propellant for aerosols, it is non-flammable, it has very low toxicity, it is almost odourless, and it is chemically stable. The only significant hazard in use is the possible thermal decomposition of the vapour by flames or red-hot surfaces to give acidic products and traces of phosgene.

The physical properties of the non-reactive blowing agents are listed in table 3-8 below.

Table 3-8 **Non-reactive blowing agents for polyurethanes**

Blowing agent	Trichloro-monofluoro-methane (CFM-11)	Dichlorodi-fluoro-methane (CFM-12)	Methylene-chloride
Molecular weight	137.38	120.92	84.94
Density at 20°C (g/ml)	1.488	1.486	1.336
Boiling point at 1 atm. (°C)	23.8	−29.8	40.1
Freezing point (°C)	−111	−160	−96.7
Threshold Limit Value (TLV*, ppm)	1000 ppm	1000 ppm	50 ppm A$_2$
Solubility (g/100 g solvent at 20°C)			
Water	Insoluble	Insoluble	2
Ethanol	∞	∞	∞
Polyethers		Adequately soluble for all applications	

See also Chapter 13, part 1, page 323, table 13-2

* From TLVs, Threshold Limit Values for Chemical Substances in the Work Environment Adopted by ACGIH for 1986-87. (A$_2$ notation indicates a substance suspect of inducing cancer in man.)

Surfactants. Surface-active materials are essential ingredients in the manufacture of most polyurethanes. Selected surfactants, or mixtures of surfactants, help in mixing incompatible components of the reaction mixture. They are particularly useful in foam making where they help to control the size of the foam cells by stabilising the gas bubbles formed during nucleation and may stabilise the rising foam by reducing stress-concentrations in the thinning cell-walls. In flexible foam manufacture, surfactants also help to control the degree of cell opening and increase the operating margin between the extremes of foam collapse, when cell opening occurs before the reaction mixture has sufficiently polymerised, and a high content of closed cells which results from cell opening being too long delayed. Early polyurethane foam systems used one or more organic, usually non-ionic, surfactants. Some polyester-based, low density, flexible foams and some semi-rigid foams are still made using organic surfactants such as substituted nonyl phenols, fatty acid ethylene oxide condensates and alkylene oxide block co-polymers. However, most flexible and rigid foams are now made using organosiloxanes or silicone-based surfactants. The first silicone polymers to be used in the production of polyurethane foams, especially in two-stage processes, were poly (dimethylsiloxanes), (PDMS) and poly (phenylmethylsiloxanes) having viscosities from about 2000 to 14000 mPa s at 25°C. These materials remain of value in some flexible and semi-rigid foam systems, but the majority of low density foams are now made using PDMS-polyether graft copolymer surfactants that have been developed specially for the purpose. To meet the surfactant needs of the polyurethane formulation and of the application technique, the surfactant structure may be modified by changing the length of the PDMS backbone, the number, length and the composition – the oxyethylene:oxypropylene ratio – of the pendent polyether chains. Silicone surfactants for rigid foams have a greater surface activity than those used for flexible foam. They have predominantly hydrophilic polyoxyethylene polyether chains pendent from the hydrophobic PDMS. Surfactants for use with in making conventional flexible polyether foams have longer grafted polyether chains with a higher polyoxypropylene content. Polyester-based flexible foams, on the other hand, require surfactants of lower activity with much shorter polyether chains.

Colouring materials. Most low density flexible foam is colour-coded during manufacture to identify the grade and the density of the foam. Specialised products, such as foams for textile laminating and for packaging, may be highly coloured to meet the requirements of the application. Rigid foams, on the other hand, being mostly made from brown-coloured polymeric MDI and sold enclosed within opaque covering materials, are often made without added colorants. The usual method of colouring is by the addition of pigment pastes to the foam reaction mixture.

Organic and inorganic pigments are both used. Dispersions of suitable pigments in polyols or in plasticisers are available from a number of specialist suppliers. It is important to use only pigments or dispersions from recommended suppliers and to make trials on a production scale when preparing new polyurethane systems. The pigments used must not react with isocyanates and must be stable at the high curing temperatures reached in the manufacture of low density foams. Dispersions must also be free from trace contaminants that can adversely affect foam formation. The most widely used colouring material is carbon black which, at levels above 0.1 part per 100 parts of the polyol used, gives some protection against surface discoloration of the foam caused by UV light. Polyurethane foam is easily coloured by dyeing or by padding with 'Procion' dyes, but dyeing is not used on a large scale because of the high cost of drying wet foam. It is, however, a very useful method for making small lots for special applications in clothing and packaging.

Polyurethanes based on aromatic isocyanates tend to yellow on exposure to daylight and uncovered items such as self-skinning foams, microcellular mouldings and RIM products must be protected to overcome the problem. Several methods are used:

- Pigmenting with sufficient dark-coloured pigment, especially with carbon black, to protect the polymer from UV light.
- Using UV light absorbing additives. ICI has developed polyether polyol systems that contain synergistic mixtures of stabilisers. Such systems are used with MDI variants to make self-skinning parts with sufficient light-fastness for most interior applications in vehicles and buildings.
- Painting the finished part. Two approaches to solving this problem are in common use. First, in order to reduce the number of operations and to simplify any difficulties associated with demoulding of the product and with paint adhesion, in-mould coating is becoming widely used, especially for priming coats. The paint is sprayed into the mould before pouring or injecting the polyurethane reaction mixture which combines with the paint to produce an integral coloured skin. Such paints are often based upon the relatively expensive aliphatic isocyanates and are stable to UV light.
  Secondly, top coats of polyurethane are applied after demoulding. These may be either flexible or rigid and they provide a high degree of abrasion resistance. One-component lacquer systems based on blocked isocyanates, and two-component systems consisting of a resin blend and an isocyanate component are each used. The former usually require stoving temperatures greater than 120°C in order to release the blocked isocyanate and polymerise the lacquer coat. Two-component systems may be cured at lower temperatures, although they have a pot life – after mixing the two components – of a few hours only.
- Using systems based on aliphatic diisocyanates. Self-skinning foam systems are available that are based on a mixture of aliphatic isocyanates. These give products with good light-fastness, even in pale colours. Such systems are much more expensive than those based on aromatic isocyanates because of the relatively small scale of manufacture.

Fillers. Particulate and fibrous fillers may each be used in most kinds of polyurethanes. There are many reasons for adding fillers. Particulate fillers are used in flexible polyurethane foams to reduce their flammability and, particularly in the USA, to increase the weight of seat cushions for furniture and to increase their resistance to compression. Fibrous fillers are reinforcing: they give increased stiffness and they increase the range of operating temperature of rigid foams, self-skinning foams and flexible RIM products. Mineral fillers are sometimes used to reduce costs and to increase the compressive strength of rigid foams used in composite building panels. Continuous fibres, nettings or scrims encapsulated in rigid foams, improve the stiffness and the heat-resistance of insulating panels and, in low density flexible foam mouldings, their use increases the tensile strength at highly stressed positions. Some fillers and their applications are listed in table 3-9.

Table 3-9 Some fillers and their application in polyurethanes

Filler	Typical applications
Calcium carbonate, (Ground chalk, ground limestone, whiting)	Flexible foams, semi-rigid foams, binder compositions, rigid self-skinning mouldings.
Barium sulphate, (Barytes)	Flexible foams, semi-rigid foams, especially for sound-absorbing.
Clays (China clay, kaolins, etc.)	Flexible systems
Expanded silicas, colloidal silicas	Flexible foams, cast elastomers
Clay balls, vermiculite, expanded mica, etc.	Rigid foams
Glass micro-spheres	Flexible, microcellular foams, RIM
Glass flakes	Elastomeric RIM
Silicates, cements	Rigid foams, sealants, grouting compounds
Short fibres, milled and chopped glass-fibre, Aramid fibres, carbon fibres, conducting fibres, (aluminium, coated glass, steel)	Elastomeric RIM, rigid foams
Glass cloths and scrims, wire mesh, organic fibres, etc.	Encapsulation in rigid foams, reinforcement of low density flexible foam mouldings.

Finely-divided fillers with a particle size ranging from a few microns up to about 100 microns are usually added as dispersions in the polyol component of the polyurethane mix. Fillers for use in polyurethanes must be dry. Some low-cost mineral fillers such as china clay, kaolins and other aluminium silicates, which contain both free and combined water, that would otherwise be satisfactory, may be difficult to dry reproducibly and economically. Fibrous fillers used for reinforcement may be dispersed in either the polyol or the isocyanate component, or both. The degree of reinforcement obtainable depends on the strength (the E-modulus) of the fibre, the concentration of the fibres, the modulus and extensibility of the polymer matrix, the interfacial adhesion and the shear strength at the fibre/polymer interface, and on the orientation of the fibres. The most important fibre used in polyurethanes is glass-fibre made from a calcium alumina borosilicate (E-glass) composition. A typical glass-fibre strand contains over 200 filaments each of 10 to 15 microns in diameter and a roving or tow will contain up to 6,000 filaments. For use in polyurethanes, glass fibres are available in the form of continuous, chopped or milled grades. The properties of some high-modulus reinforcing fibres are tabulated in table 3-10.

Reductions in the cost per unit volume are possible by filling polyurethanes with expanded fillers such as glass microspheres, expanded silicas and expanded polystyrene beads. The occlusion of air and water in expanded mineral fillers may limit their usefulness in many polyurethane systems.

Table 3-10 Some high modulus reinforcing fibres

Fibre	Young's modulus (ε)	Specific gravity (s.g.)	ε/s.g.
'E'-glass fibre	70	2.55	27
Aluminium	70	2.7	26
Aramid fibre 'Kevlar' 29	80	1.3	61
'Kevlar' 49	120	1.4	86
Carbon fibre (PAN)	250-400	1.8-1.9	140-210
Carbon fibre (Pitch)	200	1.9	105
Steel	200	7.8	26
Alumina fibres	350	4.0	88
Polyethylene fibres	30-70	0.96	31-73
Boron fibres	400	2.5	160

Flame Retardants. Polyurethanes, in common with all other organic materials, will burn given the application of sufficient heat in the presence of oxygen. The physical state of the polymer is also extremely important. Low density, open cell foams, having a large surface area and high permeability to air, burn most easily. Flame retardants are added to polyurethanes to reduce the flammability as measured by specific, often small-scale tests, that measure one aspect of flammability under a particular set of controlled standard conditions.

Flammability tests, whether on a small or large scale, should be regarded merely as a means of measuring the reproducibility of a material and of comparing one material with another under a standard set of conditions. They must not be taken as indicating the performance of the material in an actual fire, when many additional factors, including the size of the sample, the composite construction in which it is used, the heat flux from the surroundings, the amount of ventilation, and many other factors will affect flammability.

The choice of flame-retardant system in any specific circumstance will depend upon the service application of the polyurethane, the assessment of the fire hazard, and any legal requirements. Aspects of flammability that may be affected by additives include ignitability, burning rate, smoke evolution on burning, propensity to smoulder, the toxicity of the combustion gases, and the ease of extinction. The testing of uncovered foams, however, is of very limited value as most flexible and rigid foams are rarely used alone. Foams are usually enclosed within a composite article such as an upholstered chair or a refrigerator where the covering material is the most important factor determining ignitability and the primary fire hazard. Both the ignitability and the burning rate of polyurethanes may be reduced by the addition of flame retardants which operate through one or more of the following mechanisms:

- Provision of a heat sink by filling with incombustible materials to delay ignition and reduce the rate of burning.
- Provision of an energy sink and means of diluting the combustion gases by filling with substances that decompose on heating to give incombustible products such as water and carbon dioxide.
- Modification of the mechanism of burning by filling with materials, such as halogenated flame-retardants, that react with the polymer or that produce a gas that decomposes to yield free radicals.
- Inducing char formation by the use of additives.

The most widely used flame-retardants in both flexible and rigid foam systems are chlorinated phosphate esters. These have a significant effect upon the ignitability of the foams by a small heat source and they can show marked reductions in the rate of burning in small-scale tests without adverse effects upon the processability of the foam system and the properties of the product.

Many ignitability standards for furniture and vehicle seating can be met by the incorporation of 5% to 10% chlorinated phosphate esters. The presence of both chlorine and phosphorus is necessary for the optimum effect upon flammability.

The addition of aluminium trihydrate gives a further reduction in flammability and minimises the increase in smoke formation on burning, resulting from the addition of halogenated organic phosphates. Melamine is also used, together with phosphate flame retardants such as ammonium polyphosphate, in flexible foams for furniture cushions. It gives good resistance to smouldering combustion.

The rate of flame penetration or 'burn-through time' is very important in some tests of composite building panels containing rigid polyurethane foam insulation. The best results are obtained by combining foam modifications, which increase the formation of carbonaceous char when heated, with reinforcement by encapsulated glass-fibre, steel or other fibrous heat-resistant webs that prevent the cracking and disintegration of the slow burning protective char. ICI have been in the forefront in developing isocyanurate foam composites with long burn-through times.

Table 3-11 lists examples of flame-retardants in three classes: non-reactive liquid flame retardants; compounds that react with isocyanates to become bound into the polymer network; and some other materials that reduce the rate of burning. Many other flame retardants are available from specialist suppliers.

Table 3-11   **Some flame retardants for polyurethanes**

Additive	Typical application
A. Non-reactive liquids	
Tris (2-chloropropyl) phosphate, 'Daltogard' F	All polyurethane foams including polyester-based foams and microcellular elastomers.
Tris (2-chloroethyl) phosphate, (T.C.E.P.)	Polyether-based flexible and rigid foams.
Tris (2,3-dichloropropyl) phosphate, 'Fyrol' FR 2 'Celluflex' FR-2	Polyether-based rigid and flexible foams.
Tetrakis (2-chloroethyl)-2,2-bis-(chloromethyl) propylene-phosphate. 'Phosgard' 2XC20	Low volatility material for flexible and rigid foams.
Dimethyl methyl phosphonate	Rigid foams.
B. Isocyanate-reactive additives	
Tris (polyoxyalkyleneglycol)-phosphonates and phosphite esters.	Flexible and semi-rigid foams.
Tris (halogenated polyol)-phosphonates.	Flexible and rigid foams.
Dibromoneopentyl glycol. 'FR' 1138	Polyether-based polyurethanes.
Brominated polyester and polyether diols. 'Saytech' RB-79 and 42-43	Rigid polyurethane and polyisocyanurate foams.
Tetrabromobisphenol A Tetrabromophthalic anhydride	Rigid polyurethane and polyisocyanurate foams.
C. Fillers	
Ammonium salts, sulphate, polyphosphate, etc.	Together with halogenated additives in rigid polyurethanes.
Aluminium hydroxide. Melamine.	All polyurethanes but especially in low density flexible foams for ignition and smoke supression.
Calcium carbonate	Heat absorbing filler.

4   Low density flexible foams

Low density flexible foams form over 50% of the total world-wide production of polyurethanes and in Europe today such foams are produced at a rate corresponding to about 1 kg per head of the population each year. Most of this vast volume of foam is used in the furniture, bedding and vehicle industries. Foams of this type are characterised by a density in the range from about 10 kg/m^3 to about 80 kg/m^3, a high tensile strength and elongation to break, and high resilience or fast recovery from deformation.

Most flexible polyurethane foams are based upon the reaction of diisocyanates with polyether or polyester diols or triols (see Chapter 3). The earliest production of low density flexible foam was based upon slightly-branched polyester polyols (poly (diethylene adipates)) and 65:35-TDI. Polyester-based foams of this type are still produced, but over 90% of all flexible polyurethane foam production is now based upon polyether polyols and 80:20-TDI. Recent years have seen an increasing use of specially developed MDI compositions and polyether polyols to make moulded cushions for furniture and vehicle seating. MDI is also used, both alone and in combination with TDI, to make many other special-purpose, moulded, low density foams.

Low density flexible polyurethane foams are made by the simultaneous reaction of a diisocyanate with a hydroxyl-group-ended polyether or polyester and with water. The rate of the polymer-forming reaction and the rate of carbon dioxide evolution from the isocyanate/water reaction are kept sufficiently in balance by using polyols and isocyanates of appropriate reactivity, by using catalysts and foam stabilisers, and by controlling the temperature of the materials and the operation of the foam machinery. The carbon dioxide produced by the water/isocyanate reaction is thus contained within the polymerising material which expands to form a foam. Foaming or 'blowing' with carbon dioxide in this way is known as 'chemical blowing'. In addition to producing carbon dioxide, the reaction of water with a diisocyanate produces heat and forms substituted polyureas. The latter become the hard segments in the final polymer. The heat evolved has a major effect upon the course of the foam reaction and upon the final structure and the physical properties of the polymer. Although all low density flexible foams

Figure 4-1   Inspecting furniture cushions moulded in MDI-based, cold-cure foam.

Table 4-1 The major types of flexible polyurethane foams

Foam type	Typical density range produced (kg/m)	Raw materials used		Chain extending agents	Some applications
		Isocyanates	Polyols (Functionality/ Equivalent Wt.)		
Slabstock foams Polyether-based: Conventional	12 to 60	80:20-TDI	3/1000-1300	Water	1, 2, 3, 4, 5, 6, 8, 9 10, 14, 18, 19, 20
High resilience	18 to 80	80:20-TDI Modified TDI MDI/TDI mixtures Modified MDI	3/1500-2000	Water + diols	1, 2, 3, 6, 7
Filled (particularly flame retardant)	40 to 100	80:20-TDI	3/1000-2000	Water, diols, diamines.	1, 2, 3, 7, 9, 16
Semi-rigid	22 to 35	65:35-TDI 80:20-TDI	3+/1500-2000	Water + cross-linking agents	12, 14, 17, 20
Polyester-based: Technical grades	21 to 50	65:35-TDI 80:20-TDI	2+/800-1000	Water	5, 9, 10, 11, 12, 14 18, 19
Laminating grades	21 to 35	80:20-TDI	2+/1000-1200	Water	5, 8, 12, 21
Semi-rigid	22 to 35	65:35-TDI 80:20-TDI	4+/Mixture, ca 700	Water	17, 14, 18, 11
Moulded foams Polyether-based: Conventional hot-cure	22 to 50	80:20-TDI	3/800-1200	Water	1, 2, 3, 13, 20
High resilience and cold-cure	28 to 55	80:20-TDI TDI/MDI mixtures Modified TDI Modified MDI	3/1500-2000 (Also polymer polyols, polyurethane and polyurea modified polyols)	Water/diols /diamines	1, 2, 3, 4, 6, 7, 20, 22, 9
Semi-rigid	40 to 150	Polymeric MDI TDI (prepolymers)	3+/1500-2000	Water/diols /triols, etc	17, 23
Polyester-based	50 to 150	Polymeric MDI	2+/800	Water/triol	23, 17, 15, 14
'Repol' or Rebonded foams Slabstock or moulded)	60 to 300	80:20-TDI Polymeric MDI	3/1000	Water	9, 12, 13, 14, 15, 16, 17, 23, 9

See also Chapter 13, part 1, page 321

use water as the primary blowing mechanism, it is often desirable and necessary to use 'physical' or 'auxiliary' blowing as well. Physical blowing utilises some of the heat from the polymerisation and water reactions to vaporise a chemically inert liquid having a low boiling point. Many low density foams are made by a combination of chemical and physical blowing. Chemical blowing

Figure 4-2   Demoulding a complete chair of MDI-based foam moulded around steel springs

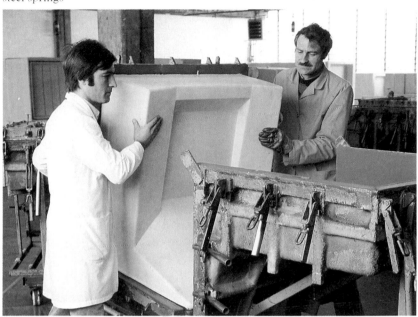

also stiffens the polymer by increasing the polyurea content of the polymer structure and the degree of secondary bonding between polymer chains. Physical blowing, on the other hand, has no effect upon the hard segment of the polymer chain but tends to reduce the stiffness of the polymer by reducing the polymerisation reaction temperature and the final curing temperature of the polymer.

As discussed in Chapter 3, flexible polyurethanes are segmented block copolymers consisting of long flexible polyester or polyether chains linked by rigid, hard block segments. By varying the type of polyester or polyether, and the length of the flexible chain, the structure and size of the hard segment (i.e. by varying the type of polyols used, the type and amount of chain extender, of isocyanate, of catalyst and other additives, and the conditions of manufacture and curing) and the amount of blowing, the foam may be tailored to meet the application requirements. The main types of flexible foam are listed in table 4-1.

The formation of flexible foam

Although it is simple in concept, the formation of flexible polyurethane foam in one stage from liquid materials is a complex interaction of both chemical and physical factors. There are no independent chemical or process variables in flexible foam manufacture. As indicated in Chapter 2, reproducible foam manufacture requires that the separate foam components are conditioned to within ± 2°C of a standard temperature, are dispensed at the required ratio, and blended together quickly before

Figure 4-3   Dispensing flexible foam chemicals between bottom and top papers.

the reaction commences. Most of the polyether polyols used in the manufacture of flexible foams are polyether copolymers containing hydrophobic and hydrophilic segments or blocks. The former are chains based on propylene oxide and the latter are chains based on ethylene oxide. Such polyols are easily mixed with water, TDI, modified MDI and the usual catalysts and additives, to form homogeneous mixtures.

In addition to mixing together the liquid chemicals to be used to make the foam, the mixer must also nucleate the reaction mixture to provide growth-points for the bubbles. Controlled nucleation is essential in order to control the cell size and the cell structure of the final product. If there are too few nucleation sites the carbon dioxide, formed by the reaction of water and isocyanate, forms a supersaturated solution. Suddenly, self-nucleation occurs. The first bubbles to be formed then grow very rapidly, the concentration of the carbon dioxide in solution falls and no further bubble nucleation occurs. The result is a foam with large cells or, even, the escape of the gas bubbles that may lead to the collapse of the foam. The nucleation of foam mixture is possible by using finely divided solids or liquids that are insoluble in the foam mixture; however almost all low density foam manufacture employs nucleation obtained by fine dispersion of a gas, usually air or nitrogen, in the reaction mixture. The formation of bubble nuclei is achieved by metering the nucleating gas into the polyol or the isocyanate streams flowing to the mixer, or by direct injection into the mixer barrel. On some foam plants, the addition of gas is unnecessary to obtain foams with a uniform medium cell size as there is sufficient dissolved gas present in the chemicals, especially in the isocyanate. This dissolved gas results from the storage of the chemicals under pressure, usually about 2 bar, in order to provide a positive feed to high speed piston metering pumps. Nucleation of the foam reaction mixture by effervescence of the dissolved gas is controlled by the pressure in the mixer and by the agitator speed. Where additional gas is needed to obtain very small cell sizes this is usually metered into the polyol stream at a rate which has been determined by trial. Mixers for continuous low density foam production usually employ simple multi-peg stirrers in plain cylindrical barrels. The nucleation of the foam mixture increases with increasing pressure drop across the mixer pins and this, in turn, increases with increasing agitator speed. The pressure drop depends also on the length and diameter of the mixer outlet nozzle and the throughput of foam mixture. Cell size adjustment during continuous foam manufacture may also be obtained by using an adjustable restriction in the outlet from the mixer.

The mixed, nucleated, foam reaction mixture is usually deposited in the mould or continuous trough less than 1 second after the start of mixing. There is then an induction period before the mixture becomes opaque with visible bubbles. This induction period is known as the 'cream time'. During this induction period the nucleation bubbles formed in the mixer become fewer in number and larger in size. The larger bubbles grow by diffusion of gas through the liquid at the expense of the smaller bubbles which have a higher internal pressure. The initial number of bubble nuclei created in the mixer is thus not the only factor influencing the cell size of the foam; other factors are:

– An increase in the *cream time* which increases the cell size
– An increase in the *foam reaction rate*, by increasing the catalyst level or temperature of the components for example, will increase the cell count (the number of cells per centimetre) i.e. it will reduce the average cell size. A secondary effect of high stirrer speeds is to increase the temperature of the reaction mixture. This reduces the cream time and, therefore, the cell size.
– *An addition of surfactants*, especially the usual polysiloxane-polyether block copolymers, which reduces the rate of nucleation bubble loss during the induction period giving a foam with a higher cell count.

During the initial period, the reactions of diisocyanate with water and with polyol proceed together. In TDI-based low density flexible foams the initial reactions produce an isocyanate-ended polyurethane, formed by the reaction of the terminal hydroxyl groups of the polyol with the more reactive 4-position isocyanate group, and a diisocyanato diphenyl-urea from the TDI/water reaction. During the early reaction and the start of the foam rise – that is until the reaction mixture reaches about 80°C – the main reactions are with the isocyanate group of TDI in the para position. There is little chain extension and no detectable formation of biuret or allophanate linkages.

All low density foams tend to be anisotropic. The small nucleation bubbles are spherical but, as they expand under the increasing pressure resulting from the generation of carbon dioxide, they tend to elongate in the direction of foam rise. The fewer and larger the bubbles, the greater will be their elongation. In low density foam formation, the volume of the expanding gas bubbles quickly becomes much larger than the volume of the polymerising liquid reaction mixture and the area of contact between adjacent bubbles becomes a membrane. The membrane is planar if the bubbles are of equal size and the line of contact between three adjacent membranes forms a rib that is roughly triangular in section. The gas pressure within the expanding foam remains low until the reaction temperature rises above about 90°C when the hindered ortho-position isocyanate groups begin to react at a significant rate. The pressure within the foam increases as gelation proceeds until the thinning wall-membranes of the cells rupture and the blowing gas is released. The production of low density flexible foams with a high permeability to air depends to a large extent on the rate at which the cell wall membranes thin and on the amount of polymer from the

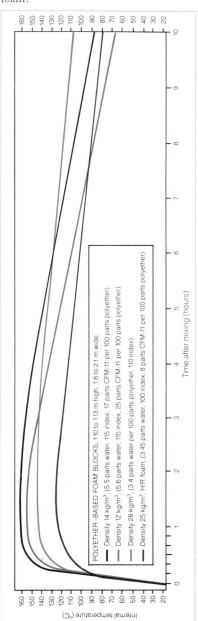

Figure 4-4 The internal temperatures recorded in large blocks of flexible foam.

POLYETHER -BASED FOAM BLOCKS. 1.10 to 1.13 m high, 1.8 to 2.1 m wide.

| Density 14 kg/m³, (5.5 parts water, 115 index, 17 parts CFM-11 per 100 parts polyether)
| Density 12 kg/m³, (5.6 parts water, 115 index, 25 parts CFM-11 per 100 parts polyether)
| Density 28 kg/m³, (3.4 parts water per 100 parts polyether, 110 index)
| Density 25 kg/m³, H/R foam, (3.45 parts water, 100 index, 8 parts CFM-11 per 100 parts polyether)

membranes that becomes incorporated into the fibrillar struts or ribs at the joint between adjacent membranes. This depends largely on the rate of polymerisation compared with the rate of blowing. That, in turn, varies with the temperature inside the reacting foam as well as with the level and type of catalysis used. The foam structure is also affected by the presence of surface-active agents that dissipate local expansion stresses and help to prevent premature cell wall rupture. The cell membranes are broken by the increasing internal gas pressure, in combination with a rapid loss in the extensibility of the polymerising material. The release of water loosely bonded to the polyether chains is also thought to play some part in cell opening by providing water for reaction to give carbon dioxide at a late stage in the foam expansion. This effect is increased by increasing the proportion of polyoxyethylene blocks in the polyether. Most water-blown TDI-based low density flexible foam systems remain homogenous until the reaction temperature reaches about 80°C when molecules of substituted and polymeric ureas begin to form a separate phase within the polymer network. This separation, which also plays a part in cell opening by creating areas of stress concentration, represents the beginning of hard block separation in the polymer.

At the time of the cell opening, the polymer strength is low, with a significant proportion of the isocyanate groups remaining unreacted. The temperature at the moment of cell opening is thought to be a measure of cell stability and is related to the chemical composition of the foam system. The temperature continues to rise after the foam has fully expanded because of the continuing reaction of isocyanate groups with water, with amine end-groups (resulting from the reaction of bound isocyanate groups with water vapour) and also, depending on the temperature, with the reactive hydrogen atoms in the urea linkages. Some bound but reactive isocyanate groups remain, even when the foam has reached its maximum reaction temperature. This maximum occurs, in large foam blocks, between about 30 minutes and 1 hour after manufacture. It may remain near to this level for 1 to 8 hours, depending on the block size, the amount of excess isocyanate present, the ambient conditions and the orientation of the foam block, because orientation affects the rate and direction of flow of convected air through the permeable foam. The 'cure' of large foam blocks, i.e. the reactions of those bound isocyanate groups remaining from the initial polymerisation and the stabilisation of labile bonds, is virtually complete within 48 hours of manufacture. The cure of small foamed mouldings may require much longer and the mechanism is different. In moulded foam cushions the level of bound but unreacted isocyanate groups falls over several days and changes in the foam properties often continue for about a week after manufacture. MDI-based foams tend to reach equilibrium much faster than those based on TDI. The differing polymerisation and curing conditions are reflected in the differing physical properties of slabstock and moulded foams. Foams made in large blocks result in stiffer, higher load-bearing polymers with higher tensile strength and lower elongation at break compared with similar foams made on a small scale from the same chemical components.

Foam formation from polyester polyols and TDI differs in detail from that using polyether polyols because the polyesters used to make low density flexible foams have predominantly primary hydroxyl end-groups and have a much higher initial viscosity than polyethers. The choice of catalysts for making polyester foams is limited and the relatively high viscosity and reactivity of the system results in less flow of the cell membranes after rupture.

Flexible foam slabstock or block-foam

See also Chapter 13, part 2, page 333

There are two main types of flexible foam slabstock
- Polyester-based foams, technical grade and high-elongation grade for laminating to textiles.
- Polyether-based foams, standard upholstery grades, high-resilience grades and flame-retardant types.

Figure 4-5   Polyether-based flexible foam leaving the production machine.

See also Chapter 13, part 1, page 327

Continuous slabstock foam

Slabstock foam is usually made continuously by metering the foam reactants – polyol(s), isocyanate(s), water, catalysts, stabilisers and other additives such as auxiliary blowing agents, flame retardants, fillers and chain extenders – in the required proportions to a mixing head. Typical formulations and component ratios are tabulated in tables 4-2 (Polyether-based) and 4-3 (Polyester-based). The metered materials are mechanically mixed and nucleated and, in the widely-used traditional process, are immediately distributed on the bottom lining of a continuously moving trough formed by a horizontal bottom paper or foil and two vertical side papers or foils, carried on a long conveyor at a controlled and adjustable speed (figure 4-3).

The distribution of the liquid foam reaction mixture may be assisted by traversing the mixing head or its outlet nozzle across the moving

Table 4-2 Formulations for making standard grades of polyether-based flexible foam

Component	Typical formulations (Parts by weight)			Formulation range (Parts by weight)		
'Daltocel' polyether, e.g. 'Daltocel' F 4801	100	100	100	100		
80:20-TDI (as Isocyanate Index)[1]	108	110	108	80	to	110
Water	2.3	4.3	4.0	2.0	to	5.0[2]
Tertiary amine, e.g. Dimethylaminoethanol	0.4	0.25	0.4	0.1	to	0.5
DABCO	–	–	–	0.05	to	0.1
Surfactant (polysiloxane/ polyoxyalkylene copolymers)	0.9	0.07	1.5	0.4	to	1.5
Stannous octoate	0.25	0.25	0.3	0.2	to	0.4
CFM-11[3]	4.0	–	25.0	0	to	25
Pigments/Dyes	–	–	–	0.001	to	10
Flame retardants/Fillers	–	–	–	1.0	to	100
Foam density (kg/m^3).	33	23	15	15	to	50

[1] Isocyanate Index $= \dfrac{\text{Actual amount of isocyanate used}}{\text{Calculated stoichiometric amount}} \times 100$
of isocyanate required to react
with all the other components

[2] The use of too much water blowing can be very dangerous, leading to foam scorching or even to fire. The maximum safe level of water depends on the type of polyol, the use of auxiliary blowing agents, the scale of manufacture, the ambient conditions and other factors affecting the maximum reaction temperature (See Chapter 10).

[3] The alternative blowing agent, methylene chloride, is also widely used.

Figure 4-6   Side-wall control for flat-topped blocks.

trough. A uniform distribution of the liquid foam reaction mixture over the bottom of the trough does not, however, result in foam blocks of the most economical rectangular cross-section. The drag and the cooling effects of the side-walls on the rising foam results in a domed section. As the final user requires foam in sheets of uniform thickness, the domed top increases the loss of foam during the cutting or "conversion" process. Much process work over the past

Figure 4-7   The Draka/Petzetakis side-wall lift for flat-topped slabstock.

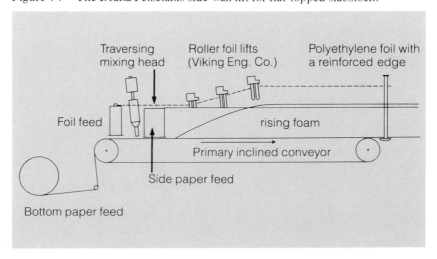

decade has been aimed at improving processes for the production of flat-topped, rectangular slabstock. There are three types of process:
- That in which the foam rise is restrained by a controlled top paper or foil (figures 4-3 and 4-8),
- A process that distributes the foam mixture as a froth onto a shaped base plate (figure 4-9) that allows the foam to expand downwards, and
- A recently introduced vertical foaming process, "Vertifoam".

An older process (figure 4-7) involved the raising of the side-wall foil linings in step with the expanding foam in order to compensate for side-wall drag. Side-wall lifting is usually practised today on 'Maxfoam' and 'Varimax' machines.
The principles of the flat-top block processes are illustrated diagrammatically in figures 4-7 to 4-11.

Figure 4-8   The 'Planibloc'/Hennecke floating top-paper for flat-topped slabstock

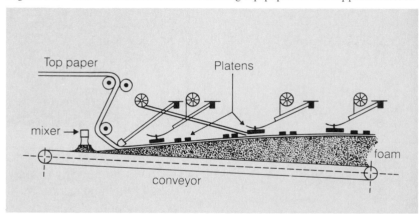

Table 4-3   **Typical formulation for making polyester-based flexible foam (slabstock)**

Component	'Technical' grade foams (parts by weight)			Laminating grade foams (parts by weight)			Formulation range		
'Daltorez' SF	100	100	100	100			100		
'Daltorez' RB4	–	–	–	–	100	100	100		
80:20-TDI (Index)[3]	–	–	105	90	90	100	85	to	100
65:35-TDI (Index)[3]	105	105	–	–	–	–	85	to	115
Water	5.0	3.0	3.7	3.6	3.6	3.8	2.0	to	5.2[1]
'Lubrol' SF2	1.0	1.0	1.0	1.0	1.0	1.0	0.7	to	1.0
N-Ethyl morpholine	–	–	2.5	1.3	1.5	2.0	1.0	to	2.5
N-Dimethlcycolohexylamine	0.8	0.6	–	0.05	0.1	–	0.01	to	0.9
N-Dimethylcetylamine	–	–	–	–	–	0.28	0.01	to	0.4
Polyurax Silicone[2] SE-232	–	0.7	–	0.9	0.8	1.0	0.6	to	1.0
Polyurax Silicone SE-236	0.8	–	1.0	–	–	–	0.6	to	1.0
Flame retardants	–	–	8.0	–	–	–	2.0	to	15
Pigments/dyes	0.05	0.05	0.05	–	0.01	0.05	0.001	to	5
Foam density (kg/m^3)	20	33	29	30	29	26.5	17	to	40

[1]   The maximum safe level of water depends on the formulation used, the scale of manufacture, the ambient conditions and other factors affecting the maximum reaction temperature (See Chapter 10).
[2]   Or equivalent materials, Niax Silicone surfactants.
[3]   The amount of TDI used is expressed as the 'TDI Index' (see page 62).

Figure 4-9 The 'Maxfoam'/'Varimax' process for flat-topped slabstock.

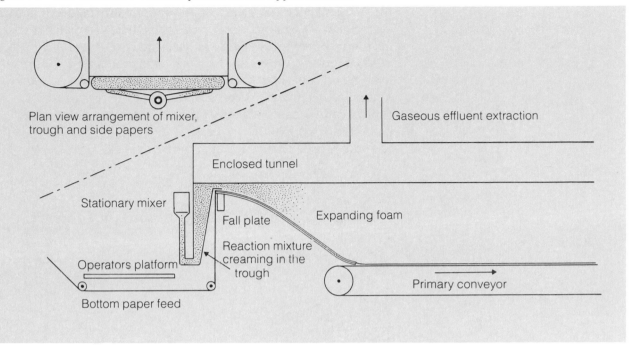

Plan view arrangement of mixer, trough and side papers

Gaseous effluent extraction

Enclosed tunnel

Stationary mixer

Fall plate

Expanding foam

Reaction mixture creaming in the trough

Operators platform

Primary conveyor

Bottom paper feed

Figure 4-10 Principle of the 'Vertifoam' process.

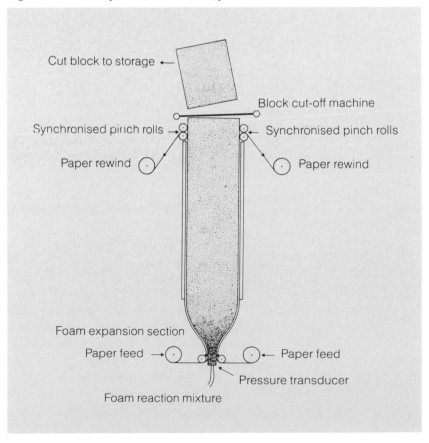

Cut block to storage

Block cut-off machine

Synchronised pinch rolls

Synchronised pinch rolls

Paper rewind

Paper rewind

Foam expansion section

Paper feed

Paper feed

Pressure transducer

Foam reaction mixture

The 'Maxfoam' and 'Varimax' processes may be operated only with foam systems that form a low viscosity, free-flowing froth in the early stages of foaming. By selecting polyethers of suitable reactivity in conjunction with selected catalysts and surfactants, the process yields the most economical slabstock foam produced by a horizontal foaming process. The 'Maxfoam'-type process, however, unlike the 'Planibloc'/'Hennecke' processes in which the rising foam is controlled by a cover of paper or foil, is unsuitable for making polyester-based foam.

Figure 4-11 Principle of round block machine.

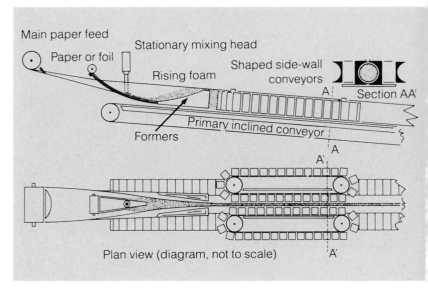

Both polyester-based and polyether-based low density flexible foams are also made continuously as cylindrical block. These blocks are converted to thin sheet foam or foil by a peeling process similar to that used in the manufacture of veneers for plywood. The detailed design of horizontal round-block foaming machines is the subject of several patents; the processes differ in the method of distributing the foam reaction mixture in the moving mould and in the means by which the rising foam is shaped into a cylinder. A typical process is shown diagrammatically in figure 4-11.

Most flexible foam slabstock is made continuously by one of the above horizontal foaming processes. A feature of continuous horizontal processes is the need to use machines with high outputs in order to obtain large blocks of foam with low wastage when the foam is converted to cushioning by cutting. High outputs require faster and longer conveyors and machines up to 100 metres long are common. The size of the foam production machine is thus dictated by the size of the foam blocks required and not by the rate of foam production needed to meet demand. Large machines are expensive to vent and heat. The 'Vertifoam' process, described in a recent European patent application, overcomes this problem.

In the 'Vertifoam' process, illustrated diagrammatically in figure 4-10 and 4-12, the foam reaction mixture is introduced at the bottom of a totally enclosed expansion chamber. This chamber is lined with a kraft paper and/or a polyethylene foil both of which are drawn upwards at a controlled rate, depending on the pressure in the chamber, the foam formulation and the rate of production. Under these conditions, the rheology of the foam reaction process, combined with the effect of gravity, ensures a stable uniform foaming front and prevents the mixing of the still liquid reacting

Figure 4-12   'Vertifoam' machine

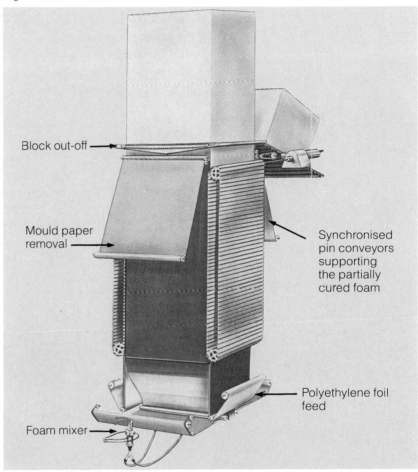

mixture with older partially gelled foam. In consequence, foam blocks with the required large cross-section may be made at relatively low outputs. The stability of the foaming front also facilitates fast colour and formulation changes. A change of foam colour shows an intermediate layer of foam that is only some 10 cm to 15 cm long. Because the machine is much smaller than horizontal machines making foam block of a similar size, there are major operational savings, especially in the rate of venting and space heating. Other advantages stem from the use of vertical foaming.

The foam blocks made in this way do not have the dense top and bottom skins produced by horizontal foaming – more of the chemicals used are converted to usable foam and the foam produced is of uniform density and structure.

The 'Vertifoam' process may, in principle, be used to produce rectangular section or cylindrical blocks of foam. It should be valuable for making blocks for peeling into thin foam foils of uniform cell orientation. The process is particularly exciting in that it may offer a significant advance in reducing the rate and improving the control of TDI vapour emissions from flexible foam production.

Discontinuous or moulded slabstock foam

The manufacture of low density flexible foam slabstock by moulding individual blocks of foam is usually carried out by mixing a batch of foam reaction mixture and transferring it rapidly into a mould. The rising foam is formed into a flat-topped block by the use of a "floating" lid. The process requires that the measured components be thoroughly mixed and poured into the mould within a few seconds – before the polymerisation reaction has caused significant increase in the viscosity of the mix. It is also important to avoid any entrainment of air during pouring, as this would cause defects in the foam structure. The floating lid system and the viscous drag of the four sides of the mould tends to result in foam having a higher density than that made by the continuous processes. This can be compensated by formulation adjustment, depending on the foam system and the size of block required. A typical modern single block process is illustrated in figure 4-14.

Figure 4-13 Demoulding a 17 kg/m^3 foam block 10 minutes after foaming.

Figure 4-14   Single block moulding machine ('Blockmatic' PTI (Viking) Ltd) Mould positioned in vented tunnel before filling.

Handling freshly manufactured slabstock foam

An essential requirement in low density flexible foam slabstock manufacture is the provision of a separate, vented storage area where freshly made blocks of hot foam must be stored until cool. This area should be monitored and equipped with fire detection and/or sprinkler systems. (See Chapter 10 for detailed recommendations and references). The foam emerging from continuous foaming machines is usually cut into predetermined lengths, weighed, marked to indicate the grade of foam, the size and weight of the block, and the date of manufacture, and transferred to the hot foam store by automatic machinery.

Conversion (cutting and fabrication of foam blocks)

Some 12 to 24 hours after manufacture, when the blocks of foam have cooled and cured, they may be cut into the sizes required for the final product. Where necessary, the first stage is usually to trim the sides and top of each block, using vertical and horizontal band-knives, to produce a completely usable rectangular block having flat, uniform faces. The amount of scrap foam removed in this initial trimming varies with the size of the block and the type of production machine used to make the foam. The trimming loss varies inversely with the block size. For that reason, many blocks are made at widths up to 2.25 metres and heights of up to 1.25 metres. Above that size the variation in density within the block tends to outweigh the advantage of any further reduction in the trimming loss. The effects of machine design and process on the trimming loss also varies with the block size, the more economic flat-top processes showing increasing gains in comparison with smaller block sizes. For example, the trimming loss in producing a block of foam with a density of 22 kg/m^3 and having usable dimensions of $2 \times 1.5 \times 1$ metres, will vary from over 15% to less than 5% depending on the foam making process adopted. The highest trim loss is that from domed-top unrestrained foam making, and such loss is reduced by the flat-topping processes in the order 'Planibloc', 'Maxfoam'/ 'Varimax', 'Vertifoam', the latter giving the least loss.

The trimmed block of foam is cut into pieces by band-knives or high speed cutting wires. Wastage at this stage depends only on the geometry of the parts required. Waste is lowest for rectangular pieces such as mattresses and plane-surfaced cushions. Curved surfaces may be cut using hot wire or vibrating blade cutters but the commonest method today is the use of high speed wire cutters. These cut the foam by means of an endless band of thin high-tensile steel wire, usually between 1.0 and 1.5 mm diameter, a method that enables cutting speeds up to 5 metres/minute to be achieved over the full width of a foam block. The cutting contour may be controlled simply by the use of templates, or by using a photoelectronic tracing system. In the latter instance, drawings of the contour shape on white paper made using black lines 1 mm thick are followed by a photoelectric cell which controls the movement of the cutting wire (figure 4-15).

Figure 4-15 Electronically-controlled contour cutting machine.

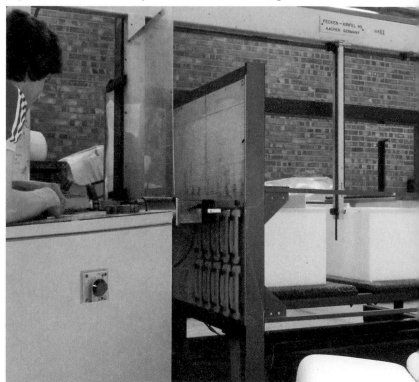

Curved shapes, including those with double curvature, may also be cut by using straight cutting blades to cut foam compressed into the desired shape. The method is usually applied when a regular three-dimensional pattern is required for foam sheets to be used for packaging or for sound absorption. The commonest method uses a modified skiving machine in which the foam sheet is compressed by patterned rollers as it is brought to the cutter. Any shape of foam cushion can by made from slabstock foam by cutting, or by a combination of cutting and fabrication of cut pieces with adhesives. Complex shapes, especially when required in large numbers, may be produced more economically by the use of moulded foam. Moulding is also capable of giving minimum variations in product size.

Aftertreatment of flexible foam

The properties of the foam may be modified by a number of methods. The most important are reticulation and impregnation.

Reticulation. This is an aftertreatment to remove residual cell membranes in order to obtain a foam with a skeletal rib structure.

70

Such reticulated foams are very efficient filters for the removal of dust and fibres from air and other gases. They allow high flow rates combined with low pressure gradients and, therefore, minimum energy consumption. Many methods of reticulation have been used, including chemical hydrolysis and the use of an explosion flame front to melt the membranes.

Impregnation. Impregnation of low density flexible foam is a process that includes the coloration of foam by dyeing and by coating with pigments in a binder or surface coating, the application of anti-fungal coatings, water resistant coatings and ion-exchange resins for special applications, but the most important treatment is impregnation with flame-retardants, especially with aluminium hydroxide. Foams are impregnated with up to 300% by weight of aluminium hydroxide usually with a synthetic latex binder, to meet the highest flame resistance requirements in upholstery for public and institutional use.

Moulded low density flexible foam

See also Chapter 13, part 1, page 331

Nearly 20% of low density flexible polyurethane foam is produced as finished products by moulding in closed moulds. Moulded foam is used for vehicle seating and interior trim, including sound absorbing trim, in furniture upholstery and bedding and in packaging. Two types of process are used: the hot moulding process, which has been established for over twenty years; and the more recent cold-cure moulding processes. Hot moulded foam is made by the reaction of TDI with polyether triols similar to those used in slabstock manufacture. External heat is applied to the mould after filling in order to obtain sufficient surface cure of the foam moulding to allow its early release from the mould. Cold-cure moulded foams are based on polyether triols with equivalent weights in the range from about 1,500 to 2,000 and isocyanate mixtures with an average functionality greater than two. Suitable isocyanates include special MDI compositions, mixtures of polymeric MDI and TDI, and TDI that has been modified to increase its effective functionality. The reactivity of the isocyanates in conjunction with suitable catalysis gives a sufficiently fast cure of the foam surface in contact with the mould, even at temperatures only slightly above room temperature. Wholly MDI-based foams give the fastest cure and mould temperatures are generally lower than those for TDI/MDI systems.

Hot moulding. Cast aluminium moulds with wall thicknesses of between 6 mm and 8 mm are used for hot moulding, or fabricated sheet metal moulds that are usually made from black steel sheet 1 mm to 2 mm thick. Moulds are usually made in two sections with provision for mechanical opening and closing of the lid. The choice of mould construction will depend on the required life of the mould,

71

on the number of similar moulds required, and on the dimensional tolerances of the moulded product. Aluminium, having over four times the heat conductivity of steel, is preferred, as it allows the construction of a stiff mould giving the minimum heating cycle and hence the minimum energy consumption. Cast aluminium moulds are also preferable to thin sheet metal moulds because they resist distortion during heat cycling and give a more uniform inner surface temperature during the heat curing stage. Moulds are usually designed to allow about 1% shrinkage of the moulded foam during manufacture but mouldings of complex shape require prototype mould trials to determine the shrinkage accurately and to find the optimum number, size and position of the vent holes that will give a product of the required high quality. It is best for the inner surface of the mould to have a coarse matt or 'scratch-brush' finish that gives good retention of the release agent and helps in obtaining moulded articles with the desired thin, highly-permeable skin.

The conventional hot-cure moulding process is based upon TDI. A predetermined amount of the foam reaction mixture is dispensed into moulds conditioned at a temperature of about 40°C. The closed mould is then passed through a curing oven at a temperature of 150° to 250°C, depending on the mould design and the production speed required, to raise the temperature of the inner surface of the mould quickly to about 120°C. This raises the temperature of the surface of the moulded foam to near the exothermic reaction temperature of the interior of the moulding and allows the cushion or other moulded article to be demoulded in 6 to 12 minutes. The hot mould then passes, in succession, to stations for cleaning, application of release agent, cooling, and back to the filling point (figure 4-16).

Figure 4-16   Hot-cure moulding

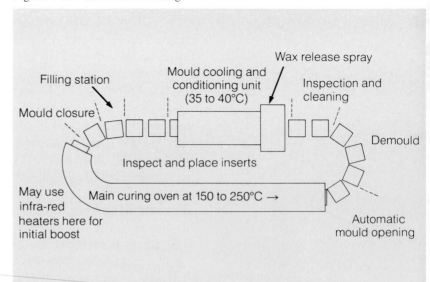

Hot cure moulding formulations are based upon specially developed polyether triols, and 80:20 TDI. The foam density and hardness depends basically on the level of water and isocyanate used for blowing, the TDI index, the shape of the mould and the degree of overpacking used, i.e. the amount of foam reaction mixture used compared with the minimum amount required to just fill the mould under the given operating conditions. The effect of water blowing on moulded foam density is shown in figure 4-17 together with a general formulation. All moulded foams show a density gradient from the core to the surface. The difference between the core density and the overall density of a moulded foam cushion is usually between 10% and 20%, but may be higher for complex shapes with a high surface-to-volume ratio. Soft foam mouldings for seat backs and pillows are usually made by reducing the isocyanate index and/or by utilising auxiliary blowing with CFM-11.

The heat cycling of the moulds from about 40°C to an outer surface temperature up to about 140°C is, however, expensive in terms of both energy and time.

In Chapter 13 CFM-11 is referred to as CFC-11

Figure 4-17   Hot-cure moulding: Formulations: Water content/Overall density

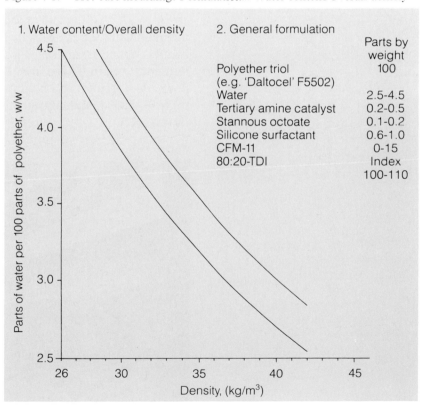

See also Chapter 13, Part 2, page 333

Cold-cure moulding. This allows moulds to be used at a relatively low and nearly constant temperature, uses much less energy and allows the use of fewer moulds and the use of alternative non-metallic mould constructions.

Cold-cure moulding systems are often specially formulated as two-component systems to meet specific applications required by foam users in the automotive and furniture industries. Typically, such two-component foam systems may be mixed and poured into moulds conditioned at 25 to 60°C, depending on the foam system adopted and on the conductivity of the material used for the mould. TDI-based systems, in general, require the highest mould temperatures whereas all MDI-based foams use mould temperatures from 25 to 45°C. The completed moulding may be demoulded a few minutes after pouring, without the need for application of external heat to complete the cure.

A wide range of machinery is used for cold-cure moulding. The simplest production facilities employ a two-component dispensing machine, with the shot weight controlled by an adjustable timer, and GRP moulds that are manually opened and closed. On the other hand, large numbers of seat cushions for vehicles are made on large oval tracks incorporating devices for automatic mould opening, filling and closing. Microprocessor control of the dispensing machine is usual on modern production lines. It gives more accurate shot weights than electrical sequence timers and is more easily and cheaply adapted to control the weight of foam dispensed into each individual mould. Requirements for smaller rates of production are met by the use of small carousels or by static moulds. The latter are temperature-controlled moulds into which foam reaction mixture is dispensed usually by a high pressure, self-cleaning, impingement mixing head. A high output cold-cure moulding line is illustrated in figure 4-18.

Figure 4-18   A typical high resilience foam moulding line

Moulds for cold-cure moulding differ from those required for hot-cure in several ways:

- Most cold-cure foam systems give the best moulding performance in 'overpacked' moulds, i.e. when using significantly more foam reaction mixture than the minimum amount required to just fill the mould. In consequence cold-cure moulds need to be designed to withstand up to 0.5 bar internal pressure. Venting of the mould and the sealing of the parting line is more critical than for hot-cure moulding.
- Most cold-cure foam systems give optimum moulding results when moulded under conditions that give a proportion of closed cells in the freshly-made foam. These closed cells are crushed soon after demould in order to avoid permanent deformation of the mouldings caused by cooling and by the diffusion of gas from the closed cells. The overall effect of the difference in the condition of the foam at the end of foam rise, combined with early crushing is that many cold-cure foam mouldings show slightly greater shrinkage from the mould and this must be allowed for in designing the mould.
- The lower operating temperatures and the greatly reduced range of the temperature cycle allow a much greater choice of materials for the construction of moulds for cold-curing foams. Non-metallic moulds are frequently used for furniture cushions and have proved very durable. Typical moulds are made from glass-fibre-reinforced epoxy resins, suitably braced to withstand the internal moulding pressure.

Chemicals for cold-cure moulding. The basic materials are polyether triols with a high ratio of primary to secondary hydroxyl groups and with a mean molecular weight from about 4,600 to 6,000. Reinforced polyols, i.e. polymer polyols, PIPA polyols and PHD polyols are also used. The isocyanate component may be a modified TDI, a modified MDI or an MDI/TDI mixture. The moulding cycle ranges from over 15 minutes for some cold-curing MDI/TDI based systems to less than 4 minutes for systems based on specially-developed MDI variants. Cold-cure moulding systems based upon MDI variants are becoming popular. They give short moulding cycles without high consumption of energy and the product is a foam with a soft permeable skin that handles like rubber latex foam. MDI-based foam systems require relatively low concentrations of reinforced polyol to produce high load-bearing foams. There are also environmental benefits resulting from the relatively low vapour pressure of modified MDI compared with that of TDI. One range of foam systems based on MDI variants is designed to make furniture cushioning with indentation hardnesses ranging from 90 N to 350 N and with core densities of between 35 to 70 kg/m^3. Systems are also available to meet the specifications for seating cushions drawn up by the main makers of vehicles. Some examples are tabulated in table 4-4.

Table 4-4 **Some MDI-based cold-cure moulding systems**

Formulation	Typical automotive system	Typical furniture system	
'Suprasec' VM 28 (pbw)	–	55.4	47.8
'Daltocel' C 971 (pbw)	–	100	–
'Suprasec' VM 25 (pbw)	63.3	–	–
'Daltocel' C 1565 (pbw)	100	–	–
'Daltocel' C 969 (pbw)	–	–	100
CFM-11 (pbw)	0 to 10	8	–
Isocyanate index	100	88	80
Processing conditions			
Polyol blend temperature (°C)	20 to 25	20 to 30	20 to 30
Isocyanate temperature (°C)	20 to 25	20 to 30	20 to 30
Mould temperature (°C)	40 to 45	ca 30	ca 30
Mould occupation time (minutes)	4 to 6	4 to 6	4 to 6
Post-cure	None	None	None
Foam properties			
Density (kg/m^3)	35 to 55	39	53
Compression set (50% compression 22 h at 70°C) - (% set)	7 to 10	4	7
Tensile strength (KPa)	145	60	80
Elongation at break (%)	140	110	105
Tear strength (N/m)	200 to 300	110	160
Indentation hardness at 40% (N) (ISO 2439)	150 to 300	165	200

MDI-based systems have many advantages over TDI-based systems for the mass production of cushions for vehicle seating. In addition to high resistance to wear and ageing, MDI-based systems give greater design freedom and economic production. Savings result from the short production cycle, the reduction in waste foam from the mould vents and the high rate of spontaneous cure of the freshly moulded cushions. The high rate of cure minimises the time required after demoulding before the cushion can be checked and despatched to the seat maker. MDI-based foam cushions can also be inserted into seat covers within a few hours of foaming, thus further reducing the necessary stock of uncovered cushions.

Another major benefit arising from the use of MDI-based flexible foam systems, is the facility with which the load bearing or indentation hardness of the foam may be adjusted over a wide range by changing the isocyanate index. All polyurethane reaction mixtures tend to give harder products when the amount of isocyanate available for reaction is increased. MDI-based flexible foam systems, however, are unusual because the reacting foam cures so quickly. Systems have been developed which allow large changes in the isocyanate index without other adjustments to the formulation

Figure 4-19   Effect of isocyanate index on indentation hardness of moulded cushions made from two MDI-based foam systems.

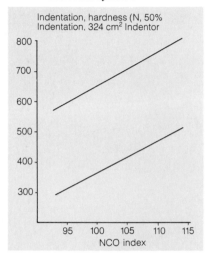

Figure 4-20   Demoulding a dual-hardness auto-seat cushion.

of the foam. The hardness of MDI-based flexible foam mouldings may therefore be adjusted over a wide range simply by changing the ratio of the isocyanate component to the polyol blend component using a relatively simple, two-component dispensing machine. In this way the hardness of the foam may be controlled accurately when moulding cushions of different sizes on the same track or the hardness of the foam may be changed from one position to another in the same mould. This allows the moulding of the *dual-hardness* cushions required for many automotive seats. These, for example, are front seat cushions with a comfortable centre and stiffer side bolsters to provide support against the lateral forces resulting from the high cornering speeds of modern vehicles. These cushions may be manufactured by expensive hand fabrication from cut slabstock foam or moulded using machines with three or more chemical component streams.

Dual-hardness cushions from MDI-based foam are also made using a two component metering machine with a single dispensing head. The foam reaction mixture is poured into an open mould in a pre-set pattern using an automatic manipulator. The isocyanate/polyol blend ratio is changed in a fraction of a second, at a pre-set position in the pour pattern. Some typical dual-hardness seat cushions made in this way are shown in figure 4-20. To show the harder side bolsters of these cushions, a pigment was injected into the foam reaction mixture during the dispensing of the high index foam.

Figure 4-21   A two component metering machine with a single dispensing head.

Because of the high rate of viscosity increase during the formation of MDI-based foams, the high- and low-index reaction mixtures do not mix together in the mould but form well-defined zones albeit without any weakness at the zone interfaces. The physical properties of typical dual-hardness seat cushions moulded from MDI-based foam are shown in table 4-5.

Figure 4-22   Comfortable seating in cars.

See also Chapter 13, part 2, page 333

Composite moulding processes

Mechanised seating is the term used to describe the manufacture of composite seats by forming the seat cover and the cushion together in one operation. Many processes have been developed, using both hot- and cold-cure moulding systems, but none have yet found favour for the mass-production of seats. Because relatively low surface temperatures occur during the cold-cure moulding process it is easy to combine fabrics, and other materials that are sensitive to high temperatures, with the foam cushion during a single moulding operation. MDI-based cold-cure foam systems, because of their rapid viscosity build-up and cure, are particularly suitable for use with textile covers. In the best-known process, the foam cushion is moulded directly into the upholstery cover. Several variations of this process are available.

Other successful systems include the manufacture of heavy duty seating for commercial and agricultural vehicles, motor cycle and outdoor furniture seats, using slush-moulded PVC plastisol covers filled with hot- or cold-moulded foam, and similar products using vacuum-formed PVC skins. Many of these products are unsuitable for domestic furniture and passenger car seating which require the appearance and comfort given by textile covering materials. Several systems are available using stretch fabrics as the top surface

of a multi-layer laminate. A typical laminate made up of three or
four layers consists of:
– a stretch fabric
– peeled flexible foam 2 to 3 mm thick
– a foil of PVC, PU or other thermoplastic, and
– an optional inner layer of peeled flexible foam about 1 mm thick.

In the 'Skinform' process such a laminate, held under controlled
tension which is varied with the cushion design, is drawn by
vacuum into specially designed moulds. When the laminate cover
has taken the shape of the mould it is filled with a cold-cure foam
reaction mixture, and the mould lid is closed until the foam has been
formed and cured sufficiently to be released from the lid. The
moulded seat cushion and the formed cover are thus bonded
together in the foam moulding process. Alternative processes such
as the 'Controform' process use laminates made up of four layers:
– PVC
– textile fabric or PVC
– foam layer
– textile fabric,

which are first shaped by hot vacuum-forming to make a shaped
cover which is then placed in a mould for foaming and filled with
cold-cure foam by foaming in place.
There are many other applications of cold-cure moulding, especially
using MDI, in processes to make composite products. A cold-cure
foam system developed by ICI is moulded into large, fitted, sound
insulation mats for use in the automobile industry. These foam-
backed mats absorb air-transmitted noise in addition to damping
bodywork vibrations (figure 4-23).

Figure 4-23   Cold-cure foam: sound insulation mats.

Controlling the properties of low density flexible foams

The properties of a flexible foam depend both upon the properties of the polyurethane/polyurea/polyether (or polyester) segmented polymer resulting from the foam reaction, and upon the geometric arrangement of that polymer to form a foam structure. The foam properties are controlled by the choice of chemical raw materials and the formulation; by the choice of processing machinery and processing conditions; and by the ambient conditions both during and after manufacture. The measurement of the physical properties of a foam is both a means of quality control during manufacture and a method of defining or describing the product in a manner that gives the skilled user an indication of its performance in a particular application. Physical tests must therefore be carried out using agreed standard methods. The recommended test methods are those of the International Organisation for Standardisation (ISO). These tests are agreed modifications of local standard test procedures, especially those developed by the American Society for Testing Materials (ASTM), Deutsche Institut für Normung (DIN), Normes Francais (NF), and the British Standards Institution (BSI). (The principal ISO test methods for flexible foam are listed in table 10-10, Chapter 10.) The most important physical properties of flexible foam used for upholstery are its hardness, i.e. its ability to support a load, and its durability. Durability is measured by repeatedly applying a load in a standard manner. Durability is related broadly to the foam density, as one might expect. For many applications, upholstery must also pass specific flammability tests. These are described in Chapter 11.

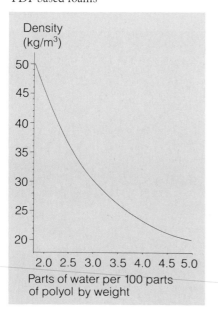

Figure 4-24   Slabstock foams: typical water/density relationship for TDI-based foams

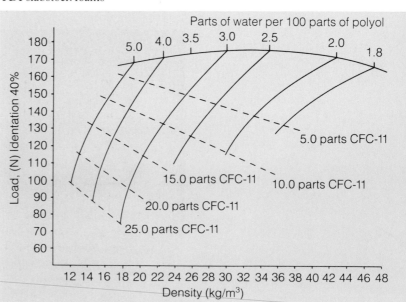

Figure 4-25   Effect of auxiliary blowing on the density and hardness of Polyether/TDI slabstock foams

Density control

The density range of most slabstock foam is from about 15 to 40 kg/m^3 and that of moulded foam from about 25 to 55 kg/m^3. Densities down to about 12 kg/m^3 and up to about 100 kg/m^3 are also made. The density of the foam is primarily controlled by the level of blowing agent, water or fluorocarbon, used in the formulation. Figure 4-24 illustrates the effect of varying the level of water blowing on the density of TDI-based slabstock foam. Blowing the foam in this way with carbon dioxide from the reaction of diisocyanate with water, also increases the polyurea hard block content of the polymer network and therefore the modulus of the polymer. The increasing stiffness of the polymer that is produced as the foam density is reduced by increased water blowing is the reason that the static load-bearing properties of water-blown foams show little change as the density is reduced over the range from about 20 to 30 kg/m^3. Auxiliary blowing with CFM-11 or with methylene chloride, however, plays no direct part in modifying the polymer structure and produces a marked reduction in foam hardness as the density is reduced. The effect of auxiliary blowing with CFM-11 is illustrated in figure 4-25 which shows the range of indentation load bearing and density normally covered by conventional polyether/80:20-TDI based slabstock flexible foams.

Although the foam density primarily depends on the level of blowing agent(s) used in its manufacture, many other chemical and process variables also affect the efficiency of blowing and the density of the foam. The effect of variations in atmospheric pressure is usually ignored in lowland factories but when foam is made at high altitudes the reduced atmospheric pressure will have a corresponding effect upon the density of the foam. Variables affecting the rate of gelation or the balance between gelation and blowing, will also affect the final density of the foam. Such factors include variations in the isocyanate index, variations in the 2,4-/2,6-isomer ratio of TDI and other isocyanate mixture variations, the initial temperature of the reactants, the catalyst type and level, the reactivity of the polyol and the scale of manufacture, all of which will affect the density of the foam. Changing the mixing conditions, i.e. changing the size of the mixer, the mixer speed, the dimensions of the outlet nozzle and other geometrical changes, will all have an effect upon the degree of mixing and nucleation of the foam reaction mixture. If other factors are constant, increasing the cell count, i.e. decreasing the cell size, causes earlier cell opening and earlier loss of the blowing gas giving a higher foam density.

The density of a block of foam is not uniform. The difference between the overall density of a block of foam and that of its core tends to decrease with increasing block size, largely because of the lower surface-to-volume ratio of large blocks. The use of high output machines to make large blocks of foam is dictated by the need to obtain the maximum yield of cut foam from the blocks, but the maximum useful block height is limited by the variation in the density and hardness of the foam from the top to the bottom of the block. The foam density is greatest at the base and falls almost linearly to a minimum about 15 to 20 cm below the top surface.

Hardness measurement and control

The standard test methods are described and discussed in Chapter 11. The load bearing properties of flexible foam for use in furniture upholstery are characterised by the indentation hardness or the indentation hardness index. These are measures of the force required to indent the foam using a circular indentor on a standard cushion, $(380 \times 380 \times 50 \, \text{mm})$, which is bigger than the indentor. Force values are quoted at 25%, 40%, and 65% indentation in order to characterise the foam. The 65% value for seating cushions is usually comparable with the force exerted by a seated adult. The hardness of small samples of foam is usually characterised by the relationship between compressive strain and stress under standard conditions. Foam samples are compressed at a standard rate between plates which are significantly larger than the foam sample. The results are quoted as the stress at 25%, 40% and 65% strain. Compression and indentation hardness measurements are sometimes confused because of the similar graphical or tabular presentation. They are different in principle and it is not possible to convert compression hardness characteristics to indentation hardness characteristics because the relationship between the two is not constant. It varies with the tensile properties of the foam because tensile forces contribute much more to the indentation load than to the compression stress. The principles of both methods are illustrated in Chapter 11 (figures 11-5 and 11-6). Physical testing must always be done under standard conditions of temperature and humidity on foam specimens which have been conditioned in a controlled environment. The stiffness of the polymer decreases with increasing temperature and with increasing humidity. The rate of indentation or compression must also be constant and reproducible. Flexible polyurethanes, in common with other organic polymers, show visco-elasticity, and in both tension and compression they exhibit stress-relaxation with time. This is attributable to the ability of the stressed polymer chains to slip past one another. When the stress is relaxed, the polymer sample has reduced stiffness. Indentation/load and compression stress/strain curves are usually drawn for complete cycles of loading and unloading. Successive stress/strain curves (figure 4-26) give a measure of the hysteresis loss i.e. the energy absorbed by the foam at the rate of working employed. The greater the area between the loading and unloading stress/strain curves the higher is the hysteresis loss and the lower is the elasticity of the foam.

The effects of varying the isocyanate composition are complex. The stiffest commercially available low density foams are made using

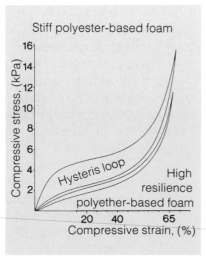

Figure 4-26   Hysteresis loops for polyether- and polyester-based foams

65:35 TDI; the use of the commoner 80:20 isomer ratio TDI yields slightly softer foams of higher resilience. The addition of a proportion of modified TDI or polymeric MDI to the TDI reduces the hardness of the foam by interfering with the alignment of the polyurea hard segments and the effective degree of hard and soft segment segregation. An important consequence is the lower melting point of high resilience foams. The melting point of the polymer falls by between 20° and 40°C i.e. significantly below the polymer decomposition point. This makes the ignition of the foam by a small flame more difficult, because the foam polymer will tend to melt and retreat from the flame rather than be decomposed into volatile, combustible materials.

The commonest method of adjusting the hardness of low density flexible foam is by adjusting the isocyanate index. This adjusts the number of bound isocyanate groups available for cross-linking at the curing stage. This method is often used in slabstock manufacture to compensate for seasonal changes in the foam hardness. A TDI index above about 108 is not, however, generally recommended for use in large scale manufacture of slabstock. The excess TDI increases the number of bound but unreacted isocyanate groups remaining at the end of foam rise. This increases the potential exothermic temperature rise during the curing stage and increases the hazard of fire in the foam block curing room (see Chapter 10 for a fuller discussion of the hazard).

The hardness of moulded flexible foam varying the isocyanate index may also be adjusted by varying the isocyanate index. The foam hardness may require adjustment from mould to mould to compensate for changes in the shape of particular cushions. Especially when using suitable MDI-based foam systems, auto seat cushions may be made with foam that is stiffer at the edges than at the centre by using automatic index changing of the dispensed foam reaction mixture during its patterned laydown in the mould.

Cell size. This also affects the hardness of flexible foams. The hardness increases with increased cell size. The effect is almost linear over the range of cell counts from about 10 to about 20 cells per centimetre for foams of uniform cell size. Fine-cell foams with a heterogeneous structure are slightly softer than those with uniform size cells and, more importantly, they have a more linear stress/strain curve. This is one reason for using excess air injection into the reaction mixture when making high resilience foams.

The position. The position in the block affects the hardness of a piece of foam. Large blocks of low density foam tend to have a higher hardness/density ratio in the core of the block where the reaction temperature is maintained for the longest time and there is limited access of atmospheric moisture.

Tensile properties

Process and chemical changes that are made in order to adjust the density and the load-bearing properties will affect all the physical properties of the foam. The tensile properties (elongation at break, tensile strength and tear strength) of most low density flexible polyurethane foams are high compared with the application requirements. Except for applications such as textile laminates and upholstery constructions using stretched foam, the measurement of elongation at break and tensile strength is done mainly for quality control purposes and also, because it varies sensitively as foam degradation occurs, in accelerated ageing tests.

Compression set

Flexible polyurethane foams have low glass transition temperatures, in the range from about −35 to about −55°C depending on the type of polyester or polyether polyol used, and they are visco-elastic at their normal service temperatures.

Thus they creep under long term stress. The rate of creep, usually measured as the rate of change of strain with time at constant stress, increases with increasing temperature and with increasing strain. Compression set tests are simple accelerated creep tests under constant strain. The standard test, 22 hours at 70°C under 50%, 75% or 90% compression, is a most useful quality control test because it is sensitive to variations in foam cure, but it often does not correlate with the creep in service at ambient temperatures.

Dynamic fatigue tests

These were developed to predict the actual performance of seating and bedding by applying an alternating load to an indentor. The cushion may be repeatedly indented to a fixed depth or repeatedly indented by a fixed load. The latter method more nearly simulates actual service because the indentor sinks deeper into the foam cushion as the foam softens because of creep, chemical changes and any mechanical breakdown. The standard test, with a circular indentor applying a force of 750 Newtons at 70 cycles/minute for 80,000 cycles, shows a high correlation with the height loss of seating cushions in service and an indication of the likely loss in foam hardness. More stringent tests involve testing under high temperature and/or high humidity when foam is to be used in tropical climates.

Physical test methods are described and discussed in more detail in Chapter 11.

5 High density flexible foams and microcellular elastomers

The products described in this chapter – semi-rigid foams, self-skinning foams and microcellular elastomers – are almost always manufactured as moulded products. The typical moulded density range for each type of foam is indicated in table 5-1.

These polyurethanes are used mainly in the automotive and footwear industries. Car parts such as crash padding, instrument panels and sound absorption mats are made from semi-rigid foam whilst integral skin foam is used for items such as steering wheels, headrests and suspension bump stops. Energy absorbing bumper mountings and auxiliary suspension units for vehicles are produced from microcellular elastomers.

The versatility of polyurethanes is illustrated by their many applications in footwear. These include microcellular and integral skin flexible, semi-rigid and rigid foams for shoe soling and for shoe platforms, injection moulded thermoplastic elastomers for heavy duty applications in studded football boots and other sports shoes, the heels and tips of ladies' shoes, and for complete ski boot shells. Low density flexible foams and laminates of foam and textiles form the basis of most house slippers.

Semi-rigid foams

These are mostly water-blown, open-cell foams with a higher stiffness or hardness and a lower resilience than flexible foams. However, unlike rigid foams, they recover – slowly but completely – from high levels of compression. Their slow recovery from compression results in high energy absorption when subjected to shock loading. Both the rate of recovery and the amount of energy absorbed can be adjusted by varying the chemical system. The foams are widely used for energy-absorbing padding in vehicles and in packaging. Their largest use is in the interior padding of vehicles, although they are also used as energy absorbers in exterior bumpers, crash helmets and protective sportswear.

The manufacture of interior trim padded with semi-rigid foam is a two-stage process. Coloured decorative skins are made first. Several processes are used: vacuum forming of sheet materials; rotational

Table 5-1 **Types of moulded high density flexible foams**

Type of foam	Moulded semi-rigid foams	Self-skinning foams		Microcellular mouldings
		For light duty	For heavy duty	
Property Typical moulded density range (kg/m^3) Overall Core Skin Cell structure	50 to 120 30 to 100 90% open cell	225 to 450 100 to 300 500 to 800 Core 90% open cells	500 to 700 400 to 650 800 to 1000 Core of closed cells	500 to 700 400 to 600 600 to 800 Closed cells
Additives Blowing agent (parts per 100 parts of polyol) CFM-11/Methylene chloride Water Cross-linking agent Chain-extender (1,4-Butane diol or ethylene glycol, etc)	1.5 to 4.5 2.0 to 10+	10 to 30 0.15 to 0.25 0 to 8 0 to 30	3 to 8 0.10 to 0.25 1.0 to 3.0 5 to 15	0 to 1.0 0.3 to 0.7 5 to 15
Applications	**Vehicle trim** 'Crash padding' Instrument panels Arm rest filling Post and screen pillar padding Roof lining pads Sound absorbing pads Seat-back pads Energy absorbing filling for bumpers **Packaging** Energy/absorbing pads	**Vehicle trim** Head rests Steering wheels Arm rests Knee rails Post and pillar pads Radio consoles Switch gear Control knobs **Furniture** Decorative padding Arm rests Metal cladding **Bicycle trim** Saddles.	**Vehicle parts** Suspension bump stops Puncture-proof tyres for light cycles and toys Boat fenders Shoe soling Moulded soles, both cast and injected. Bouncing balls	**Shoe soling** Cast and injected shoe soling Vehicle suspension bump-stops Boat-fenders Energy absorbing bumper (fender) mountings Tyres for trolleys

moulding of PVC plastisols, (slush moulding), and blow moulding of a variety of thermoplastics. In the second stage, the preformed skin is placed in a suitably shaped mould, often together with a support armature carrying the attachment devices required for assembly in the vehicle, and the semi-rigid foam reaction mixture is poured or injected onto the skin. The expanding foam reaction mixture fills the complex and often narrow cavity between the support armature and the pre-formed skin. Semi-rigid foam systems based on polyether polyols and polymeric MDI have been specially developed for moulding in this way. The expanding foam reaction mixture must flow easily between the skin and the armature and must produce a foam consisting largely of open cells, that adheres well to both the skin material and the insert material. Foams containing closed cells may shrink and cause the assembly to distort. Good adhesion is essential for a satisfactory service appearance over

Figure 5-1 Renault 25 dashboard of semi-rigid polyurethane foam in a thermoformed PVC skin.

the wide temperature range encountered during the life of a vehicle interior.

Moulds for foaming semi-rigid foam-filled items may be made from GRP reinforced with steel tubes or backing struts, but the most durable moulds are made from cast aluminium faced with epoxy resin. The mould lid may be treated with a release agent but, depending on the mould shape, the use of a thin polyethylene release foil is often more convenient and avoids the possible contamination of the cover by the foam. Mould lids have also been coated with a silicone rubber containing a silicone oil in order to reduce the time required for mould cleaning. Mouldings may be demoulded after 4 to 10 minutes depending on their size and complexity, the foam system and the mould construction and temperature.

Semi-rigid foams may be made from either polyester or polyether based polyols reacted with either TDI or MDI but the most popular and economic foam systems are based upon high molecular weight polyether polyols and polymeric MDI. The general formulation is:

Table 5-2

Ethylene oxide tipped polyether triol, (Mol. wt. 4500 – 6000)	100	parts by weight
Water	1.5 to 4.5	parts by weight
Tertiary amine catalyst	0.1 to 1.0	parts by weight
Cross-linking/curing agent	2.0 to 10.0	parts by weight
Surfactant	0 to 1.0	parts by weight
Silicone oil	0 to 0.5	parts by weight
Polymeric MDI	80 to 105	Index.

There are two advantages in the use of pre-formed skins filled with semi-rigid foam for energy-absorbing padding. The process allows a wide choice of skin materials with the freedom to use colour-matched, printed and embossed sheet materials – such as PVC, PVC/ABS or TPR – having a high lightfastness. The skin material also acts as a load spreading device to increase the deceleration of an impacting body. This may allow the use of a softer foam to meet a

Figure 5-2 Ford Sierra fascia/crash pad with MDI-based semi-rigid foam core.

Figure 5-3   Car bumper core of moulded, MDI-based semi-rigid foam.

specific low speed impact performance requirement or the use of a lighter backing plate or attachment armature. The latter is usually designed to be the main energy absorbing device in a high speed head impact and may be made of perforated steel sheet, compressed wood pulp fibre or injection-moulded thermoplastics. A typical weight-saving construction is that used in the fascia for the Ford Sierra which uses a compressed fibre backing armature, see figure 5-2. The combination of a flexible skin and an MDI-based semi-rigid foam is also used to make light-weight, but damage resistant and energy absorbing, exterior bumpers, see figure 5-3.

Integral skin foams

See also Chapter 13, part 1, page 332

These are made by using physical blowing instead of water blowing and by overpacking the mould. Integral skin or self-skinning foam parts can be made from flexible, semi-rigid or rigid foam, and depending on the density, with an open or closed cell foam core (Self-skinning rigid or structural foams are discussed further in Chapter 7).

The combination of a cellular core and a skin having a much higher density results from the high temperature gradient across the reacting material adjacent to the relatively cool mould surface, together with the use of sufficient reaction mixture to give some excess pressure in the closed mould. The cooling of the surface layer, under a pressure greater than atmospheric pressure, ensures that the liquid blowing agent present in the skin remains liquid until the polymerisation is substantially complete. The process requires fast reacting exothermic foam systems and low levels of chemical blowing or of air entrainment.

Lower density self-skinning foams with an open cell core are used to make automotive interior trim – steering wheels, arm-rests, headrests, gear lever knobs, complete instrument panels, and many other interior and exterior items such as mirror surrounds and wheel trims. They are also used in furniture, especially office furniture; for application in toys; and for bicycle and motorcycle seats. The higher density self-skinning foams have a microcellular, closed-cell foamed core and are used for items such as heavy duty shoe soles, bump stops and protective fenders. The properties of the moulded articles are matched to their application requirements by selection of the basic polyol, the chain extender or cross-linking agent and especially of the functionality of the isocyanate type, together with control of the density of the foam and of the integral skin. The isocyanates used for integral-skin foam manufacture are usually special polymeric MDI compositions or modified liquid forms of pure MDI. Some typical formulations and product properties are listed in table 5-3.

Table 5-3 Self-skinning foams: typical formulations and physical properties

Formulation	parts by weight			
'Daltocel' C972	100	100	100	100
'Suprasec' VM 30	–	44	49	54
'Suprasec' DND (or DNR)	45.3	–	–	–
CFM-11	18	18	18	5
'Dabco' 33 LV	–	–	–	0.4
Isocyanate index.	100	90	100	110

Processing parameters				
Process type (pressure)	Low	Low	Low	High
Chemicals processing temperature (°C)	20 to 25	20 to 25	20 to 25	20 to 25
Mould temperature (°C)	30 to 35	35 to 40	35 to 40	40 to 45
Mould occupation time (minutes)	3 to 4	3 to 5	3 to 5	2

Physical properties				
Overall density (kg/m^3)	380	380	380	850 to 950
Skin density (kg/m^3)	900	900	900	1,100
Skin thickness (mm)	1.5 to 2.0	1.5 to 2.0	1.5 to 2.0	1.0 to 1.5
Surface indentation hardness (Degrees, Shore A)	50 to 55	25 to 30	60 to 65	88 to 90
Skin properties (measured on a 4 mm thick section of skin and core)				
Average density (kg/m^3)	600 to 750	600 to 750	600 to 750	1,000
Tensile strength (ASTM D412, MN/m^2)	1.8 to 2.3	1.4 to 1.9	2.5 to 3.0	5.5 to 7.0
Elongation at break (ASTM D412, %)	55 to 70	105 to 120	70 to 90	70 to 85
Trouser tear strength (DIN 53507, kN/m)	0.7 to 0.9	1.2 to 1.5	1.7 to 2.2	5 to 8

Integral skin foams in the automotive industry

Self-skinning foams for vehicle trim are based upon polyether polyols with equivalent weights up to about 2,000, low molecular weight chain extenders and MDI variants or polymeric MDI compositions. They are usually formulated as two-component systems but sometimes the blowing agent is separated as a third stream to permit adjustment of the density of different products on the same track. Fully-formulated systems are available for the production of moulded items with core density from about 100 to 500 kg/m³, and with overall densities, depending on the shape and size of the moulding, from about 200 to 800 kg/m³. A wide range of MDI variants allows the mechanical properties required for each application to be obtained at the minimum cost. Mouldings such as steering wheels and arm-rests require a tough, abrasion-resistant, flexible skin. This is obtained by the use of a low functionality, modified MDI (such as 'Suprasec' VM10, VM021 or VM30), depending on the detailed specification.

Decorative trim such as fascia trim, instrument panels and steering wheel centre pads, may often be made at a lower cost using other special grades of polymeric MDI. Systems manufacturers can usually advise on the optimum system for each application or suggest how to meet the individual specification set by some of the major car makers.

Figure 5-4   Colour-matched steering wheel in self-skinning foam based on MDI.

The moulds for self-skinning foam vehicle trim are usually made directly from hand-made patterns. The reproduction of fine surface detail is obtained by the use of metal moulds with electroformed copper/nickel shell linings. These linings are made by the deposition of copper, followed by nickel, directly onto the surface of the pattern. The shell so formed is then reinforced and mounted in a steel mould. Alternative mould constructions include sprayed metal linings, which give accurate design reproduction with limited mould durability, and engraved steel moulds. Elastomeric moulds, made from cast polyurethane or cast silicone elastomers, are also used to make integral-skin products. They are easily and quickly made and give easy demoulding of undercut parts, but they have a limited life of up to 200 mouldings depending on the type and the cure rate of the integral-skin system being moulded. For the mass-production of auto trim it is usual to avoid designs with severe undercuts in order to allow the use of durable metal moulds. Metal moulds have the further advantage of allowing accurate temperature control of the mould. This is essential for the uniform and reproducible production of the integral skin. Mould temperatures in the range from 35 to 55°C are usual. As all the foam in contact with the mould surface produces a skin, low density, flexible integral-skin mouldings usually need to be punctured on removal from the mould to allow the open cell foam core to breathe. Otherwise, the reduction in the internal pressure as the gas in the cells cools to room temperature may cause the moulding to distort. Self-skinning parts always show 2% to 3% shrinkage from the dimensions of the mould. When dimensional accuracy is important, the shrinkage should be

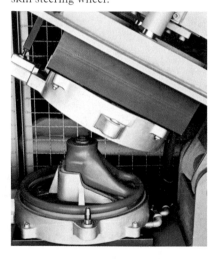

Figure 5-5 Demoulding an integral-skin steering wheel.

Figure 5-6 Cold-cure polyurethane bonded air filter.

ascertained using a prototype mould so that the production moulds can be adjusted to compensate for any asymmetrical shrinkage. Self-skinning foams, in common with all polyurethanes based upon aromatic diisocyanates, tend to discolour in sunlight. Mouldings are often coloured with carbon black or other dark-coloured pigments to minimise the effect of sunlight. Light-coloured trim has, for many years, been obtained by painting, either directly or by in-mould transfer coating. Painting is expensive because it adds extra steps to the production line and in-mould transfer coating increases the mould-cycle time and the minimum number of moulds required. Transfer coating also requires high quality metal moulds and may restrict the design of mouldings to allow a uniform spray coating of the interior of the mould. For these reasons, constructions made up from slush moulded PVC skins filled with polyurethane foam are sometimes still preferred for interior auto trim, especially as they allow precise colour matching of PVC-coated upholstery fabrics. The high energy usage of slush moulding processes, combined with the trend away from PVC-coated fabrics for car seating, has encouraged the development of light-resistant self-skinning foams. Light-coloured, self-skinning polyurethane products with a very high light fastness for special applications may be made with available systems based on aliphatic diisocyanates. These isocyanates are less reactive and more expensive than aromatic diisocyanates such as MDI. Satisfactory self-skinning foam systems have, however, been developed to reduce mould occupation times to about two minutes. This type of system has been in use since 1980 to make coloured steering wheels, especially for the American market.

The first low-cost system for coloured integral skin parts was introduced by ICI in 1978. This system was based upon selected MDI variants and polyol systems containing a synergistic mixture of agents to give protection against UV light. These products gave a satisfactory lightfastness (5 to 7 on the blue scale) for car interior trim in many of the colours required. They allowed the manufacture of colour matched trim at the same rate as that of black pigmented trim with little increase in cost. Pigment dispersions containing matched UV light protective agents are now available for use in self-skinning foams.

Self-skinning foams are also used in exterior fittings and body panels for vehicles and also under the bonnet. Applications under the bonnet utilise the wide operating temperature range and the resistance to petrol and oil of polyurethanes. Almost all European cars use engine air intake filters that are held together by moulded, MDI-based, cold-cure, self-skinning or high density foam that also forms an air-tight gasket when the filter is positioned on the vehicle. Polyurethanes have replaced PVC plastisols which required oven curing. The result is a saving in energy and an improved factory

91

Figure 5-7 In-mould painted, self-skinning foam sunroof surround.

environment during the filter-making process. The polyurethane seal also has better resistance to petrol and oil. Another under-bonnet application is the use of self-skinning, semi-rigid foam to insulate the fuel system and reduce the risk of the fuel vaporising at the high temperatures found under the bonnets of high performance cars. Self-skinning foams are also used to make external mirror surrounds because they simplify the design requirements needed to meet international safety standards.

The combination of a high density skin and an open-cell core makes an efficient sound absorption system. This property is used in the application of 2 to 3 cm thick, semi-rigid, self-skinning mouldings for engine house covers in forward drive vehicles and boats.

The resilience and durability of a low density, self-skinning flexible foam is used in the large mouldings which form the surround and sealing medium for the glass sunroofs fitted to many popular cars. These self-skinning foams are pigmented and are also painted in a matching colour. The painting is done by in-mould transfer coating from the mould surface. This system allows the removal of thin mould flash without further finishing. If required the coated parts are easily finished to match the car body colour.

Integral skin foams in footwear

The production of polyurethane microcellular elastomers and polyurethane self-skinning foams for shoe-soling is now at a rate of about 150,000 tonnes/year and represents about 5% of the shoe soling materials used worldwide, and over 10% of those used in Western Europe.

The advantages of using polyurethane materials in shoe-soling are their high abrasion resistance and durability in actual use combined with high flexibility, low weight, high thermal insulation and cushioning which together give high comfort.

Other useful properties of polyurethane shoe-soling include good grip on wet surfaces and resistance to oil, petrol and many common solvents. Most polyurethane shoe-soling is made from specially developed, fully compounded chemical systems. A wide range of systems is available to make polyurethanes suitable for the whole range of footwear from industrial and safety boots to fashion shoes, sports shoes and sandals.

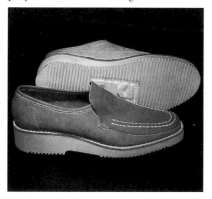

Figure 5-8 An example of moulded polyurethane shoe-soling.

From the point of view of the shoe makers, polyurethanes offer low capital and labour costs per unit, with a high production rate per mould and a low energy demand. Shoe soles and shoe platforms may be moulded by pouring the polyurethane reaction mixture into an open mould or by injection into a closed mould. Moulded shoe-sole units are often made for subsequent attachment to the shoe uppers by adhesive or by sewing, but the direct moulding of the sole onto the shoe uppers is becoming more popular. In this method of manufacture the top half of the mould is replaced by a last carrying the shoe upper and inner sole. The reacting polyurethane elastomer adheres strongly to the shoe upper without the use of adhesives or adhesive primers. The direct moulding of a cushioning mid-sole between the shoe upper and a high density, wear resistant, outer sole is also becoming very widely used for example, for tennis shoes. A thin, almost solid, outer sole is first cast or injected in the base of the mould. About 2 minutes later the mould is opened, the upper half of the mould replaced by a last carrying the shoe upper, (figure 5-9), and a foam reaction mixture is then injected or cast to form a lower density, softer microcellular foam between the outer sole and the shoe upper and in-sole. This provides cushioning for comfort, good adhesion between the shoe upper and the sole, economy in production and great design freedom.

Microcellular polyurethanes now find extensive use in shoes for leisure wear, for fashion and for work.

Figure 5-9   Two-colour/density shoe-soling process.

The sequence of twin-colour casting in the mould:

1. First colour casting.
2. Second colour casting.
3. Shoe upper brought into contact with the cast foam.
4. Demoulding of the finished shoe.

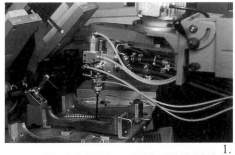

1.

2.

3.

4.

Figure 5-10   PFI (Pirmasens) Abrasion test machine with heated abrasive disc.

Polyester-based polyurethane shoe-soling. Systems based on polyester polyols are now the polyurethanes most widely used in Europe for this application. Polyester-based microcellular polyurethanes have higher abrasion resistance, tear strength and elongation at break compared with polyether-based polyurethanes. Because some laboratory abrasion tests tend to exaggerate the superior performance of polyurethanes relative to traditional shoe-soling materials, special abrasion tests have been developed by the Prüf- und Forschungsinstitut für die Schuhherstellung, (PFI), Pirmasens, West Germany, and by the Shoe and Allied Trades Research Association, (SATRA), Shawbury, England. These

Figure 5-11   Abrasion loss of shoe-soling materials by the PFI test.

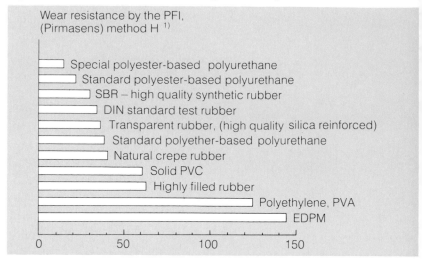

Wear resistance by the PFI,
(Pirmasens) method H [1]

Special polyester-based polyurethane
Standard polyester-based polyurethane
SBR – high quality synthetic rubber
DIN standard test rubber
Transparent rubber, (high quality silica reinforced)
Standard polyether-based polyurethane
Natural crepe rubber
Solid PVC
Highly filled rubber
Polyethylene, PVA
EDPM

0   50   100   150

[1] Prüf und Forschungsinstitut für die Schuhherstellung. (PFI), Pirmasens, West Germany.

special abrasion tests simulate the abrasion and shearing action of normal walking on a hard surface and give good correlation with actual user walking trials. Figure 5-10 shows the PFI apparatus. The application of the PFI abrasion test to a range of polymeric shoe-soles gives a realistic measure of the wear resistance of polyurethanes, (figure 5-11).

The PFI abrasion test machine has recently been modified to permit the testing of materials at elevated temperatures. This is important in evaluating materials for use in some types of sports shoes, such as tennis shoes and squash shoes, where transient high surface temperatures may arise from friction between the sole of the shoe and the surface of the games court. The abrasive disc which is slowly rotated against the rocking test specimen (figure 5-10) is electrically heated to a pre-set temperature. The abrasion loss increases with temperature (figure 5-13) but the effect may be reduced and the shoe performance enhanced, by the development of modified formulations.

A very wide range of two-component polyurethane systems is available. In addition to general purpose systems there are systems formulated to meet the requirements of specialised applications such as football, tennis and safety shoes, (table 5-4). Additionally, many systems are available that have been developed or modified for use with particular dispensing and mixing equipment. High quality polyester-based polyurethane shoe-soling is made from two-component systems consisting of an isocyanate-terminated polyester prepolymer and a second component which is a mixture of one or more polyester polyols and chain extenders, catalysts, surfactants, blowing agents, pigments and stabilisers.

The polyester prepolymers are often waxy solids, with melting

Figure 5-12   A sports shoe with polyurethane sole.

Figure 5-13   Effect of temperature on abrasion loss by the PFI test.

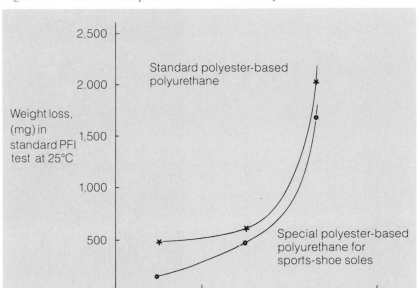

Table 5-4 Systems for shoe-soling: typical properties

System	Polyether based	Polyester based
Chemicals processing temperature (°C)	20 to 25	30 to 35
Mould temperature (°C)	40 to 50	40 to 50
Mould occupation time (minutes)	4	3 to 4
Moulded density (kg/m^3)	500 to 600	400 to 600
Tensile strength (DIN 53504, MPa)	5.5 to 6.5	8 to 10
Elongation at break (DIN 53504, %)	380 to 410	450 to 550
Trouser tear strength (DIN 53507, kN/m)	12 to 14	15 to 18
Ross flex test (Cut growth in mm/1,000 cycles) at −5°C	0.001	–
at −18°	–	nil

points up to about 45°C, which require melting out in ventilated air ovens at temperatures up to 70°C. It is important to follow the manufacturer's recommendations for handling these reactive materials. The dispensing and mixing machines used with these systems are thermostatically controlled to maintain the chemical components in the liquid state. Polyester-based polyurethane systems are also available that are liquid or are easily melted at or near to normal room temperatures to yield relatively low viscosity liquids. These systems do not usually yield the highest levels of abrasion resistance and other physical properties but are satisfactory for some footwear markets.

Figure 5-14   Testing flex fatigue resistance (DIN 53507).

A high resistance to splitting or cracking after repeated bending is a critical requirement of shoe soling materials and this resistance to cracking must be maintained even in very cold climates. The best "flex-cracking" resistance at low temperatures results from a combination of a suitable, usually polyester-based, microcellular polyurethane in a design of shoe that avoids points of stress concentration during normal wear and flexing.

The moulded density of microcellular shoe soles will depend upon the design and the thickness of the soling as well as upon the filler and pigment content and the level of water used for blowing. Densities from about 400 kg/m^3 to about 800 kg/m^3 are used. The density across a microcellular polyurethane moulding is not uniform but tends to increase towards the surface of the moulding. Testing should, therefore, be done on complete moulded sections of shoe soles, or moulded test pieces, and not on slices from which the surface layers have been removed.

Polyether-based shoe-soling. These systems are based upon slightly branched polyether polyols and modified pure MDIs such as 'Suprasec' VM021 and similar low functionality MDI variants which have been specially developed to meet the needs of the shoe industry. With these systems, moulded flexible shoesoles may be produced with overall densities from about 400 kg/m^3 to over 1,000 kg/m^3.

Systems are also available to make rigid and semi-rigid shoe platforms in foam densities from about 250 kg/m³ to over 500 kg/m³. Preblended systems are thus available to meet a very wide range of requirements but the main applications of polyether-based shoe soling is for fashion shoes and lightweight summer shoes and sandals. Fully formulated systems are available for the manufacture of synthetic crepe and transparent coloured soles. As polyether based systems employ low viscosity liquid components they may, in principle, be processed on either low pressure or high pressure dispensing machines. The choice of machinery will depend upon the range of shot weights required and upon the reaction rate of the chemical system. High pressure dispensers have some limitations of minimum shot weight compared with the best low pressure shoe system dispensers but are more suitable for use with highly reactive systems.

Some polyether-based shoe soles have an integral skin structure with a relatively low density cellular core and a high density, abrasion resistant skin. Integral skin moulded soles are made in a similar way to microcellular soles but using physical blowing, with CFM-11 or a mixture of CFM-11 and methylene chloride, instead of chemical blowing with water. The skin thickness is controlled by the mould temperature and the degree of overpacking of the mould, i.e. by the pressure generated in the mould. The density of the skin and its flexibility may also be controlled by precise adjustment of the low level of water in the polyol blend to give a controlled microcellular structure to the skin layer. The adjustment of the water content of the polyol blend is a useful method of varying the surface hardness and the appearance of an integral skin part.

The abrasion resistance of an integral skin shoe sole is directly related to the density of the elastomer. Integral skins give good resistance to wear although it has been suggested that the long term resistance to flex cracking at low temperatures may be reduced because of increased stress concentration in the outer skin of the shoe sole. At ordinary temperatures, integral skin soles give excellent wear. The extension of the foam into cleats helps to spread the stress at the base of the cleats and the compressibility of the foam is believed to increase the durability of the skin by spreading the wear. Integral skin systems are also used to mould thin soles and as the wear surface for inflexible shoe platforms. Thin mouldings tend to be of a uniform high density with good flexibility and durability.

Machinery. There are many types of metering and dispensing machines in use for making microcellular and self-skinning parts but almost all polyurethane shoe sole production and direct soling onto the shoe upper is done using low pressure metering with small, high shear, mechanical mixing heads. A popular machine is that made by Desma-Werke GmbH in Germany. This uses a horizontal,

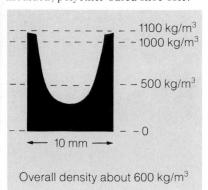

Figure 5-15   Typical density profile a moulded, polyether-based shoe-sole.

1100 kg/m³
1000 kg/m³
500 kg/m³
0
←— 10 mm —→

Overall density about 600 kg/m³

Figure 5-16   Reaction-injection moulding polyurethane shoe soles.

high speed, (22,000 rpm), reverse scroll mixer with a patented self-cleaning action. This mixer, originally developed by Semperit AG in Austria, is manufactured under licence from ICI Polyurethanes. The current Desma machines for polyurethane shoe soling have a variable speed metering pump drive with electronic speed display for accurate component ratio control and are available for two, three and four components to enable rapid colour changes to be made. Unlike earlier Desma machines, the components are recirculated to the mixing head and, by the use of two polyol blends, the density and properties of the moulded microcellular polyurethane may be varied from shot to shot. In this way two-colour shoe soling may be manufactured using a single dispensing machine.

The design of the mixer is particularly critical in the manufacture of small, low density, self-skinning parts because the production of a smooth skin of uniform colour, free from striations, requires the elimination of entrained, unevenly dispersed air bubbles.

Mechanical high shear agitators with minimum clearance in the mixer barrel, are operated at high speeds to expel the air from the mixer in front of the reaction mixture at the start of each dispensing shot.

High pressure metering of low viscosity systems in conjunction with hydraulic piston type self-cleaning, impingement mixers can give satisfactory results for all but the smaller shot weights. Uniform gas nucleation, essential to avoid streaks of light-coloured bubbles in the skin of the moulded part, is best obtained by using material feed tanks that are pressurised by nitrogen or dry air pressure, rather than the boost pumps and air injection systems that are satisfactory for moulding low density foam cushions. A typical Desma machine, used in many factories, is shown in figure 5-16.

"Hold-up" mixing machines are sometimes used for making both microcellular and self-skinning parts. In this type of machine, all the reaction mixture required for a moulding is held in a mixing vessel and mixed for a predetermined time before its discharge into the mould. "Hold-up" mixing avoids the problems of component metering surge and of variations in gas nucleation which, with machines which dispense continously from the beginning to the end of each shot, may cause defects in the moulding. "Hold-up" mixing is, in principle, a simple method of producing a homogeneous reaction mixture and a uniform moulded part. The main limitations are that any one size of "hold-up" mixer is only effective over a limited range of shot weights and that the method cannot be used for fast reacting foam systems.

The production of pairs of right and left polyurethane solings in a range of sizes requires up to about 20 moulds for each design. Shoe sole moulds are therefore usually carried on an indexing track or a carousel, (figure 2-12 and figure 5-16). As there is a wide spread in the number of mouldings required from each size of mould, it is also essential that moulds are easily removed from, or replaced on, the track. Moulds are usually made from cast aluminium or Kirksite and are provided with a means of controlling the mould temperature to within ± 5°C of the optimum. This is typically between 40 and 50°C.

Shoe soles may be produced by casting, i.e. by pouring the reaction mixture into an open mould and then closing the mould, or by injection into a closed mould. Two colour soles moulded directly onto a shoe upper are often made by a combination of casting the outer sole and injecting the mid-sole.

Quality standards and control testing.

In addition to the National and ISO test methods which are applicable to polyurethane shoes there are minimum quality standards recommended by trade associations such as SATRA and PFI. There are also test methods and standard performance requirements required to meet the RAL Quality Mark of the Gütegemeinschaft für Schuhteile aus Polyurethan e. V., in W. Germany. These are:

Table 5-5

Tensile strength (DIN 53504)	Equal or greater than 3.5 MPa.
Elongation at break,	Not less than 380%
Flexing life test (DIN 53507)	30,000 flexes, maximum widening of puncture, 6 mm.
Abrasion test, (DIN 53516)	Maximum loss 300 mg.
Tear propagation test, (DIN 53507)	Not less than 7 kN/m.
Values of properties after temperature/humidity test (DIN 53508, 168 hours at 70°C and 100% R.H.)	The values for tensile strength and elongation at break must not drop below the minimum original values specified above.

Figure 5-17   Abrasion testing to DIN 53516.

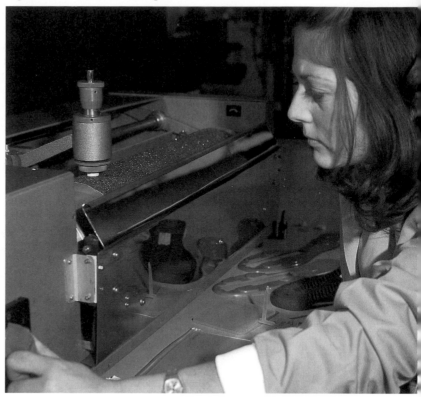

The right to use the RAL Quality Mark is granted only after application to the Gütegemeinschaft für Schuhteile aus Polyurethan e.V., in Pirmasens, W. Germany, who carry out random testing of the applicant's production. The user of the quality mark is required to maintain approved quality control procedures using the DIN test procedures listed above.

In addition to the test listed above, a most useful routine control is to check the density or weight of each moulded part. The DIN 53516 abrasion test is a very quick test which is useful for quality control purposes but does not necessarily correlate with service performance. The special PFI abrasion test should be used in development and research work on shoe soling systems.

6   Reaction injection moulding

The Reaction Injection Moulding (RIM) process is a method of making moulded articles from liquid components at production rates that are competitive with conventional injection moulding. The polyurethane RIM process involves the rapid injection of a mixture of reactive components into a closed mould through a self-cleaning mixing head mounted directly on the mould.

The use of polyurethane RIM has become widespread over the last few years. The reasons for this rapid increase in the use of RIM may be summarised as follows:

Figure 6-1   Demoulding a car wing/RIM auto part.

Large parts may be produced. Parts up to about 10 kg are made with standard machinery. Larger parts involve special high output machines but only a marginal increase in capital costs.

Low energy consumption as the polyurethane RIM process uses much less process energy than any alternative moulding process.

Low equipment costs since, even for large parts, the process needs only relatively light moulds and low clamping pressures, lightweight mould presses are used. RIM cquipment costs less than equivalent thermoplastic injection moulding plant.

Tough, impact resistant polyurethane RIM elastomers can be produced over a wide range of stiffness.

Thin sections are easily moulded to meet part performance specifications at the minimum material cost.

Excellent, paintable surface finish of RIM mouldings allow an automotive Class "A" surface finish to be achieved. Surfaces are free from swirl marks.

Precise reproduction of intricate parts and of intricate patterns. The low viscosity of the reaction mixture and the expansion to produce a microcellular core ensures good contact in complex moulds and with patterned surfaces.

Minimum sink marks; RIM mouldings show much less tendency than thermoplastic injection mouldings to show surface defects above strengthening ribs and inserts.

Inserts and reinforcements are easily incorporated during RIM moulding. Inserts placed in the mould are wetted by the low viscosity liquid reaction mixture and become an integral part of the moulded article.

Stress-free, lightweight RIM parts, both large and small, are obtained free from moulding stresses without the use of expensive hot-runner moulding systems.

Economic production, since the RIM moulding process may be operated at high production rates. These are especially attractive for large parts.

In-mould painting yields a paint finish that is chemically bonded to the RIM part and is extremely resistant to damage.

Early RIM processes were used to make low density, self-skinning parts in both flexible and rigid foam. RIM is now used to produce a wide range of integral skin parts in flexible, rigid and semi-rigid foam, as well as for making almost solid polyurethane parts. The process has also been adapted for the fast moulding of seat cushions. The wide range of machinery now available enables RIM to be used to make mouldings of all sizes from a few grams to an upper limit which depends largely on the capacity of the metering units. Metering capacities of 300 kg/minute are available from standard dispensing machines and are used to make mouldings weighing over 10 kg. Many car bumpers and other external parts for vehicles are being produced with part weights in the range from 2 to 10 kg. The development of systems for stiff, impact resistant, RIM polyurethanes has resulted in lighter weight auto bumpers and other exterior parts with a high resistance to damage. (Figure 6-2.) Polyurethane RIM processes give economical production with short demould times. The minimum demould time is a function of the size, thickness and complexity of the part and of the design of the mould, since the maximum usable reaction rate of the polyurethane system is governed by the need to obtain a smooth flow of the reaction mixture into the mould. This limits the speed at which the mould can be filled satisfactorily. The reaction rate of the polyurethane system may be adjusted to yield the minimum demould time consistent with the speed of filling. Large elastomeric polyurethane RIM parts having thin sections are usually demoulded in 30 to 60 seconds. The use of amine-terminated polyethers and aromatic diamine chain-extenders in high output machines, yields polyurea RIM parts which are demoulded in about 10 seconds depending on their size and shape.

Although the MDI based RIM process is comparable with thermoplastic injection moulding in the rate of production and the degree of automation that is possible, RIM requires a relatively low investment in plant. The cost advantage is greatest for large parts as these can be RIM without the need for high mould clamping pressures. The surface finish of the RIM moulding will exactly mirror that of the mould surface so that a very wide range of surface finishes may be obtained. The thickness of the RIM moulding may be varied with the minimal risk of surface sink marks. In addition to the other advantages of the RIM process listed earlier, the range of

Figure 6-2   Impact resistant RIM bumper.

applications has been increased by the development of RIM systems containing short fibres and/or flaked glass, (RRIM), which gives higher stiffness, higher dimensional stability and a reduced coefficient of linear thermal expansion.

Figure 6-3   RIM polyurethane soft front.

Figure 6-5   RRIM polyurethane wheel-arch extensions.

The RIM process

Figure 6-4   RIM integral skin rear spoiler.

Polyurethane RIM parts are normally produced using two liquid chemical components, a polyol blend and an MDI-based isocyanate mixture. The stages in the RIM process are:

- Preconditioning the two liquid chemical components, i.e. adjusting the temperature within the optimum range, removing entrained air or gas and adjusting the level of dissolved nucleating gas.
- Metering the components in the required ratio by positive displacement pumps.
- Mixing the required quantity of the metered components in a small, self-cleaning, impingement mixing head to form the reaction mixture.

- Injecting the reaction mixture into the prepared and temperature-controlled mould through a well designed gate to prevent turbulence and avoid air entrapment.
- A delay period to allow the reaction mixture to polymerise and to cure sufficiently to allow demoulding.
- Demoulding the finished part.
- Allowing the moulded part to cure and cool until it can be handled without the risk of permanent distortion.
- Trimming and finishing as required.

Figure 6-6 shows a plan view of a typical RIM dispensing machine connected to a single mould. In practice, one dispensing machine will usually serve several mixing heads and moulds in turn. The in-line machine consists of two tanks containing the liquid chemical components, two metering pumps and at least one mixing head together with the ancilliary hydraulic systems, electronic or electrical controls and the temperature control units for both the chemicals and the mould.

Figure 6-6   Plan view of a RIM dispensing machine connected to one mould.

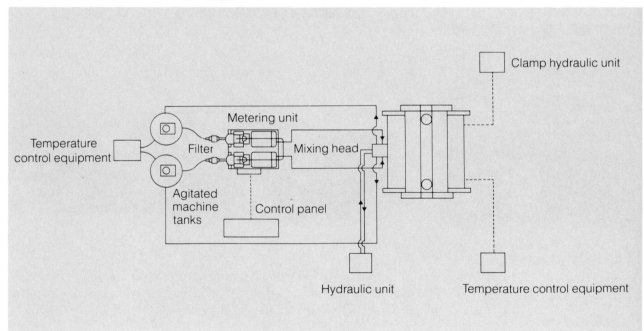

The machine tanks are closed vessels from which the components can be fed under pressure to the metering pumps. The tanks are pressurised using dry air or nitrogen. They are also provided with a temperature-conditioning system and with mechanical agitators to prevent separation of the components and to maintain a pre-set temperature. For continuous production the tanks are usually equipped with automatic recharging to maintain a constant level of the component liquids.

Figure 6-7   Piston dosing cylinder for RRIM.

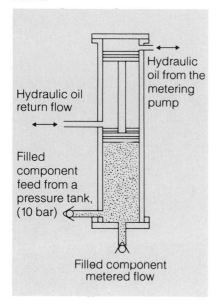

Hydraulic oil from the metering pump

Hydraulic oil return flow

Filled component feed from a pressure tank, (10 bar)

Filled component metered flow

RIM employs high pressure metering of the isocyanate and the polyol components to small, opposed nozzles, in order to obtain the high velocity impinging jets of liquid giving high turbulence in the small mixing head. High pressure metering is most commonly done with either axial piston or radial piston pumps because these give the required accuracy of metering with stepless variable output. Such pumps are satisfactory for mixing with opposed nozzles in simple mixing chambers when used with low viscosity systems, especially with two component systems requiring a component ratio near to 1:1. The need to incorporate fillers, both particulate and fibrous and the introduction of more viscous systems, has led to the development of alternative metering systems using single action dosing cylinders having, usually hydraulically driven, pistons or displacement lances. Displacement lances, especially when designed for use with highly-filled components, are usually driven at an accurately controlled speed by a monitoring system with feed-back to the hydraulic motor, or alternatively may be mechanically driven by electronically controlled motors in order to ensure accurate control of the metering ratio. Displacement lance metering machines with total outputs up to about 8 kg/second are used in RRIM.

Mixer development has met the need for recirculation of the components between dispense shots in order to improve the temperature control and to avoid any settling of insoluble additives,

Figure 6-8   Displacement lance metering.

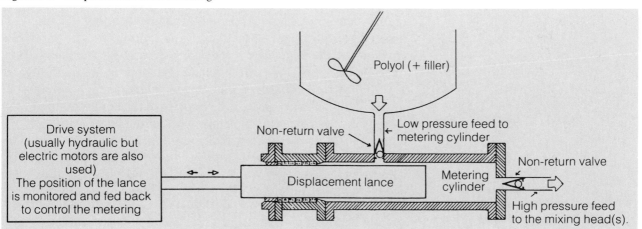

Polyol (+ filler)

Drive system (usually hydraulic but electric motors are also used)
The position of the lance is monitored and fed back to control the metering

Non-return valve

Low pressure feed to metering cylinder

Non-return valve

Displacement lance

Metering cylinder

High pressure feed to the mixing head(s).

and the need to eliminate the use of solvents and/or compressed air to clean the mixer. At the same time material wastage has been reduced and, for some types of parts, sprueless mouldings may be made from low viscosity polyurethane systems. Several satisfactory self-cleaning impingement mixers are available. The use of a hydraulically operated plunger to eject the residual reaction mixture from the mixer at the end of the dispense cycle is common practice, but the valving arrangements and their control, and the mixer

105

geometries differ in the machines from different suppliers. Some widely used mixing heads are illustrated diagrammatically in figures 6-9 and 6-10.

Figure 6-9   Principle of the Krauss-Maffei mixing head.

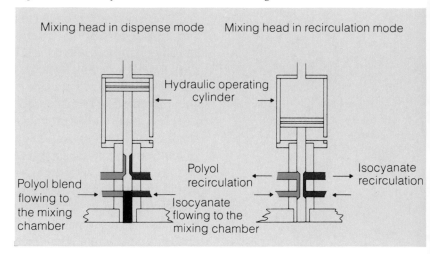

Self-cleaning mixers are available in sizes designed for specific flow-rates depending on the rheological properties of the materials to be dispensed. The polyurethane system suppliers should be consulted about the suitability of equipment for use with their materials. The efficiency of small impingement mixers depends upon the creation of high turbulence inside the mixing chamber. The degree of turbulence in turn depends upon the kinetic energy of the impinging liquids and this is a function of the flow rate, the nozzle size and the rheological properties of the components. Satisfactory impingement mixing requires the use of high pressure metering. At a fixed flow rate, mixing efficiency will deteriorate with increasing nozzle size and with increasing viscosity of the components. It follows that mixing efficiency falls with falling component temperatures. It also tends to deteriorate as the required ratio of the chemical components increases from the optimum ratio of 1:1 by volume. RIM and RRIM systems based upon MDI variants, when used in equipment with available metered feed pressures in the range from about 100 to 200 Bar, require aftermixers between the impingement mixer and the mould gate.

A typical aftermixer (figure 6-11) has two main functions; it divides and recombines the stream of reaction mixture via several small impingement mixers in order to complete the mixing of the components and it collects the material from the start of the injection cycle which is likely to be outside the desired component ratio, especially in high ratio mixes or where the components differ widely in viscosity.

The reaction mixture must enter and fill the mould with laminar or non-turbulent flow to avoid trapping air bubbles and to ensure a

Figure 6-10   The principle of piston-ejector mixers with recirculation valves.

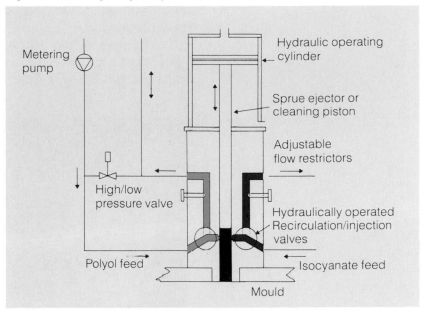

high quality, void-free moulded surface. The creation of laminar flow is the function of the mould gate. Several types of gate arrangement are used. The most useful for elastomeric polyurethane RIM is a film gate of sufficient width to ensure a flow velocity just below that at which turbulence occurs and begins to entrap air bubbles. The critical velocity is between 2 and about 4 m/s depending upon the part geometry, the viscosity of the RIM system and the temperature, surface finish and thermal conductivity of the mould. The conductivity of the material of mould construction affects the rate of heat conduction to the reaction mixture and

Figure 6-11   Typical aftermixer and rod film gate.

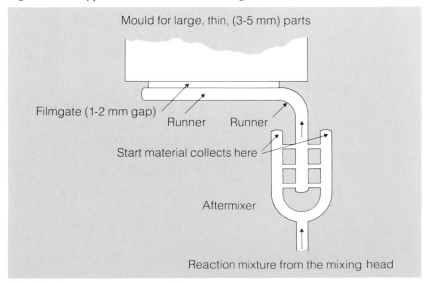

therefore, its viscosity and rate of reaction. Film gates always lie on the parting line of the mould and the film of sprue that is removed from the mould with the moulding is usually thin enough (1-2 mm), to be trimmed without spoiling the appearance of the part. Film gates may be fan-shaped (figure 6-12) but for large; thin parts a rod or ribbon gate (figure 6-11) is more satisfactory.

Figure 6-12   RRIM part showing the fan gate.

Small parts, on the other hand, may be moulded using multicavity moulds with simple cylindrical injection ports. If the mould cavities in a multicavity mould are all of equal volume, they are best arranged symmetrically around the single injection port so that the flow path to each cavity is identical. Small parts having different volumes may also be made in a multicavity mould by balancing the entry gates to ensure that all the mould cavities fill at similar rates.

Mould construction

A well-designed mould is of critical importance in the economic production of RIM parts. The most important design consideration are discussed below.

Materials
The choice of material for mould construction will depend on the number of moulded parts required, the type and quality of the surface finish required, the required production rate per mould and on the possible need for future modifications to the mould.
Metals are the most durable materials for mould construction and, because of their high thermal conductivity, are best for use with

highly reactive RIM systems. Epoxy resin moulds, or metal-filled epoxy moulds, in GRP shells, can easily be made strong enough for RIM and they are easily modified. Their use is restricted to prototype work or for small production runs. This is mainly because the surface quality and the ease and speed of release of RIM mouldings is a direct function of the quality of the mould surface and the control of the mould surface temperature. The latter is particularly important with high modulus RIM systems which have a high exothermic heat of reaction. This heat must be removed from the surface of the moulding before it can be cleanly released. Epoxy resin mould surfaces are inferior to metal surfaces in quality, durability and the ease of temperature control. The poor durability of epoxy resin surfaces on heat cycling, combined with their low thermal conductivity, often results in mould-release problems. Metal moulds are made in steel, aluminium, zinc alloys, beryllium/copper and nickel/copper shell construction. Machined tool steel, finished by polishing or by chromium or nickel plating, is the most durable mould material and gives good release. A highly-finished mould surface will allow some polyurea RIM parts to be moulded without the application of release agents to the mould surface. Steel moulds used in mass production have given from 100,000 to 200,000 demoulds. Machined, forged aluminium is less durable but it gives good release and easier temperature control than other metals, with the advantage of light weight. The low density of aluminium may significantly reduce the cost of handling very large moulds. Kirksite (zinc alloy) moulds are heavy and because of the relatively low thermal conductivity of the material they are usually cast around closely pitched heat exchanger pipes. They give good release and surface finish and, with reasonable care, have a useful life of over 50,000 mouldings. Kirksite moulds are easier to repair or refurbish than other metal moulds. Cast aluminium moulds are relatively cheap but have been found to be unsatisfactory for RIM moulding because of surface porosity and surface roughness. Cast aluminium is, however, satisfactory for RIM low density cushioning foams and for other products which are covered before use. Detailed surface patterns are often reproduced by using electroformed copper-backed nickel shells which are made by electroplating, or sometimes by vacuum depositing, nickel onto a master pattern. The thin shells, typically 50 to 90 mil of nickel backed with up to 4 mm of copper, are built into a GRP body which is usually reinforced with a metal frame. This type of mould is used for the manufacture of low density integral-skin RIM parts. Mould durability is good and over 100,000 mouldings have been obtained from a mould of this type.

Other mould making techniques include the use of sprayed metal shells backed by GRP or metal and the use of silicone rubbers and polyurethane elastomers. All three techniques will give an accurate reproduction of intricately patterned surfaces at low cost but the mould surface durability is limited. The life of these moulds depends to some extent on the reactivity of the RIM chemical system, especially on the rate of surface cure. In general, sprayed metal, silicone rubbers and polyurethane elastomers are not suitable for moulding stiff RIM elastomers and especially polyurea RIM, because the thermal shock from the high exotherm causes breakdown of the mould surface. Sprayed metal moulds may give from several hundred to a few thousand self-skinning foam parts, depending to a large extent on the care taken in demoulding and in cleaning the mould as the sprayed metal surface is rather soft and easily damaged. Silicone rubber moulds are usually satisfactory for between 100 and 200 parts made from flexible self-skinning foam. Mould linings cast from polyurethane elastomers are more durable and may allow up to 1,000 parts to be made.

Temperature control

Temperature control of the mould is important in obtaining reproducible mouldings with uniform properties, a high surface quality and good release from the mould. Both halves of the mould should be constructed with tubes or ducts for the circulation of a heat transfer fluid – usually water. Moulds are usually filled at a controlled temperature between 50 to 70°C depending on the RIM system used, its filler content and the conductivity of the mould. Temperatures above 70°C may be needed with some polyurea RIM systems. Integral-skin parts are usually made with mould temperatures in the range from 40°C to 50°C. The surface temperature of the mould should be uniform to about ± 2°C. The RIM process is exothermic; a typical system generates about 50 kcal/kg of material, which especially in thin parts, is mostly conducted into the mould. Uniformity of mould cooling is therefore most important in order to avoid the development of hot spots during production. The rate of heat removal from the mould will influence the demould time and the rate of production. Fast, uniform cooling is especially important with polyurea RIM.

Mould venting

Adequate mould venting is essential to allow the air in the mould to escape as the mould is filled, or almost filled, with reaction mixture. Injection times vary from about one second to several seconds, depending on the part size and the chemical system but typical injection times for high modulus RIM are between one and two seconds. Moulds are best vented via the parting line at the highest point – remote from the filling gate. This method allows sufficient venting to avoid excessive air pressure in the mould and minimises the spew of polyurethane through the vent because the reaction mixture has become a viscous polymer by the time it reaches the uppermost parting line. The split line near to the runner and film gate must remain sealed to contain the fast flowing, low viscosity reaction mixture during the injection phase. Narrow elastomeric

Figure 6-13   Cross-section through a RIM tool showing the coolant channels.

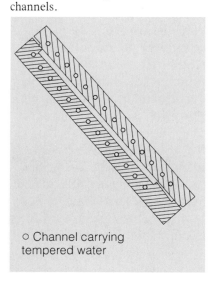

○ Channel carrying tempered water

seals are sometimes incorporated in this area of the split line. These may be formed from the RIM material itself by using sealing grooves in the mould split flanges which are filled when the first moulding is made. In many cases, careful choice of the mould split line, combined with programmed mould tilting, will allow venting entirely through the upper split line. When this is not possible because air is trapped in essential protrusions above the tilted split line of the mould, additional vents will be required. A small hole (ca. 2 mm), may be all that is needed but it will have to be cleaned after every shot. Vents combined with ejector pins are used in moulds for the reaction injection moulding of rigid self-skinning foam parts. Small area ejector pins are unsuitable for the rapid demoulding of thin section elastomeric RIM parts. Such parts may be demoulded by the use of compressed air together with large-head ejectors. These cannot be used on the show face of the moulding and may be too large to be used as vents for trapped air. One solution is to include a cavity in the mould to contain the air trapped in ribs or bosses. The cavities may be positioned on the parts of the moulding which are hidden when the moulded part is in use or can be trimmed after demould. This technique is used in some automotive trim manufacture because it avoids increases in the mould cycle time for vent cleaning (figure 6-14).

Figure 6-14 Demoulding a complex RIM soft front.
This slatted grill design incorporates cavities to control flashing and eliminate bubbling.

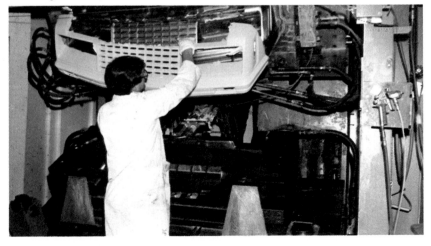

Mould strength and mould manipulation
Moulds for RIM and RRIM are designed for relatively low moulding pressures compared to those required for the injection moulding of thermoplastics. The maximum pressure developed in polyurethane RIM is usually less than 4 bar, although point pressures of about 12 bar have been recorded when making large area, glass-fibre reinforced mouldings in the laboratories of ICI

Polyurethanes. Small moulds and prototype moulds may be mounted in a rigid frame employing hinges and toggle clamps for opening and closing the mould. Toggle clamps and other lever type closing devices are not satisfactory for production use, except with very small moulds, unless they are backed up by a pneumatic or hydraulic support to take up wear and to ensure an even closing force. The best method is to mount the mould in a suitable hydraulic or pneumatic press or mould clamp. Mould clamps designed for use with moulds for RIM are often provided with a means of tilting the mould in one or two planes, either manually or automatically, during or immediately after filling in order to assist mould venting and filling. Large mould clamps are usually arranged to present the open mould to the operator in a fashion that assists demoulding of the part, the examination of the upper and lower mould surfaces, and the application of release agent to the mould surface, (figure 6-15).

Figure 6-15   Demoulding RIM sheet from an open mould in a hydraulic clamp.

Mould clamps may be designed for use with one particular mould or type of mould, for example, moulds for the manufacture of bumpers or soft front and rear ends for cars, or may be designed for use with any type of RIM moulding. Universal mould clamps are usually designed to tilt in two planes and with adjustable opening and closing speeds. Large mould clamps are invariably hydraulically operated and usually use proportioning valves to obtain a combination of high opening and closing speed with a gentle closure of the mould to avoid shock loads on the mould flanges. The mould locking forces required for RIM and RRIM are very low compared with those required for the injection moulding of thermoplastics, and light presses or mould clamps designed to exert a maximum closing pressure of 25 to 50 tonnes per square metre of platen area are used. Moulds should be designed to withstand both internal and external pressures of about 10 bar without significant distortion. Flat, thin section, large area moulds are usually reinforced and mounted on the press platens, using shims if necessary, so that the mould is effectively supported by the clamping pressure.

Mould release

It is usually necessary to apply a mould release agent to the mould in order to prevent the moulding sticking to the mould surface. The release agent (sometimes called a 'parting agent') is best applied by spray coating the open mould. The usual release agents are micro crystalline waxes dispersed in solvents. The operator must be protected from breathing any aerosol of the release agent or any solvent vapour produced during the application of the release agent. As the mould carrier installation will already include the positive exhaust ventilation that is required when handling isocyanates (discussed in detail in Chapter 10), it is usually only necessary to check that the system also removes any vapours arising from the application of mould release agent. Otherwise, protective clothing must include a suitable breathing mask.

In addition to mycro crystalline and/or polyethylene wax, some proprietory release agent compositions contain other additives including silicone oils. These are valuable in assisting the dispersion of a thin layer of release agent on the mould surface and in helping to prevent the polyurethane reaction mixture from wetting the surface of the mould. Silicone-based release agents are valuable in obtaining a glossy surface on pigmented mouldings but they should never be used when RIM parts are to be painted. This is because a small part of the release agent coating is transferred to the surface of the RIM moulding. Wax-based release agents may be removed from the part before painting, by aqueous detergent washing or by a conventional degreasing solvent, but silicone oils are very difficult to remove and will interfere with the paint coating.

A great deal of work has been done to try to avoid the need for the use of release agents or at least to avoid the need to apply a fresh coating for each RIM moulding. The objective is not only to speed up the process and avoid a rather unpleasant job, but also to improve the uniformity and the surface quality of the moulded parts. The build up of release agent on a RIM mould can adversely affect the reproduction of the surface detail and may impose frequent breaks in production to allow mould cleaning. There is no universally applicable alternative to the use of wax release agents, but several techniques are available to avoid the application or removal of release agents in particular product areas.

Mould liners. Disposable mould liners of thin vacuum-formed polystyrene sheet or polypropylene sheet have been used in the manufacture of flexible self-skinning foam. Such liners are only suitable for making parts with plain surfaces.

Transfer coating. A coat of paint is transferred from the mould surface to the moulding. This technique was first used for RIM integral-skin foam parts, both flexible (Chapter 5) and rigid (Chapter 7), and is now being applied to external automotive trim made from RIM and RRIM polyurethane elastomers. The transferred paint coating becomes chemically bonded to the RIM polyurethane, giving excellent resistance to impact damage.

Durable release coatings. These are polymeric coatings which reduce the mould surface treatment to one application per shift. They find some limited application. For example, thin silicone elastomer coatings are used in some self-skinning foam moulding and for RIM cold-cure flexible foams.

Internal release agents. These are materials incorporated in the chemical components. They have found some useful applications in rigid integral-skin RIM, where they greatly reduce the need for conventional release agents.

Multiple release chemical systems. These are fast reacting polyurea RIM systems and polyurethane systems containing additives and curing agents to accelerate the cure of the surface of the moulding. They are used in polished steel moulds with a surface temperature of about 70°C and allow several demoulds from one coat of a conventional wax release agent. Over 1,000 demoulds are possible from one application of release agent depending on the reactivity of the chemical system and upon the shape and surface finish of the (metal) mould.

RIM elastomers

See also Chapter 13, part 2, page 338

Although the RIM process is applicable to the manufacture of most types of polyurethanes, from low density flexible and rigid foams (Chapters 4 and 7), semi-rigid and self-skinning foams (Chapters 5 and 7), to non-foamed elastomers, it is the latter area which has seen much recent development. A wide range of RIM elastomers can be made by varying the formulation of the components. The components used in this technology consist of a mixture of a polyether polyol and/or an amine-terminated polyether, chain extenders, catalysts and optional additives, such as surfactants pigments and fillers. The isocyanate component is usually a low functionality modified pure MDI or, when a lower performance and cost are acceptable, a specially developed polymeric MDI

Figure 6-16   Demoulding a RRIM wheel-arch extension.

composition. The polyols are linear or slightly branched polyether polyols with an equivalent weight in the range from about 2,000 to 3,000. Modified polyether polyols such as polymer polyols and amine-terminated polyethers such as the 'Jeffamines' are also used. The chain extenders are usually low molecular weight diols or diamines although hydroxy-amines have also been used. Mixtures of chain extenders may be used but, because mixtures often give hard block segments which do not easily form dense domains, they tend to yield elastomers with lower levels of physical properties. Many RIM elastomer systems are based upon simple diols because these give adequate properties for many applications together with sufficient processing latitude for the production of both small and large parts. Also, for many applications diol-based systems are satisfactory when post-cured at room temperature. Aromatic diamine chain-extenders are useful in obtaining very fast reacting RIM systems with the minimum mould occupation times but, because the high rate of reaction is accompanied by a rapid rise in the viscosity of the mixture, the production of large parts requires very high output RIM machines. Outputs up to 500 kg/minute are used for some polyurea RIM parts. The use of diamine chain-extenders also has some practical advantages however, as systems containing aromatic diamines often require a higher ratio of the polyol/isocyanate component compared with systems using diol chain extenders. This higher ratio is advantageous when high levels of fillers such as glass-fibre or glass flakes are added. This is because in practice reinforcing glass fillers can only be added to the polyol component and a high polyol/isocyanate component therefore reduces the concentration of filler required for a given degree of reinforcement. Blends of polyethers and diamines are also less likely to separate than blends containing simple glycols. Diamine-based RIM elastomers, on the other hand, may require a post-cure above room temperature to develop the optimum physical properties and the elastomers produced have a greater tendency to discolour in sunlight than those based on glycol chain extenders.

Pre-blended, two component RIM systems are available which cover the product range from soft elastomers to hard materials with high stiffness or flexural modulus. These products effectively fill the gap between soft rubbers and rigid plastics or metals. High modulus RIM elastomers have sufficient resilience and toughness to resist impact damage, but have sufficient rigidity to replace metal sheet in many applications where impact and corrosion resistance, allied to light weight, are important.

The hardness and stiffness of elastomeric RIM parts depends not only upon the properties of the elastomer itself, but also on the thickness of the moulding and its geometry. RIM systems are always formulated to give some expansion in the mould to ensure good surface definition. The resulting moulding consists of two high density skins surrounding a lower density, slightly blown, core. The

115

overall density obtained from most RIM formulations depends upon the thickness of the moulding because, as the thickness is reduced, the proportion of high density skin material increases. Thin section mouldings are thus harder and the higher overall density tends to compensate for the loss in stiffness that would be expected to result from a reduction in part thickness.

The wide range of elastomeric RIM materials results largely from the development of several MDI variants and special polymeric MDI compositions for use as the isocyanate component with matched polyol compositions. High quality elastomers are based upon MDI variants such as 'Suprasec' VM10, a low viscosity derivative of pure MDI, which is used with a range of polyol compositions to produce a range of RIM elastomers. These are sometimes characterised by measuring their flexural moduli (a measure of stiffness) at 23°C, and also, to define the effect of temperature, at −30°C and at +70°C. Some typical moduli for RIM sheets with a thickness of 4 mm, derived from 'Suprasec' VM10 are tabulated in table 6-1.

Table 6-1 **Typical flexural moduli of a range of polyurethane RIM elastomers**
(all samples cut from moulded sheet having a nominal thickness of 4 mm.)

Polyol component	'Daltoflex' 1478	'Daltoflex' 1619	'Daltoflex' 1613	'Daltoflex' 1591
Isocyanate component	'Suprasec' VM 10	'Suprasec' VM 10	'Suprasec' VM 10	'Suprasec' VM 10
Mean density (kg/m³)	1100	1100	1100	1100
Flexural modulus (Mpa)				
at 23°C	300	520	760	940
at −30°C	810	1300	1590	1680
at +70°C	190	275	380	400
Ratio Modulus at −30°C / Modulus at +70°C	4.3	4.7	4.2	4.2

Elastomeric RIM polyurethanes are characterised by excellent tensile properties and good resistance to tearing and abrasion. Tensile strength changes little with changes in flexural modulus. Typical moulded parts have strengths from about 23-24 MPa when made from flexible low modulus materials increasing to about 30 MPa when made from high modulus elastomers. The elongation at break of these materials varies from about 300% for the most flexible products to 50 to 90% for very stiff materials. Tear strength tends to correlate with the tensile strength and is usually within a range from about 90 to 150 kN/m, the higher levels corresponding to the higher modulus materials.

The combination of rigidity with good impact resistance over a wide temperature range, combined with the good surface obtainable from relatively low cost tools, has led to many applications for RIM in American, Japanese and European cars and trucks. Typical

Table 6-2 **The effect of thickness on the flexural modulus of RIM mouldings**

(System: 'Daltoflex' 1478/'Suprasec' VM 10.)

Sample thickness (mm)	2.7	4.1	7.3
Mean density (kg/m^3)	1100	1100	1100
Flexural modulus (MPa)			
at + 23°C	350	300	170
at − 30°C	860	810	390
at − 70°C	240	190	120
Ratio Modulus at − 30°C / Modulus at + 70°C	3.6	4.3	3.3

elastomeric RIM parts include complete soft fronts and rear ends, bumpers, wheel hub covers, bumper over-riders, exterior driving mirror surrounds, spoilers, wheel-arch liners and extensions. Additionally there are many RIM parts in production for commercial vehicles especially heavy duty trucks and buses.

Polyurethane RIM parts are either pigmented with carbon black or painted in a separate operation, and are then attached to the vehicle after it has passed through the body painting stage. This is necessary because polyurethane RIM parts, in common with many other elastomeric materials, are likely to distort at the high baking temperatures used for car paint finishes. High modulus, polyurea RIM systems are used for parts to be painted on-line at 165°C. All elastomers and plastics expand much more than metals on heating, therefore unreinforced RIM materials are normally unsuitable for use as vehicle body panels. The properties of RIM elastomers may, however, be modified by the incorporation of fillers. Particulate fillers such as barium sulphate are used in sound-absorbing covers for diesel engines and gear-boxes in trucks and other commercial vehicles but the most important fillers used in RIM elastomers are short fibres and laminar fillers especially milled or chopped glass-fibres and glass flakes. Such fillers increase the stiffness, the dimensional stability and the distortion temperature of RIM elastomers. They also reduce the coefficient of linear thermal expansion and make the material more compatible with the metal vehicle structure. Work in this area has resulted in materials that will survive the high paint oven temperatures used in stoving car bodies without distortion. These are typically about 150° C for up to 4 cycles of up to 30 minutes.
Existing RRIM high modulus elastomers will withstand typical "touch-up" oven temperatures up to 120°C for 30 minutes.

Reinforced RIM

Glass-fibre reinforced RIM systems (RRIM) are used in many automotive applications such as lightweight abrasion-resistant

117

rubbing strips, boot lids, door panels and wings because RRIM parts meet the requirements of the automotive industry. These have been summarised as follows:

- RRIM parts must retain a "Class A" automotive finish.
- RRIM parts must not distort when in position on the car body during its exposure to paint oven temperatures.
- RRIM parts must retain good impact resistance from − 29°C to + 70°C.
- RRIM parts must be paintable with standard body paints.
- RRIM parts must have a sufficiently low coefficient of thermal expansion to be compatible with metal body frames.

Figure 6-17   Truck with RIM mudflaps and mudguards

All these objectives may be attained by the optimum combinations of short-fibre and particulate reinforcement in a RIM elastomer*. The reinforcing fibres are dispersed in the polyol component which is then metered to the impingement mixer by a dosing unit (figure 6-7). The most popular type of glass-fibre used in polyurethane RRIM is milled glass-fibre consisting of 15 μ diameter filaments milled through a 1/16", (1.6 mm) screen. Typical material has a mean filament length of 0.1 to 0.3 mm depending on the surface treatment, and a length distribution from about 0.01 to about 1.0 mm. This mean filament length is, however, less than the critical length below which the interfacial shear stress under tension will cause the fibre to pull out of the matrix in which it is embedded. Above the critical length, failure under tension will occur in the fibre rather than in the elastomer. As the fibre length is increased above the critical length the reinforcing effect becomes more consistent. This is because variations in interfacial adhesion become less significant. A lot of work has been done to determine the optimum combinations of polyurethane chemical system, fibre type and surface treatment, and the method of processing. As the fibre length increases, the free volume occupied by the fibres and hence the viscosity of the fibre dispersion in the liquid polyurethane components, also increases. Metering, pumping and mixing of the suspensions becomes much more difficult. However, much less fibrous reinforcement is required to obtain a given degree of reinforcement when the fibre length is above the critical length. ICI Polyurethanes, working closely with the Research Department of Fibreglass (UK) Ltd., developed a 1.5 mm long chopped glass strand with a surface treatment which ensures easy dispersion in the polyol component of polyurethane systems. This material gives more reinforcment than over twice its weight of typical milled glass-fibre fillers.

*   The use of standard bodypaints may reduce the impact resistance of RRIM parts at low temperatures because cracks in the paint layer can propagate cracks in the elastomer.

The effect of glass-fibre reinforcement on flexural modulus or part stiffness

The effects of typical milled glass-fibres on the flexural modulus of RRIM are compared with those of 1.5 mm chopped glass-fibre strand in table 6-3 below. In both cases the rate of change of modulus with temperature is not significantly affected by glass-fibre reinforcement. In practice, it may be found that the best combination of reproducible processing, good surface

Figure 6-18   The effect of milled glass-fibre loading on the flexural modulus of three RRIM elastomers, measured in the direction of mould filling

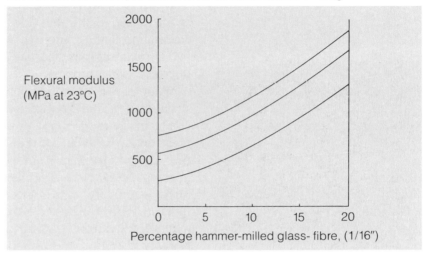

(Samples moulded using a Krauss-Maffei 40/80 machine having a polyol component piston dosing unit and a 4 stream impingement mixing head to fill a $750 \times 800 \times 3$ to 4 mm mould held in an automated mould press under a closing force of 25 tonnes.)

Table 6-3 **The flexural modulus of RRIM made with milled and chopped glass-fibre**

Chemical system									
Isocyanate component	'Suprasec' VM 10								
Polyol component	'Daltoflex' 1478			'Daltoflex' 1619			'Daltoflex' 1613		
Type of glass-fibre	–	1.	2.	–	1.	2.	–	1.	2.
Percent glass fibre weight	0	6	20	0	6	20	0	6	20
Flexural modulus (MPa) at + 23°C	300	650	1050	520	800	1300	760	1250	1850
at – 30°C	810	1240	1300	1300	1610	2100	1590	2050	2750
at + 70°C	190	325	700	275	380	570	380	480	720
Ratio of moduli – 30 + 70°C	4.3	3.8	3.1	4.7	4.2	3.7	4.2	4.3	3.8

Notes

Glass-fibre type: 1. = 1.5 mm Chopped strand, Fibreglass Ltd., type WX 6461.
                2. = 1/16″ milled glass, Owens Corning Fibreglass type 737 AA.
Measurement: All moduli were measured parallel to the direction of flow in the mould.

reproduction, glass-fibre reinforcement, the minimum wear of the machine and cost, may be obtained by using mixtures of chopped and milled glass-fibre in suitable systems.

Glass-fibre reinforcement has a marked effect on the coefficient of linear thermal expansion. The effect is greatest in the direction of the liquid flow during the filling of the mould, because the fibres tend to orientate in the flow direction. In tables 6-1, 6-2 and 6-3, all measurements were made parallel to the flow direction.

Anisotropy of RRIM parts

The glass-fibres dispersed in the reaction mixture become orientated in the direction of liquid flow during mould filling. The degree of orientation tends to increase with the average length of the fibres and is greatest for thin parts. RRIM parts are thus anisotropic. The degree of anisotropy increases with the increasing length and orientation of the reinforcing fibres, and it is also proportional to the strength of the bond between the fibres and the polymer matrix. The effects of the reinforcement, i.e. the increase in stiffness and the reduction in heat sag, the reduced coefficient of thermal expansion, the reduced shrinkage from the mould dimensions and the reduction in elongation at break, are all greatest in the direction of liquid flow during mould filling. The effect on flexural modulus is illustrated in figure 6-19 for the rather extreme case of a large, thin sheet of material filled from one end of the mould with an elastomer reinforced with 1.5 mm chopped strand glass-fibre dispersed in the polyol component. The glass-fibre had been pre-treated to ensure good bonding to the polymer.

Figure 6-19   Anistropy of RRIM Sheet reinforced with 5% chopped glass-fibre. Effect on the flexural modulus, (MPa), of sample orientation.

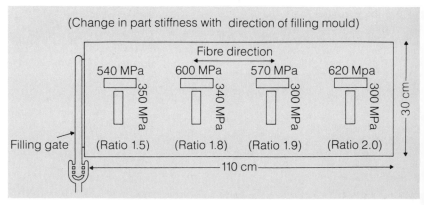

The anistropy of RRIM parts is an important factor in both the design of the part and the design of the mould. It influences the position of the filling gate and the position of stiffening profiles or ribs. As the ex-mould shrinkage is anisotropic the use of prototype moulds for the final part design is particularly important for RRIM parts. Whereas the shrinkage of RIM parts is typically from 1.0 to

120

1.2%, that of RRIM may vary from as little as 0.3% to about 1.0% depending upon the orientation of the glass-fibre reinforcement. Anisotropic shrinkage is reduced by the partial replacement of glass-fibre reinforcement with laminar fillers such as flaked glass and mica.

The tensile properties of RRIM elastomers

The tensile strength of elastomeric RRIM materials is usually in the range from about 23 to 29 MPa at 20 to 25°C, depending on the system – high modulus materials having the higher tensile strengths. The effect of glass-fibre reinforcement on tensile strength is small. Both milled and chopped glass-fibre give a reduction in tensile strength of 5 to 10% at the maximum levels that can be added in the polyol stream, compared with the base RIM material without reinforcement. The elongation at break, on the other hand, is greatly reduced by the addition of glass-fibre reinforcement, especially in systems having good glass to elastomer adhesion. The reduction in elongation at break is greatest in the direction of liquid flow during mould filling and is roughly in inverse proportion to the increase in flexural modulus when there is significant bonding between the glass and the polyurethane elastomer. In the absence of bonding the reduction in elongation is much less. Some typical effects are listed in table 6-4.

Table 6-4 **Typical effects of glass-fibre reinforcement on the elongation at break of RRIM metered at 3 to 4 mm thickness**

Amount of glass, (% by weight.)	None	5	25	20
Glass/Polyurethane bond strength	None	Good	Good	Poor
Type of glass-fibre	None	1.5 mm chopped strand	Milled of mean length 0.1 mm	Milled of mean length 0.1 mm
Elongation at break, (As % of that of the basic elastomer)[1].				
Direction of measurement				
Parallel to flow	150	20% to 50%	5% to 20%	60% to 70%
Perpendicular to flow	150	50% to 80%	30% to 50%	70% to 80%

[1] Elongation at break of the unreinforced RIM elastomer will vary from about 150% to about 300% depending on the system and the density and stiffness of the product.

Resistance to impact damage

The resistance to impact damage of both RIM and RRIM polyurethane elastomers is high. Impact testing of RIM and RRIM parts is usually done using falling dart or pendulum type impactors (see Chapter 11.). These are of many types, ranging from small laboratory bench machines to the large machines which are used to test automotive bumpers and the complete front and rear ends of

Table 6-5 **RRIM: The effect of glass-fibre loading on the thermal expansion of RRIM polyurethanes elastomers.**
(Typical values for large, thin mouldings).

Type of fibre		Chopped strand						Milled fibres	
Mean length		1.5 mm						0.1 mm	0.2 mm
Percent glass-fibre the final polymer	0	1	3	5	7	9	25	25
Overall density (kg/m^3)	1200	1300	1300	1300	1300	1300	1300	1300
Flexural modulus at 20°C (MPa)	300	390	420	540	720	850	900	1100
Coefficient of linear thermal expansion[1] (x 10^{-6}/°C)	126	100	70	55	60	45	45	35

[1] Values measured parallel to the direction of flow in the mould for systems giving good adhesion between the glass and the elastomer. Values measured perpendicular to the flow direction will be higher by a factor varying from 0 to about 2.5 depending on the degree of orientation, the concentration and the length of the glass-fibres. For perfectly aligned, thin test plaques, the coefficient of linear thermal expansion in the direction perpendicular to the direction of flow, is in the range from 125 to 90 (x 10^{-6}/°C). For real parts, reinforced with 5% chopped strand or 15% milled glass-fibre, the value is usually in the range from about 55 to 75 (x 10^{-4}/°C).

cars. Any of these tests may be used to demonstrate the high resistance of RIM and RRIM polyurethane elastomers to low speed impact damage. Impact forces that would cause severe distortion of sheet steel or aluminium parts and the perforation of GRP or SMC

Figure 6-20   Falling dart impact on RRIM, steel, aluminium and SMC sheet.

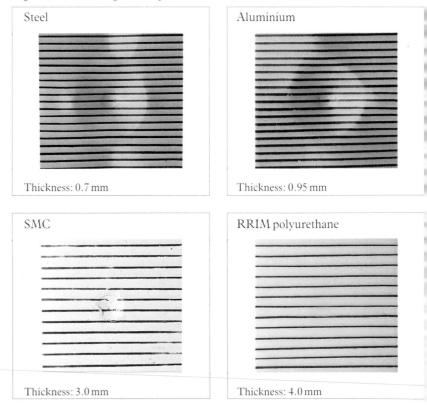

Steel — Thickness: 0.7 mm

Aluminium — Thickness: 0.95 mm

SMC — Thickness: 3.0 mm

RRIM polyurethane — Thickness: 4.0 mm

Figure 6-21   Falling dart impact: force/displacement traces.

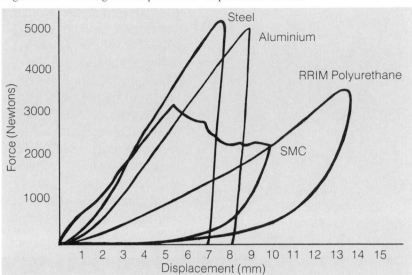

sheets, are absorbed by RIM and RRIM without damage. The photographs reproduced in figure 6-20 show the effects of dropping a dart having a weight of 5 kg and an impact speed approaching 15 km/h onto steel (thickness: 0.7 mm), aluminium sheet (0.95 mm), SMC sheet (3 mm) and RRIM sheet (3.9 mm). The falling dart had a hemispherical head of diameter 25 mm and was used in an instrumented machine to obtain both force/displacement and energy/time traces. Figure 6-21 shows the force/displacement graphs obtained which demonstrate the complete recovery of the RRIM sheet compared with the permanent deformation of the metal sheets and the cracking of the SMC sheet.

Trends in the use of RIM and RRIM

The rapid growth, especially in the use of RRIM, confirms the many manufacturing advantages of the process compared with the alternative traditional methods. Moulds and presses are of light weight compared with those required for injection moulding thermoplastics or forming sheet metals. Prototype work is easily, quickly and inexpensively carried out. Large and complex parts in RRIM are free from built-in stress. Surface sink marks above internal supporting ribs, which are a problem in all thermoplastic moulding, are not significant on many RIM mouldings. Sink marking is reduced by dispersing pigments or gases in the chemical components. Heavily filled RRIM shows only minimal sink marking which is usually easily corrected by design or surface modifications. Compared with thermoplastic injection mouldings, RRIM and RIM mouldings are unique in that they do not show any part weakness at the junction of the flow lines which are often unavoidable in complex parts.

123

An important factor in the growth of RRIM moulding is the low energy requirement of the process. It has been calculated that the total energy usage of a RRIM moulding plant producing large RRIM mouldings is 600 MJ/h. This figure includes venting and factory air heating. At the maximum production rate of a single dispensing machine this equates to a process energy usage of about 300 J/cm^3 for mouldings weighing about 8 kg each. This is only a small fraction of the energy required by the alternative manufacturing processes. It has also been shown that the overall energy consumption of RRIM from crude oil to finished part is lower than that of the alternative materials. Even with recent developments in the use of thinner, high strength steel sheet, a simple press-formed steel part uses almost 50% more energy than that required to make the part in RRIM.

Figure 6-22   Reliant 'Scimitar-SS' with RRIM front and rear wings and bumpers

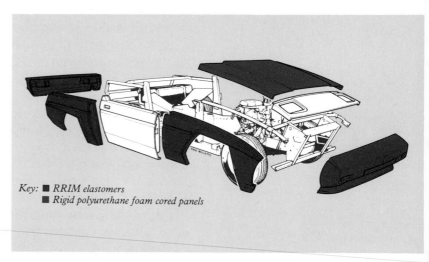

Key: ■ *RRIM elastomers*
■ *Rigid polyurethane foam cored panels*

Trends in the use of RRIM are illustrated by two automotive applications. The ease with which RRIM polyurethane moulds can be modified or re-tooled permits a modular approach to design with several body styles carried on the same basic structure. RRIM mouldings may be used to modify a mass produced car body in order to give a distinctive appearance to a relatively small volume version of the vehicle.

Figure 6-24   The metal space-frame of the 'Fiero'.

Figure 6-25   The 'Fiero's plastic body panels.

The second, and the most revolutionary application, is the use of RRIM moulded polyurethane panels to replace steel panels. General Motor's Pontiac Ficro, the first mid-engined car produced in large numbers by General Motors, consists of a metal space frame (figure 6-24) clad with plastic outer panels. SMC is used for the roof, bonnet and boot lid, but polyurethane RRIM elastomer is used for the front and rear panels, the wings and the door panels because of its greatly superior resistance to impact damage. Two types of reinforcement have been used, 10% milled glass-fibre or 22% of a specially developed grade of flaked glass. To avoid any tendency to sag, the horizontal front end is reinforced with milled glass-fibre, but the other, vertical, parts are reinforced with glass flakes. The glass flakes orientate in the direction of liquid flow as the mould is filled, to form laminar reinforcement. This improves the stiffness and reduces the thermal coefficient of expansion of the part without excessive anisotopy in the plane of the panel. This helps to ensure the high standard of panel fit which is a feature of the car and by reducing anisotropic shrinkage compared with fibrous reinforcement, gives a stress-free moulding of improved surface quality. Since the development of the 'Fiero', polyurea RIM and RRIM has been used to make large numbers of front and rear bumper covers and panels in the USA. The fast reaction of low-functionality MDI variants with amine-terminated polyethers and aromatic diamines, even without

Figure 6-26 Motorcycle with RRIM fairings.

the addition of catalysts, yields production rates of about 60 parts per mould cavity every hour. A second, important benefit is the high stiffness and high temperature resistance resulting from alignment of the polyurea hard segments.

Although the use of RIM and RRIM polyurethane elastomers has been pioneered by the automotive industry, they are now finding use in many other applications requiring a tough, damage resistant material. Examples include the use of RRIM hoods and body panels for lawn mowers and snow-mobiles and the development of RIM parts for heavy duty office furniture.

7   Rigid polyurethane foams

See also Chapter 13,
part 1, page 324

Within the wide range of polyurethane materials manufactured today, rigid foams are second only to flexible foams in terms of the tonnage and volume produced. The quantity of rigid foam produced has grown at an annual rate of about 10 percent – about twice the growth in flexible foam production – for many years. Like flexible foams rigid polyurethane foams can be made in a range of densities from less than 10 kg/m^3 to about 1100 kg/m^3 – almost solid polyurethane plastics. However, the major proportion of rigid polyurethane foam production consists of lightweight foams for thermal insulation with densities from about 28 kg/m^3 to about 50 kg/m^3.

Low density rigid polyurethane foam

Figure 7-1   The thermal conductivity of various building materials.

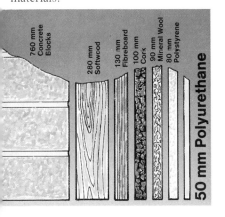

Please note that CFM-11 is now referred to as CFC-11

Low density rigid polyurethane foam is a highly effective thermal insulant (see figure 7-1) that is made by mixing together easily transportable, liquid components. The foam also has a high strength to weight ratio and it is extremely adhesive during the short time – usually about 1 minute – between mixing and surface curing. The adhesion of rigid polyurethane foams to most of the facing materials used in building construction is high. In fact, the strength of the bond between the foams and the facing material is almost always higher than the tensile or shear strength of the foam. This means that it is impossible to peel away the facing material without tearing the foam itself. The ease of manufacture together with the excellent adhesive properties, facilitates the combination of low density rigid polyurethane foam with other materials to make extremely cost-effective, multi-layer structures.

The manufacture of low density rigid polyurethane foam does, of course, require the use of a blowing agent to create the foam structure. The blowing agent may be carbon dioxide formed by the reaction of a measured amount of water with the isocyanate component. Most rigid foam, however, is made using physical blowing with a volatile liquid chlorofluoromethane, CFM-11. The more volatile CFM-12 is added in the frothing process where the mixed foam components are dispensed as a froth rather than as a liquid mixture. The CFM gas has a much lower thermal conductivity than air and, because it is retained within the closed

cell structure, it yields a rigid polyurethane foam with a much lower thermal conductivity than foams containing air or other gases.
A typical low density rigid polyurethane foam having a density of 30 kg/m³ consists of 97 percent gas, mostly CFM-11, entrapped within the closed cells of the foam. This explains the unusually high insulation value of rigid polyurethane foam in comparison with insulating materials which rely upon air-filled interstices to reduce the rate of heat loss.

High density rigid polyurethane foams

Figure 7-2   A structural foam housing for a computer terminal.

High density rigid polyurethane foams with densities from about 250 to over 1,000 kg/m³ are strong, rigid materials. They are used alone as moulded structural materials or, together with other materials, to form composite articles having high stiffness and a high strength to weight ratio. Structural foams with densities from about 250 to over 800 kg/m³ with an impervious integral skin, are used to make moulded cabinets, housings for computer terminals and for other specialised equipment. The material combines high stiffness with high heat distortion temperature and a high resistance to impact damage. Examples of the use of these foams for reinforcement include their use in aircraft structures and in many types of sports equipment.

Rigid polyurethane foams are also used in the furnishing industry to mould decorative, simulated wood articles and furniture trim.
Complete chair shells are moulded from rigid polyurethane foams having densities in the range from about 40 to 250 kg/m³, depending on the thickness and design of the chair shell.

Table 7-1 **A comparison of some common thermal insulants**

Material	Density (kg/m³)	λ-value (W/mK)	Thickness required for equivalent thermal barrier (mm)
Rigid polyurethane foam (Impermeable facings)	32	0.017	20
Rigid polyurethane foam (Permeable facings)	32	0.022	27
Flexible polyurethane foam	16	0.035	44
Polystyrene foam	16	0.035	44
Kapok between paper	16	0.035	44
Corkboard: low density	86	0.036	45
Rockwool: low density	100	0.037	46
Hairfelt	190	0.037	46
Glasswool	65 – 160	0.041	51
Rockwool: high density	300	0.041	51
Expanded diatomaceous earth	160	0.045	56
Corkboard: high density	220	0.049	61
Fibre insulating board	220 – 250	0.049	61
Timber: white pine	350 – 500	0.112	140+

The main applications for low density rigid polyurethane foams are:

Figure 7-3   Some applications of rigid polyurethane foam.

In construction. Rigid polyurethane and polyisocyanurate (see glossary) foam-cored, factory made, composite boards are used for the insulation of the envelope of a building (roof, walls and floor). When the composite has rigid facings, such as steel or aluminium for example, it can be made as prefabricated sections of walls and roofs that reduce erection time dramatically and result in a building that is well insulated at the initial construction stage. Cut rigid polyurethane and polyisocyanurate foam slabstock can also be used as an insulant. Boards with thin impervious facings (such as aluminium foil) can be used as partial cavity lining boards, while boards with a facing such as plasterboard can be used to line and insulate walls internally – particularly in high-rise flats and old buildings with solid walls.

Both kinds of foam can also be applied directly by spraying to insulate and seal roof structures, the inside walls of cow sheds and other animal houses and the exterior walls of dwellings. Sprayed foam must always be covered to shield the foam from UV light and inclement weather and – especially inside buildings – to reduce the rate of flame-spread in a fire. Coverings include suitable sprayed membranes, sheet materials and foils.

By pouring or frothing the foam reaction mixture they can be used for cavity insulation and also in the manufacture of foam-insulated and reinforced doors.

Packaging and crash padding. Low density rigid polyurethane foam and isocyanate-derived polyurea foams are used for the pour-in-place packaging of delicate objects. Rigid polyurethane foam is used together with energy-absorbing, semi-rigid foam for crash-padding in aircraft.

In the refrigeration industry. Polyurethane foam is nowadays used in almost all domestic refrigerators and freezers. It is also the most widely used insulant for commercial freezers, cold storage installations, refrigerated vehicles and container transport.

In plumbing and chemical plant. Hot water tanks and pipes are most effectively insulated with rigid polyurethane foam. Domestic hot water tanks now come factory-insulated with rigid polyurethane foam. Many oil storage tanks and pipelines are insulated with rigid polyurethane foams applied in situ either by spraying or cladding with cut, profiled slabstock. Pipework for use at higher temperatures, such as that installed in steam district heating systems, is insulated with polyisocyanurate foam which may be operated at surface temperatures up to 140°C.

For wall, soil and mine stabilisation. The bonding properties of rigid polyurethane are used in replacement of wall ties, renovation of sewers and mine and soil stabilisation.

In ships and boats. Puncture-proof buoyancy, lightweight stiffness and impact resistance is added to ships, and other items such as lifebuoys, surfboards and boats, where rigid polyurethane foam fills a cavity to form a sandwich structure.

Making low density rigid polyurethane foams

The earliest rigid polyurethane foams were made using TDI. The manufacture of blocks of rigid foam from 65:35-TDI or 80:20-TDI was a two stage operation via an intermediate prepolymer stage. The two stages were necessary in order to reduce the exothermic temperature rise during foam production. Single stage or 'one-shot' systems were then developed using modified TDI or 'crude' TDI, (TDI containing undistilled phosgenation products), i.e. TDI variants having modified reactivity and functionality. However, the rapid growth in the use of rigid polyurethane foams dates from the pioneering work done by ICI to develop and manufacture the first commercially available polymeric MDI. It thus became possible to make rigid foams by a simple one-shot process from two low-viscosity liquid components. The lower volatility of polymeric MDI, and therefore the lower hazard involved in handling it compared with TDI, was also a major factor in the development of processes for making foams, and of applications for their use. Most rigid polyurethane foams are now made by mixing polymeric MDI with a polyol blend component containing catalysts, surfactants and blowing agents.

Rigid polyurethane foam is easily made in the laboratory from a wide range of polyols and polyisocyanates. Pre-weighed components, a polyol blend and polymeric MDI, are quickly blended together with the blowing agent. The reaction of the polymeric MDI with the polyfunctional polyol gives a rapid increase in both the temperature and the viscosity of the mixture, so vaporising and trapping the blowing agent. If the rate of polymerisation is fast enough, the blown polymer becomes self-supporting and the blowing agent is retained within the cellular structure. There are two major benefits from blowing with CFM-11. The first advantage is a very low thermal conductivity (measured by the λ-value) that results from the trapped CFM-11 and the second is a reduction in the foam reaction temperature attributable to the heat of vaporisation of the blowing agent. The latter is an important factor in avoiding excessively-high reaction temperatures in some systems. Rigid foams, unlike flexible foams, are rarely blown only by chemical blowing using carbon dioxide from the reaction of polyisocyanate and water. Carbon dioxide has a relatively high thermal conductivity and rapidly passes through the cell walls to be replaced by air. The use of some water-blowing as a minor additional aid may be advantageous in modifying the reaction profile and the flow behaviour of the rising foam. Water blowing is rarely used in polyisocyanurate foam systems because the amine end-groups produced by the initial reaction of the water and isocyanate react very quickly with further isocyanate and tend to inhibit the isocyanate trimerisation reaction.

It is customary to use polyols and polyol blends with average functionalities from 3 to 5 and with mean molecular weights from about 300 to 800 to make rigid polyurethane foams. The development and formulation of rigid foam systems for industrial use requires skilled research and development that must include the testing of candidate systems under actual conditions of application. A satisfactory foam system must give a constant, specified product and the system must allow sufficient latitude in processing for use under actual conditions of application.

As discussed in Chapter 4, the bubble formation in a polyurethane foam reaction mixture must be nucleated in order to obtain a fine, uniform cell structure and, because the number of nucleation bubbles falls with increasing cream time, the final cell size is related to the reactivity of the foam system.

Polyisocyanurate foams. All rigid foams made from polyisocyanates are often loosely described as polyurethane, or urethane, rigid foams, although some systems actually contain relatively few urethane groups. The principal linkages formed during the polymerisation may be urethane, isocyanurate, urea or carbodiimide groups, depending on the foam system. Rigid polyisocyanurate foams are particularly important because of their resistance to high temperatures and their relatively low combustibility. In these foams, most of the isocyanate groups are polymerised to produce isocyanurate ring structures, which are thermally stable. Complete polymerisation to isocyanurate unfortunately results in products that are extremely brittle and of no practical interest. Therefore, almost all of the established polyisocyanurate foam systems, such as the 'Hexafoam' systems, are actually polyurethane-modified polyisocyanurate systems. The purpose of the modification, made by incorporating a small amount of a selected polyol, often a diol, is mainly to control the degree of cross-linking of the polymer network in order to optimise the properties for the intended application.

Special polymeric MDIs. Although most rigid polyurethane and polyisocyanurate foam is made from standard grades of polymeric MDI, such as 'Suprasec' DNR and 'Suprasec' DND, polymeric MDI compositions with a higher functionality and viscosity are produced especially for some types of rigid foam manufacture (table 7-2). The higher viscosity

makes it easier to control the uniform distribution of the reacting foam mixture in the production of many types of foam laminates and of continuous slabstock foam. Compared with standard grades of polymeric MDI these products give a foam reaction mixture with a higher initial viscosity and a faster rate of increase in viscosity. These factors give foam with cells that are less elongated and, therefore, with higher strength and dimensional stability. The use of MDI with a higher functionality has also been found to be useful in making fire resistant polyurethane and polyisocyanurate foams.

131

Table 7-2 Polymeric MDIs for making rigid polyurethane and polyisocyanurate foams.

Product	Description	NCO value	Viscosity (mPa s at 25°C	Average functio- nality	Typical applications
'Hexacal' LN	Modified polymeric MDI	27.0	1300	2.70	Isocyanurate foam as contin- uous laminate.
'Hexacal' SN	Modified polymeric MDI	26.0	2700	2.70	Isocyanurate foam as contin- uous slabstock.
'Hexacal' F	Polymeric MDI	30.7	230	2.70	Isocyanurate foam moulding.
'Suprasec' VM80	High functionality polymeric MDI	30.6	430	2.80	Rigid foam of modified properties.
'Suprasec' VM85 HF	High functionality polymeric MDI	30.4	550	2.9	Rigid foam of modified properties.
'Suprasec' VM90 HF	High functionality polymeric MDI	30.2	900	3.0	Rigid foam in continuous lamination and slabstock with improved properties.

Manufacturing processes for low density rigid polyurethane foams

The free-flowing expansion of a reacting rigid polyurethane foam and its strong adhesion to most materials provides the basis of the many manufacturing processes in which the low density foam becomes part of a composite product such as boards and panels. But there is, nevertheless, substantial manufacture of rigid foam slabstock i.e. unfaced foam which is converted, using modified wood-working machinery, into insulation boards, pipe sections and many other shapes. Foam may also be used to fill complicated cavities such as refrigerator cabinets.

Slabstock production

Rigid foam slabstock is made both by discontinuous and continuous production methods. Discontinuous, or batch production, methods range from simple 'hand' mixing to those using high-output machines for mixing and dispensing.

Discontinuous manufacture

The simplest method of discontinuous manufacture is to weigh and hand-mix the components of a slow reacting system and pour the reaction mixture into a wooden or cardboard mould of the required

132

The minimum demould time depends on several factors:
- The reactivity of the foam system.
- The density of the foam.
- The temperature reached inside the block of foam.
- The size of the block of foam.
- The degree of overpack in the mould. This depends upon the amount of restraint applied to the rising foam by the floating lid.

Figure 7-4   Batchwise slabstock production using a floating lid.

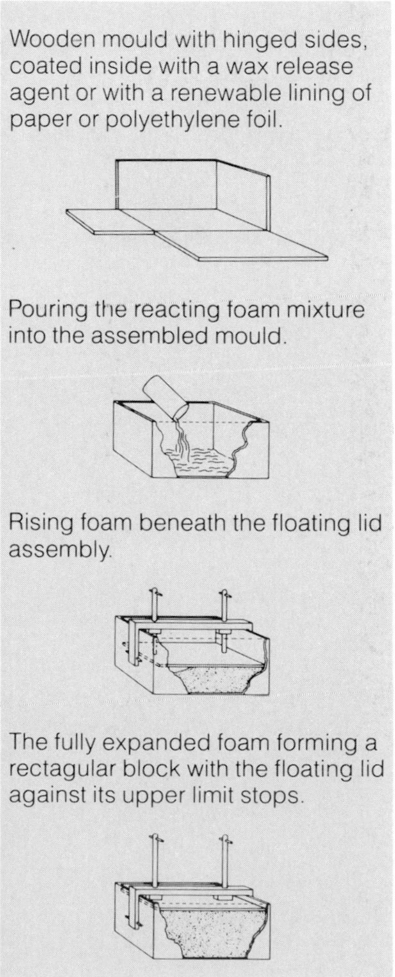

Wooden mould with hinged sides, coated inside with a wax release agent or with a renewable lining of paper or polyethylene foil.

Pouring the reacting foam mixture into the assembled mould.

Rising foam beneath the floating lid assembly.

The fully expanded foam forming a rectangular block with the floating lid against its upper limit stops.

dimensions. Small quantities of a slow reacting foam mixture may be mixed by hand using a simple disposable stirrer but any quantity above about 500 g is best mixed with a mechanical stirrer. There are many suitable designs of propeller and turbine type agitators available from equipment suppliers depending on the batch size and the viscosity of the foam reaction mixture. Foams with a wide range of properties can be obtained by adjusting the formulation in order to vary the density and hardness of the foam. However, the method is restricted to chemical systems that can be formulated to give a long cream time at ordinary temperatures. It is usual to flatten the top of the rising foam by using a floating lid. A convenient form is shown, diagrammatically, in figure 7-4. This simple slabstock process can provide about two blocks of foam per mould per hour, as each block must be left in the mould for at least 10 to 15 minutes after the end of foam rise to avoid any distortion of the foam.

It is usual to use sufficient foam mixture to give 3% to 5% overpacking of the mould. This is usually sufficient to give an almost flat-topped block and a more uniform foam of reduced anistropy compared with that from unrestrained foaming. The weight of the floating lid assembly should be sufficient to apply the required restraint to the rising foam, but the lid should be so designed that accidental over-filling of the mould will lift the whole of the floating lid assembly rather than exert excess pressure on the mould.

Machine dispensing and mixing. Some of the disadvantages of batch mixing are the limited choice of chemical systems, the high labour cost, the loss of material remaining in the mixing vessel and the increased hazards associated with handling the chemicals compared with those when machine dispensing is used. These disadvantages are reduced or eliminated by the use of machine mixing and dispensing. The capacity of the dispensing machine should be high enough to dispense the required weight of material into the mould within the cream time of the foam system. It should be borne in mind that, because of the energy input during the mixing operation, a foam system dispensed through a mechanical mixer, or a high-pressure impingement mixer, will show a significantly reduced cream time compared with that obtained by batch mixing. The manufacture of large, homogeneous blocks of foam from highly reactive foam systems requires a dispensing machine with a high capacity. For example, to make a foam of density 30 kg/m^3 in a mould with dimensions 2m × 1m × 1m requires about 66 kg of foam reaction mixture. To dispense this within 20 seconds requires a machine output of 200 kg/minute. The capacity and the cost of the machine required are reduced when the ICI mini-slabstock process is used.

Mini-slabstock process. This is a method of making blocks of foam by using a traversing dispense tube to distribute the foam mixture in

133

This diagrammatic representation of the process does not show the safety and venting equipment required (see Chapter 10)

Figure 7-5   ICI rigid foam mini-slabstock process using a traversing dispense tube.

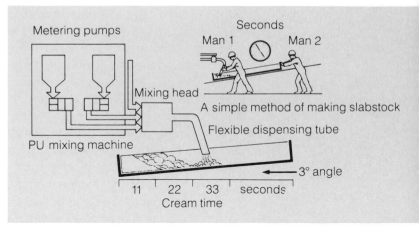

an inclined mould, figure 7-5.

The method may be used both with rigid polyurethane foam systems and rigid polyisocyanurate foam systems, the latter having relatively short cream times. Low level frothing with CFM-12 increases the robustness of the polyisocyanurate foam process by helping to prevent newly-deposited foam reaction mixture from flowing

Table 7-3   **Typical physical properties of rigid polyurethane and rigid polyisocyanurate slabstock foam**

	Foam type	
	Polyurethane	**Polyisocyanurate**
Reaction profile (Mechanical mixing at 20°C)		
Cream time	24 seconds	13 seconds
Rise time	90 seconds	55 seconds
Tack-free time	150 seconds	50 seconds
Foam properties		
Density (kg/m³)	33	34
Apparent closed cell content (BS 4370:Part 2:1973)	92	92
Thermal conductivity (λ value, W/mK)		
Initial value at 0°C	0.018	0.018
Aged value at 0°C	0.023	0.024
Compression strength at 10% strain, (kPa) (BS 9370)		
Parallel to block height	225	170
Parallel to block length	180	150
Parallel to block width	130	90
Tensile strength (BS 4370:Part 2:1973) (kPa)		
Parallel to block height	340	260
Parallel to block length	240	220
Friability (BS 4370:Part 3:1974) (% weight loss/minute)	5	30

Some foam systems: 'Daltolac' 51 with 'Suprasec' DNR/DND
'Daltolac' K135/17 with 'Suprasec' DNR/DND
'Hexacal' SN with 'Hexacal' Activator S4

Figure 7-6 Cutting slabstock foam

underneath that already reacting. The mini-slabstock process will produce blocks several metres long with a cross-section of about 100 cm × 50 cm from a machine with an output as low as 50 kg/minute. ICI systems are available that produce rigid polyurethane and polyisocyanurate foams with densities from 30 to 200 kg/m^3. Typical products are described in table 7-3.

Continuous manufacture

This is the most economical method of making large quantities of slabstock foam. It also tends to give a product of higher quality, as it is easier to control the cell size, the cell structure and the uniformity of the foam in continuous production. The foam reaction mixture is dispensed continuously into a trough, formed by paper or polyethylene film, on a moving conveyor belt. The machines used for making rigid foam slabstock are similar in principle and appearance to the conventional inclined conveyor machines used to make flexible foam slabstock which are described in Chapter 4, (figures 4-6 and 4-8). Rigid foam machines, however, must be designed to accommodate the foam pressure on the sidewalls, pressure that occurs just after the foam has risen completely. Most machines for making rigid foam slabstock continuously have sidewalls consisting of vertical conveyors which are driven synchronously with the base conveyor. It is also possible to use fixed sidewalls faced with closely-pitched vertical rollers to reduce the friction between the sidewalls and the moving paper or polyethylene film covering the rigid foam block. A flat-topped block is obtained by the use of a top conveyor that is adjustable in height, or by a modified Planibloc/Hennecke type of top platen arrangement, see figure 4-8.

The rise profile of the foam is an important factor in obtaining high quality foam. The ideal system combines a short cream time with a controllable, extended rise time and an early tack-free time. The advantages of using reactive chemical systems with short cream and tack-free times are, first, that a foam with a fine uniform cell size and optimum properties can be obtained more easily and, second, that the length of expensive, driven sidewall conveyor required can be kept to a minimum. Another reason for using fast curing foam systems is to allow early cutting of the foam block into lengths that permit easier transport of a material which, unlike flexible foam, cannot be bent to negotiate curves and changes in conveyor angle. The problems associated with inclined foaming conveyors may be overcome by using an inclined section beneath the dispensing head only.

Production of boards and panels

A very wide selection of foam-cored panels and insulation boards is available in which the rigid polyurethane foam is bonded to the facing materials to form a strong, rigid composite. Some laminates

135

are made by the adhesive bonding of skins to foam boards cut from slabstock foam, but most are made by the injection or pouring of a foam reaction mixture into pre-assembled panels or by continuous manufacture. Large quantities of laminates for buildings are made by continuous processes. The facings may be flexible (e.g., aluminium foil or coated paper) on both sides, or may be made with a rigid material such as plaster-board on one side of the laminate. Some building panels are also made by spraying foam onto the backs of tiles or bricks assembled in a jig.

Adhesive bonding

Rigid foam boards may be faced with sheet materials using conventional adhesives such as solvent-based contact adhesives, hot-melt adhesives and chemical bonding agents based on epoxy resins, unsaturated polyesters and polyurethanes. The foam boards should be vacuum-cleaned to remove cutting dust before allowing them to contact the adhesive.

Another method of construction is by laying up glass-fibre reinforced polyester or epoxy skins onto a cut or moulded rigid foam. This gives very rigid laminates and is useful for making curved building panels, and in making boats and sporting equipment.

Continuous lamination processes

The continuous production of laminated boards is a major application for rigid polyurethane foams. This success results from the wide range of functional laminates that save labour and energy both in their manufacture and in their applications in the building industry.

The first commercially successful continuous laminate production was established by 1963. Since that time there has been continuous improvement in the versatility and economy of the process. Laminated rigid foam boards are produced with the range of properties required to meet the diverse structural and statutory requirements of most countries. The biggest single application in Europe is foam cored laminate for roofing and roof insulation. These may be made from rigid polyurethane foam or from polyisocyanurate foam. The market is expanding as new laminate constructions are produced that not only supply the required heat barrier but also simplify roof construction and reduce labour costs. Several laminate systems, such as that illustrated in figure 7-7, retain the traditional tiled or shingle-covered roof.

There are two principal types of continuous laminating machines; the continuous horizontal laminator and the inverse laminator. The former uses facing materials that are flexible enough to be supplied in long lengths on rolls. The inverse laminator is a variation developed by ICI, that enables one face to be a rigid facing supplied

Figure 7-7   Roofing laminate with traditional tiling.

in sheet form. There is a third process that produces laminate faced with metal sheet on both sides.

Figure 7-8 The continuous manufacture of rigid foam laminate for roof construction and insulation.

Horizontal lamination

In horizontal continuous lamination, the upper and lower facing materials are drawn horizontally from rolls. The temperature-conditioned bottom facing is coated with a uniform layer of foam reaction mixture from a continuous metering, mixing and dispensing machine. The edges of the bottom facing may be folded to form a faced square edge to the laminate. Alternatively, a continuous flexible, or a discontinuous rigid, edging strip may be introduced as the foam rises. In most processes the foam rises completely or almost completely before the upper facing contacts the foam. The soft, newly-formed laminate then passes into a temperature-controlled heated conveyor press where the foam reaction is completed with good contact and adhesion of the foam to the facings. When the laminate leaves the conveyor mould press, the edges of board made without edging strips are trimmed before it is cut into pre-set lengths, stacked horizontally and allowed to cure and cool.

Inverse lamination. Rigid facings such as reinforced-cement boards, particle boards, plywood and plasterboard may be used to form one face of the foam sandwich by the use of automatic or manual indexing to insert boards from a stack. Laminates with one flexible and one rigid facing may be made on most types of horizontal laminator but the ICI 'inverse laminator' is specially

137

Figure 7-9   The ICI inverse laminator.

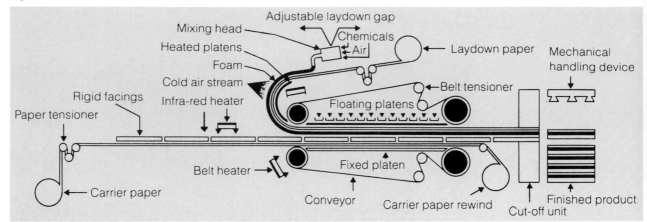

Production rate: (single shift of 5 or 6 workers), about 550,000 m/year or higher depending on the thickness and the number or different products manufactured.

Maximum production speed	7m/min
Maximum board width	1.25 m
Minimum board length	1.25 m
Board thickness range	12.5/65 mm
Foam density range	32/50 kg/m^3

designed for this job. The foam mixture is applied to what will be the inner surface of the upper, flexible facing which is then inverted as the foam rises. The freshly risen foam, whilst still tacky, is pressed into contact with the rigid boards carried on the lower conveyor through an ICI/Viking floating platen press (figure 7-9).

Double metal-faced lamination. Laminate faced on both sides with metal sheet is made on specially designed laminators. Aluminium or steel sheet, primed on one side to ensure good adhesion of the foam and often with a coloured coating of PVC or a galvanised finish on the other side, is fed from coils and formed into stiff facings by profiling rolls. The continuous profiled sheet is then conditioned to a pre-set temperature to ensure satisfactory foam adhesion and filling at the lamination stage. A typical production unit (figure 7-10), is designed to make foam-cored panels with profiled-metal facings, at rates up to 10 metres/minute. The high stiffness that results from the longitudinally-profiled metal makes these panels very popular for the construction of large, insulated roofs.

Figure 7-10   Double metal-faced lamination.

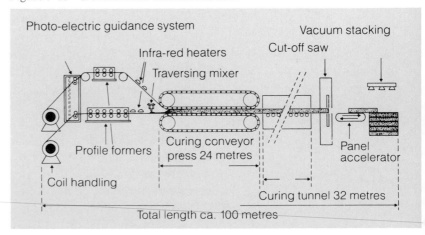

Lamination machines

There are many designs of horizontal continuous lamination machines. The main functional differences are in the method of metering the foam chemicals, the method of distributing the foam reaction mixture on the bottom facing and in the function and operation of the conveyorised press section (see figures 7-11 to 7-13).

Metering and mixing the components. Several methods are available, such as:
– High pressure metering with impingement turbulence mixing.
– Low pressure metering and mechanical mixing.
– Low pressure metering with a static, air-assisted, turbulence mixer.

Most new machines have metering systems that are suitable for operation at both high and low pressures. Most of the mixers designed for use with rigid polyurethane foams are two-component mixers but two-, three- and four-component metering systems are often used in series with batch or in-line blending systems.

Foam mix distribution. Distribution onto the bottom facing material is obtained by:
– Traversing a mixing head with a spray outlet or with divided
– stream liquid pouring, or
– fixed or traversing mixing head(s) combined with a spreading device or a calendar roll under which the top facing passes.

Figure 7-11   The original ICI/Viking floating platen laminator.

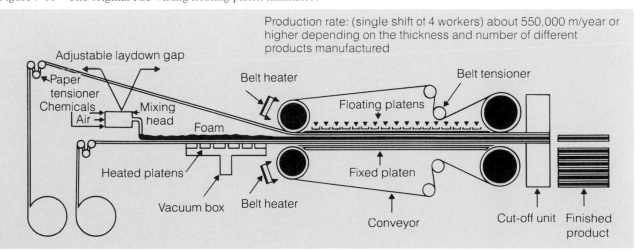

Production rate: (single shift of 4 workers) about 550,000 m/year or higher depending on the thickness and number of different products manufactured

Production speed	4 to 8 m/min
Maximum board width	1.25 m
Minimum auto-cut board length	1.25 m
Board thickness range	6/100 mm
Foam density range	32/90 kg/m^3

Conveyor presses. There are three main types of continuous moulding press used in continuous lamination:
– The ICI/Viking floating platen press which uses rubber belt conveyors over flat platens. The lower platens are fixed and heated, but the upper platens are articulated and floating and designed to apply the minimum stress to the foam (figure 7-11).

139

The ICI floating platen laminator, (figure 7-11), is used for many lightweight facings such as:
Polyethylene-coated kraft liner board
Kraft liner board/aluminium foil laminate
Chipboard liners
Glass tissues and polyethylene coated glass tissues
Glass tissue/aluminium foil laminates
Bitumenised papers
Kraft papers.

The floating platens apply a low and adjustable stress to the freshly-risen foam. This gives the minimum distortion of the foam cell structure resulting in the optimum strength/density ratio of the foam core. This is particularly important with lightweight facings.

– Fixed-gap conveyors which are designed to ensure a flat uniform product of the required dimensions by restraining the expansion of the foam. The pressures likely to be generated require the use of strong conveyors which are usually faced with steel slats (figure 7-12).

Figure 7-12   Fixed gap laminator.

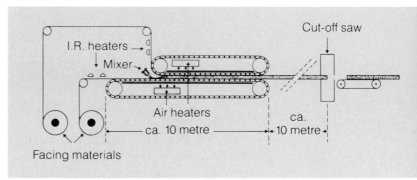

– The third type of curing press uses upper and lower flat metal press plates in an enclosed tunnel. There are no moving belts or slats. The laminate is drawn through the press tunnel on a film of heated air which is continuously supplied under pressure through angled nozzles in the press plates (figure 7-13).

Figure 7-13   High speed continuous lamination: The Kornylak 'Process tunnel'.

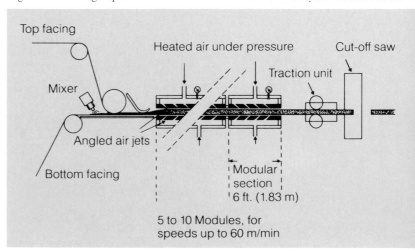

The choice of machinery for continuous lamination depends upon the type and the range of facings to be laminated and upon the production rate required. Conventional laminating machines

140

operating at production rates from about 4 metres/minute to about 10 metres/minute depending on the foam thickness, will handle a wide range of flexible facings.

High speed lamination. High speed laminators may use extended floating platen or fixed gap conveyors. In general, to allow some scope to vary the foam system, 25 mm thick foam-cored laminate needs to remain in the press for about one minute. A conveyor press 30 metres in length, for example, will allow production rates up to about 30 metres/minute. The alternatives are free-rise tunnels, which have limited application, and the Kornylak 'Process tunnel', (figure 7-13).

The main problem in high-speed lamination lies in obtaining an even distribution of the foam reaction mixture on the bottom facing. The single traversing mixing head is limited to production rates of lamination below 9 to 10 metres/minute at widths up to 1.25 metres. At higher speeds the linear traverse speed and the high rate of traverse speed reversal that is necessary to avoid excess foam mixture deposition at the edges of the laminate, imposes mechanical forces that are too high for reliable operation. One solution is to use more than one mixer carried on one or more traverses. This is not popular because of difficulty in maintaining a matched performance from separate mixing and distribution units. The preferred method is to use a single mixer, usually traversed at production rates up to about 10 metres/minute. For manufacture at higher speeds, the mixer is often fixed centrally and is operated in conjunction with a calendering roll or other spreading device. Any spreading device should operate with the minimum reservoir of reaction mixture. Too large a quantity of reaction mixture in front of the unit tends to give a spread layer containing reaction mixture of varying age, which results in the uneven onset of foaming. This, particularly with fast rise, low-viscosity foam systems such as isocyanurates, may cause redistribution of the foaming reaction mixture and the production of a non-uniform product. Calendering rolls and other fixed-gap spreading devices are most effective with high linear facing speeds, i.e. at high production rates. The sensitivity of the high speed process to temperature and to changes in foam reaction rates, especially with isocyanurate foam systems, is reduced by the use of high viscosity grades of MDI and by the incorporation of continuous glass-fibre reinforcement. Both factors help to reduce the effects of uneven creaming of the foam reaction mix, of uneven heat transfer and of uneven facing stresses acting on the foaming material.

Glass-fibre reinforced laminates. The properties of rigid foam-cored panels and boards may be further improved by the incorporation of reinforcing fibres and meshes. Many types of reinforcement have been found useful, including steel wire meshes, natural fibres, and various synthetic textile fibres, but by far

The glass is bonded to the foam core and improves the following properties:
- Dimensional stability, especially at high temperatures.
- Fire resistance, especially of polyisocyanurate foams.
- Stiffness.
- Impact resistance.
- Resistance to thermal shock from hot bitumen and asphalt in built-up roof insulation.

the most popular reinforcement is glass-fibre. Special grades of glass-fibre mesh have been developed for use in continuous lamination.

The most useful effect of the incorporation of glass-fibre is the improvement in the fire-resistance of polyisocyanurate foam laminates. The glass-fibre embedded in the foam prevents the development of deep fissures in the protective carbonaceous char that is formed when polyisocyanurate foam is exposed to high temperature flames. The effect may be demonstrated in the 2-foot corner test of ASTM D3894 (figure 7-14).

Figure 7-14   Glass-fibre reinforced laminate faced with aluminium foil.

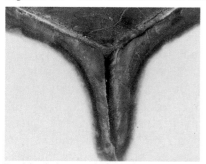

Residual char – reinforced

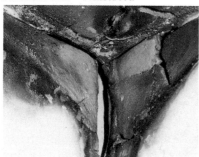

Residual char – unreinforced

Glass-fibre reinforced insulation boards may be made continuously on a conventional horizontal laminator equipped with a suitable unit to feed the glass into the foam reaction mixture at the foam lay-down position. The form of the glass-fibre and its surface treatment are important. Many systems and processes for doing this are patented.

Discontinuous production

Foam-cored panels are made, singly or in batches, by several discontinuous methods. These are often preferred to continuous manufacture for low rates of production, for very thick or large panels and for panels of complex construction. Two production methods, the 'Press Injection Process' and the 'Vertical Lamination Process', were pioneered in the laboratories of ICI Polyurethanes. In the Press Injection Process, a foam reaction mixture is injected into the hollow interior of a panel held horizontally in a press or jig. The foam is overpacked by from 10% to 15% to ensure a complete and stable foamed core. In the Vertical Lamination Process a panel with an open top edge is held vertically in a jig and filled with foam by successive traverses of a mixing head. The foam is not overpacked and it is possible to achieve a lower uniform density than can be obtained by press injection.

The press injection process. This is a very popular process because of its economy and versatility. A typical production arrangement is illustrated diagrammatically in figure 7-15.

Figure 7-15   The press injection process.

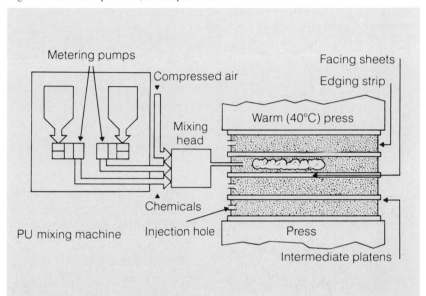

The facings may be rigid or flexible, flat or profiled. The edges may be complex extrusions designed to simplify jointing and fixing the panels on a building. Alternatively, a moulding frame may be used, made up of pieces of wood that are removed after moulding is completed and re-used. The sections of such a frame can be profiled to provide tongue-and-groove (male and female) ends to the foamed panel, to simplify the formation of joints and to eliminate cold bridges. Polyurethane or polyisocyanurate foam systems, liquid or frothed, may be used. Internal reinforcement may be incorporated. Large panels such as those forming the side of a house or the side of a refrigerated vehicle or container, may be made by sub-dividing the interior of the hollow panel into sections of a size suitable for foam injection with the equipment available (figure 7-16).

The choice of press or jig depends upon the required production rate. Multi-daylight presses can be used to hold a stack of up to about eight pre-assembled panels. Light presses capable of applying a clamping force of about 1.0 kg/cm^2 (about 1 bar or 100 kPa) are satisfactory for manufacturing of most types of panel. Temperature control is important. To ensure a reproducible product of high

143

Figure 7-16   The press injection process: filling a large sub-divided panel.

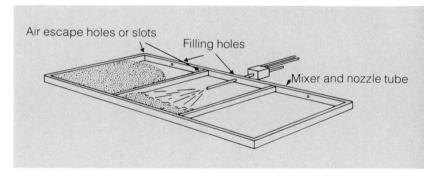

Table 7-4 Typical cycle for filling a panel with a 50 mm thick polyurethane foam

Operation	Minutes
1. Load a stack of pre-assembled panels in the press and close the press.	1
2. Inject the panels with foam reaction mixture, say, 8 × 10 seconds injection and nozzle manipulation.	2
3. Foam rise and cure.	10
4. Open press and remove stack of panels. Contingency time	1 1
Typical press cycle time	15

Figure 7-17   Using a low pressure machine to inject foam into panels held in a press.

quality, and with good adhesion of the foam facings to the foamed core, the facings should be at a temperature of between 30 and 40°C at the time of injection of the foam, the optimum temperature depending on the conductivity of the facing materials and the reactivity of the foam system. Stacks of pre-assembled panels may be warmed by circulating hot air. The rate of production depends upon the time taken to assemble the panels and also on the dwell-time in the press, which depends largely upon the thickness of the foamed core and the degree of overpack. Each centimetre of foam thickness requires a dwell time in the press of about two to three minutes in order to avoid distortion of the panel on its release from the press. Rigid polyurethane foam systems that allow somewhat shorter cycle times are available. The precise minimum dwell time in the press depends on a number of factors; the temperature of the panel facings, the temperature and the reactivity of the foam chemicals, the density of foam produced, the distribution of the foam mixture in the panel, the degree of overpacking, i.e. the amount of foam mixture injected above the minimum amount required to just fill the panel with foam, and upon the type of dispensing and mixing equipment used. The minimum dwell time is therefore found by experiment. A sufficient time margin should be allowed for day-to-day variations in the ambient conditions.

Very long panels are filled using a long dispensing tube or, preferably, a specially-designed flat mixing head which is inserted into the panel and withdrawn at a controlled speed during the dispensing cycle. Foam reaction mixture is thus distributed evenly from the back to the front of the panel and foaming proceeds smoothly from the back of the panel, so minimising the number of air escape holes required (figure 7-18). Panels up to 12 metres in length may be filled with the standard equipment.
Alternative methods of filling very long panels include simultaneous injection from two or more machines through holes in the long edge of the panel, or, using one machine only, by successive injections through carefully spaced holes on the long edge. The latter method requires precise positioning of the filling holes and accurate timing

Figure 7-18   The press injection process: filling long panels using a thin mixer and lance.

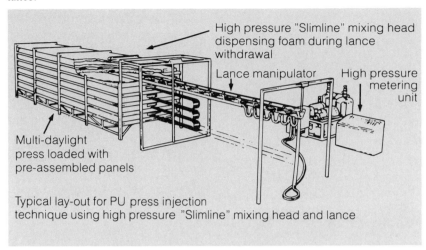

of the consecutive injections in order to obtain a reproducible and uniform foamed core. The use of a flat mixer and lance is preferred, especially one with an arrangement for automated withdrawal.

The vertical lamination process. This process was developed for the production of large foam-cored panels without the need for high output foam machinery or for sub-division of the panel. It has the major advantage that thick, large panels may be filled with highly-reactive foam systems having short cream times. Panels are produced in long lengths with a range of rigid facings including deeply profiled coated steel and aluminium sheet. The equipment required is a vertical clamp or jig to support the panel faces and a dispensing machine with a low output capacity (from about 4 kg/min to about 15 kg/min, depending on the panel length and thickness), feeding a lightweight mixing head capable of producing a reactive foam mixture frothed with CFM-12. The froth from the mixing head is laid down in the panel using a traversing mechanism of constant speed. Typically, the mixing head is traversed from end to end of the panel at a rate about 30 cm/sec. The output is chosen, depending on the thickness of the panel and the density of the foam, to give a foam layer height of about 15 cm for each passage of the traversing head. A moulded top edge may be obtained by a suitable arrangement of top edge lids that are closed in sequence following the final traverse of the mixer. The pressure generated by the expanding foam is much lower than that exerted in the press injection process except when the final layer of foam is restrained by top closures. The design of the jig, the use of manual or automatic loading and locking mechanisms, and the method of loading will depend upon the required production rate and the type and size of facing materials to be used. Advice on suitable equipment may be obtained from machinery suppliers.

145

Other lamination processes. There are many other methods of making laminates with a core consisting wholly or partly of polyurethane foam. In one widely-used discontinuous method the foam reaction mixture is distributed by an airless spray into a tray-like bottom facing. A top closing sheet is applied as the foam rises.

Such processes may be operated on a production line using a conveyor press of the fixed-gap type. Similar processes use the adhesive and gap-filling properties of rigid polyurethane foam reaction mixtures to bond together insulating fillers such as expanded clays and hollow glass spheres to form the cores of sandwich panels. Such panel constructions can have a high compressive strength combined with improved fire resistance.

Cavity filling

Polyurethane foam reaction mixtures can be formulated to flow easily during the foam expansion in order to fill complicated cavities. One of the most successful applications of this property is in the insulation of refrigerators and freezers. There are several advantages in using rigid polyurethane foams in these appliances. The most important is that, because of the low λ-value of the the foam, it is possible to use much thinner insulation panels than is possible when mineral wool or glass-wool insulant is used. Consequently, for a typical domestic appliance, the useful refrigerated storage space is greater by up to 30%. Other advantages include:

– Reduced labour costs in production line assembly of the appliances.
– The excellent bond between the foam and the interior and exterior casings gives a stable, impact-resistant sandwich structure using lightweight thermoplastic liners and thin sheet-metal or foil outer casings.
– The core of rigid polyurethane foam forms a gap-free insulating layer which, unlike mineral wool and similar materials, is unaffected by the vibration of the refrigeration unit.

Refrigerators

Almost all refrigerators and freezers are now made using rigid polyurethane foam. The pre-assembled inner liner and outer casing are loaded into a temperature-controlled supporting jig, foam reaction mixture is injected into the cavity between the inner and outer casing and, after an interval of time for curing, the completed cabinet is removed from the jig. There are two different arrangements of the production line. In the first, a single mixing head fed by a dispenser operating at either low or high pressure is used. The cabinets, in their jigs, are carried by carousels or conveyor tracks through the foam filling and curing stations. The second

146

arrangement uses stationary jigs each with its own mixing head. A single high-pressure metering unit with re-circulation lines can be used to feed up to 20 dispensing heads, depending on the shot times and the frequency of operation (figure 7-19).

Figure 7-19   Static jig refrigerator/freezer production line.

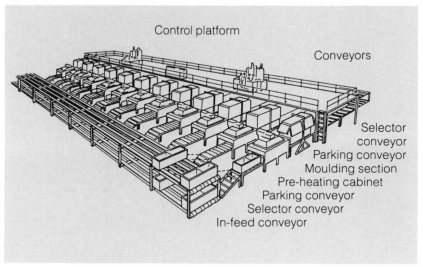

The feeding, loading and foam injection of the cabinets may be controlled by a microprocessor, thus permitting simultaneous production of different sizes of cabinet, each with its own optimum production cycle time.

Whatever the arrangement of the production line, it is important that the design of the cabinet, the method of joining the inner and outer casings, the position of the filling point and the jig design are matched with a suitable foam system. It is suggested that an experienced foam system supplier be consulted at the design stage. Refrigerators are always filled with the door opening facing either downward or upward so that the expanding foam can flow to fill the cavity without developing excessive pressure. A typical jig is shown in figure 7-20. The jig is shown in the open position so that the core plugs for the refrigerator and freezer compartments can be seen. The combined unit is mounted with the door openings uppermost. This is the preferred orientation as the liquid foam reaction mixture, initially of low viscosity, will flow around the back of the assembly. The foam rises evenly between the inner and outer casings and reaches the joint between the two late in the foam reaction when the foam viscosity is high and leakage less likely. Where the construction of the appliance requires foam filling with the door opening facing downward careful sealing of the front joint is necessary. The use of a frothed foam system may be advantageous. It reduces the risk of leakage when filling with the door opening facing downward but increases the number of possible variables in the foaming process.

147

Figure 7-20   Jig for foam-filling a typical refrigerator/freezer unit.

Figure 7-21   Pouring rigid foam reaction mixture into a refrigerator door panel.

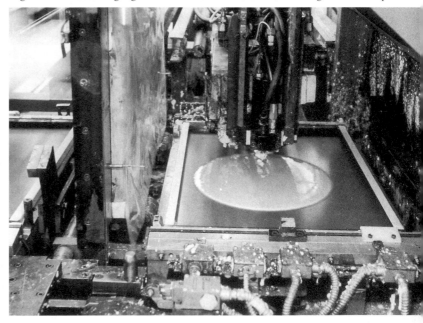

Refrigerators may also be insulated using a separately moulded foam layer or by moulding rigid polyurethane foam in contact with either the outer cabinet or the inner casing only. These three methods of construction give a product having lower strength than that obtained by injecting foam to produce a rigid foam-cored sandwich and the absence of an impermeable layer on one face of the foam means that the equilibrium λ-value will be higher. A better method of making units without an outer rigid casing, is to mould the foam around the inner refrigerator-liner in a mould lined with an impermeable foil which becomes the outer surface of the foam sandwich. This construction is often used to make units for installation into custom-made casings and kitchen units.

The doors or lids of refrigerators and freezers are also insulated with polyurethane foam, usually by a foam-injection technique similar to that used to make sandwich panels. The lid is assembled from an inner and outer shell and placed in a jig. The loaded jig is carried on a carousel or conveyor through the conditioning, foam filling and curing stations. Refrigerator doors are also manufactured by pouring the foam reaction mixture into the outer case of the door which is held on the lower platen of a jig (figure 7-21). The inner liner is then clamped over the rising foam to form the complete door. High production rates for making simple doors and freezer lids may be obtained more economically by using a fixed-gap conveyor press to restrain the lid assembly during the rise and the initial cure of the foam core.

Other cavity filling applications

There are many other applications of rigid foam where the foam mixture fills a cavity to form a rigid sandwich structure. Rigid polyurethane foam is widely used in boat-building and other marine applications where the closed-cell foam core gives puncture-proof buoyancy in addition to lightweight stiffness and impact resistance. The mechanical strength and lightness of rigid foam sandwich structures is also utilised in the construction of cars, commercial vehicle cabs, buses and trains (figure 7-22 and 7-23).

Figure 7-23   Ambulance constructed by injecting polyurethane foam between the inner and outer skins of GRP.

Figure 7-22   Sports car incorporates rigid foam sandwich structure.

A wide range of formulations for foams is available, depending upon the type of cavity to be filled, the requirements of economic production and the relative importance of the thermal and mechanical properties of the foam in a particular application. In many instances, such as foaming between a pipe and an outer protective pipe or sheath, for example, the optimum thermal insulation with adequate reinforcement is obtained with rigid polyurethane foam of a density between 32 and 40 kg/m³. In other applications a high compression strength is most important and foams having densities from 500 to 600 kg/m³ may be used. Engineering applications include foam-lined roadways, structural foam cores for skis, aeroplane propellers, windmill vanes and linings in tunnels and sewers.

Rigid polyurethane foams are also very widely used in lifebuoys, surfboards, sailing boards and boats. A core of rigid polyurethane foam provides the required buoyancy and also greatly strengthens the composite structure. Several methods of construction are used:

– Rigid polyurethane foam reaction mixture is injected or poured into the preformed shell of a sail board, surf board or boat, or into the cavity between the inner and outer hulls of a boat. See figures 7-24 and 7-25.

Figure 7-24   The polyethylene skin of a sailboard is held in a press and injected with rigid polyurethane foam.

Figure 7-25   The complete, foam-cored sailboard is removed from the press less than three minutes after injecting the liquid, foam reaction mixture.

Figure 7-26   A typical sailboard with a rigid polyurethane foam core to provide lightweight strength and buoyancy.

– A block of rigid polyurethane foam is shaped and then covered with glass-fibre scrim and polyester resin to form the outer shell of a boat or sail board.

– The inner and outer hulls of a boat which have been preformed in glass-fibre reinforced polyester or vacuum formed thermoplastic sheet material, are stiffened before assembly by a layer of rigid polyurethane foam. The rigid polyurethane foam is applied by spraying to the inner surface of the outer hull and to the outer surface of the inner hull. Once the inner and outer hulls are assembled together then rigid polyurethane foam may be injected to fill the cavity and bond together the foam coated inner and outer hulls. See figure 7-27.

Aircraft propellers, 3 metres in diameter, made with carbon-fibre reinforced foam cores show a saving in weight of up to 60 kg compared with those made from conventional aluminium alloy.

Figure 7-27   Glass-reinforced polyester boats with rigid polyurethane foam cored hulls.

On-site foaming

Although most rigid polyurethane and polyisocyanurate foam is made in factories producing composite products, two-component foam systems matched to portable dispensing machines are widely used in the construction industry to apply insulation by spray coating and cavity filling. In Europe, possibly because of the problem of spray drift in the wind and the need for protective respirators, the application by spray of rigid polyurethane and polyisocyanurate foams is largely confined to industrial uses. In North America, Japan and elsewhere, extensive use has been made of on-site spraying for the insulation of domestic buildings in addition to that of storage tanks, industrial roofs and other structures.

Sprayed rigid polyurethane and polyisocyanurate foams. Roofs, storage tanks, ducting and pipework (whether above or below the ground) are all commonly insulated by the spray-application of rigid polyurethane and polyisocyanurate foams. Most of the work is carried out using a hand-held, high-pressure, spray gun. The two components are metered separately under pressure, through non-return valves into a small chamber in the spray gun where the two streams impinge and are ejected through suitable deflector plates as a fine spray. Low-pressure impingement mixing, using compressed air both to increase the mixing turbulence and to spray the reaction mixture, gives a foam of high quality. The use of excess air, however, will produce a fine aerosol and excessive overspray. Airless high-pressure spray guns are preferred because the spray pattern is more uniform and there is less overspray than with low-pressure guns. Airborne spray drift is minimal. To spray rigid foam it is necessary for the metering machines to have outputs from about 1 to 5 kg/minute, and small portable machines are available with self-cleaning mixers and with heated hoses. Machines such as the Gusmer portable unit, which was developed for use outdoors in both winter and summer in North America, allow the spray application of suitable foam systems at temperatures down to $-10°C$.

Figure 7-28   Hand-held spraying with rigid polyurethane foam.

Foam systems for spraying must be highly reactive, especially if foam is to be sprayed onto vertical surfaces. In general a foam thickness of up to about 30 mm can be applied in one layer, depending on the foam system, the temperature and the thermal conductivity of the substrate. Where thick foam coatings are required the best results are obtained by repeated application of layers about 15 to 20 mm thick. The spray is applied at right-angles to the substrate from a distance of about 1 metre. Spray operators must wear protective clothing including a full air-fed hood or an approved respirator to avoid inhalation of aerosols (figure 7-28). Sprayed foam must be given a protective coating or covered by a protective sheath to resist weathering and spread of flame in the event of fire. Sprayed rigid polyurethane or polyisocyanurate foam is a particularly economical method of insulating curved and contoured objects with a continuous layer of a low-permeability, closed-cell foam. Gaps and open joints are sealed during the process and sprayed foam is, therefore, increasingly used to seal and insulate old, leaking roofs provided the basic roof structure is sound. (figure 7-29).

Each layer of sprayed foam consists of a uniform, fine-celled, homogeneous foam with a thin, dense, foamed skin. Average densities are usually in the range from about 35 to 45 kg/m^3 and, therefore, sprayed foam has a high mechanical strength. With a suitable top protective coating it easily withstands normal traffic during roof maintenance. The continuous layer obtained by spray application eliminates local heat bridges and can be important in reducing the danger of spread of flame through open joints.

Figure 7-29 Insulating a large, profiled roof with sprayed rigid polyurethane foam.

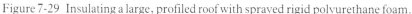

Foam injection and foam pouring. There are many applications for rigid polyurethane and polyisocyanurate foam in which the foam reaction mixture is poured, injected or sprayed into cavities. Most of these applications use both the insulating and the bonding properties of rigid polyurethane foams but some, such as the use of foam to consolidate friable strata in mining operations and in sewer renewal, utilise the gap-filling nature and bonding strength of specially-developed foam systems.

Foam reaction mixtures may be dispensed directly into the gap between the inner and outer wall facings of a building during construction. Multiple passes from a small portable dispenser are used to fill and insulate even large cavities. The technique is

153

applicable to most methods of construction from wooden-framed buildings to reinforced slab construction. Similar foaming methods are used to insulate pipe-lines laid in trenches. Foam is also used to stabilise the ground around pipe-lines to prevent erosion by wind and water.

An example of the dual role of rigid polyurethane foam is a special rigid polyurethane foam for bonding together brick cavity walls when the original galvanised steel wall ties have corroded. This obviates the need to demolish the outer brick wall and rebuild it using new stainless steel or polypropylene wall-ties. The rigid polyurethane foam bonds together the inner and outer walls and enhances the thermal insulation of the old construction to meet the highest modern standards (figure 7-30).

Figure 7-30   Insulating and strengthening houses using polyurethane foam injected into the wall cavity

Figure 7-31   To stop subsidence at a hospital, rigid polyurethane foam was injected into all cavities between the ward floors and a new concrete base.

Another engineering use of rigid polyurethane foam is in the renovation of large sewers. One example is the relining of an old, corroded, wrought-iron sewer pipe 2.6 metres in diameter. It was cleaned, primed and lined with interlocking GRP lining segments sealed with a polyurethane sealant. An annular gap of about 65 mm between the old corroded sewer pipe (which was no longer of uniform diameter) and the new lining was filled with a fast-curing foam by spray application using heated chemical components. The result was a rigid, impervious and corrosion-resistant composite lining, firmly bonded to the old construction.

High-density rigid polyurethane foam is also used in coal and ore mines to consolidate faulty ground strata so that weak areas can be safely held by supports. A two-component rigid polyurethane foam system is injected into fissures, or through holes bored into friable roof structures, to consolidate the faulty rock and prevent crumbling and formation of gaps behind the roof or face supports (see Chapter 9).

Figure 7-32   Applying a seal of one-component frothed foam.

One-component frothed foams (OCFs), were invented within ICI in 1969 and since that time they have found many applications in the building industry. The main advantages of OCF systems are their portability and ease of application. OCFs are MDI-based prepolymer compositions containing dissolved CFM-12. They are supplied, both to the building industry and to the do-it-yourself market, in pressure cylinders or aerosol cans fitted with dispensing nozzles. The nozzle is designed to extrude a bead or strip of liquid material. As the viscous, liquid prepolymer leaves the nozzle it is expanded to a low density froth by the vaporisation of the CFM-12. The density of the froth foam may be adjusted to suit the application conditions by changing the amount of CFM-12 dissolved in the prepolymer. Once exposed to the air, the froth foam cures by reaction with atmospheric moisture. Full cure of the foam may take several days depending on the temperature, the relative humidity of the atmosphere and the reactivity of the froth foam system. OCF systems are usually formulated to become 'tack-free' in one hour or less under typical application conditions. After it becomes dry to the touch, the foam continues to cure internally as traces of moisture diffuse through the outer skin. During this process a small amount of carbon dioxide gas is also formed. This ensures that the foam tends to expand slightly as it cures and so maintains some gap-sealing pressure. OCFs adhere strongly to most building materials and may be applied to damp surfaces. They form self-adhesive, insulating fillers and gap sealants.

Major applications include the sealing in place of door and window frames, the joining of insulated panels, roofing boards and pipe insulation and for general draught proofing around pipe and cable runs. The use of OCFs to seal and fix door and window frames not only gives good thermal insulation and sealing against draughts but also gives much improved sound insulation compared with traditional methods of fixing.

Moulded rigid foams

Low density structural foam Both rigid polyurethane and polyisocyanurate foams may be moulded either by pouring into an open mould or by using reaction injection moulding (RIM) techniques. Low density foams with thin or cellular skins are moulded to form complex rigid structural cores for articles such as chair shells. The use of moulded structural foam allows great freedom of design because complex shapes and surfaces having double curvature can be produced economically. Low density polyurethane foam chair shells are usually reinforced by the incorporation of glass-fibres, polypropylene-fibre mesh or expanded metal during the moulding process. It is also common practice to bond strips of wood into the foam during the moulding bond strips of wood into the foam during the moulding process to form tacking strips for the upholsterer to attach the covering fabric. Self-skinning rigid polyurethane foam is moulded either to replace wood or to supplement it in both decorative and structural furniture components (figure 7-33).

Figure 7-33   Demoulding rigid polyurethane foam mouldings for furniture.

High density structural foams. These foams are made from specially-formulated two-component foam systems. The mouldings have a dense outer skin and a low density cellular core forming a stiff sandwich structure. The average density of moulded, self-skinning, structural foam articles may be varied from about 250 kg/m^3 to over 800 kg/m^3, depending on the foam formulation and the surface/volume ratio of the item.

Many articles in structural foam are made at densities from 400 to 600 kg/m^3, usually by a RIM process. The low viscosity of the foam reaction mixture and the excellent flow of the expanding foam allows large and complex mouldings to be made with low in-mould pressures. Polyurethane structural foam mouldings are free from sink marks even where reinforcing ribs and inserts are moulded in-situ. Because of the low pressure developed by the rising foam, prototype parts and small production runs may be made using GRP moulds lined with epoxy resin, epoxy resin filled with metal powder, or with sprayed metal. The latter allows the exact reproduction of textured surfaces by spraying a metal having a low melting point onto a hand-made pattern and backing the metal skins with GRP or aluminium filled epoxy resin (figure 7-34). Fine surface detail is easily reproduced in self-skinning, polyurethane structural foam. The moulding of flat uniform surfaces requires a smooth tool surface free from blemishes. For long production runs, moulds made from tool steel with ground surfaces are the most durable.

Figure 7-34   Resin-backed, sprayed metal mould for structural foam moulding.

The structural foam process is particularly suitable for the production of large items such as casings for computers, office machines and other specialised equipment (figure 7-35 and 7-36). Structural foams with a high heat-distortion temperature and with low combustibility (USA Underwriters Laboratory test, Subject 94, Classification V - O) have been developed for use in the manufacture of such equipment as colour video display units where high temperatures occur.

Table 7-5 **Polymeric MDIs for making polyurethane and polyisocyanurate structural foams.**

Product	Description	NCO value	Viscosity at 25°C (mPa s)	Average function-ality	Typical application
'Suprasec' VM30	Low functionality modified poly-meric MDI	28.6	200	2.30	Structural foam moulding, flexible and semi-rigid self-skinning moulding.
'Suprasec' VM50	Low functionality polymeric MDI	30.6	130	2.49	Structural foam moulding.
'Suprasec' DND	Polymeric MDI	30.7	230	2.70	Rigid and semi-rigid foams.
'Suprasec' DNR	Polymeric MDI	30.7	230	2.70	Rigid and semi-rigid foams.

Figure 7-35 Moulding an X-ray body scanner in polyurethane structural foam.

Because polyurethane structural foams slowly darken in colour on exposure to daylight, they must be coated with a suitable protective finish or pigmented with carbon black or some other stable, dark-coloured pigment. Advice on the many finishing systems available can be obtained from the suppliers of structural foam systems. In-mould coating is often used to apply priming coats to those faces of large mouldings that will be visible when the article is in use. Structural foams may be made with standard grades of polymeric MDI ('Suprasec' DNR or 'Suprasec' DND). Improved compatibility with some polyol components, improved flow in the mould and a moulding with enhanced physical properties are obtained by using systems having 'Suprasec' VM50 or 'Suprasec' VM30 as the isocyanate component (table 7-5). Mixtures of 'Suprasec' VM30 with a standard grade of polymeric MDI are also sometimes used for RIM structural mouldings. The RIM process using static mould clamps (see Chapter 6) allows the production of both small and large items using programmed injection from a single high-pressure dispensing machine. Moulds are temperature-conditioned by circulating water at controlled temperatures to maintain the optimum mould-filling temperature of $55 \pm 5°C$ and to remove the heat generated by the foam reaction. The mould occupation time depends upon the chemical system and especially upon the thickness of the moulded item, and lies within the range from less than 2 minutes to about 10 minutes.

The mechanical properties of a structural foam moulding depend upon its shape and thickness as well as upon the formulation and the density of the foam. For purposes of comparison and design, physical properties are determined from samples cut from flat moulded sheets, usually about 10 mm thick. Typically, such samples have a cellular core of almost uniform density. The skins

Figure 7-36 Structural foam computer cabinet housing.

158

are between 1 mm and 2 mm in thickness and the outer surfaces are almost solid. This structure allows a combination of high stiffness and high heat-distortion temperature with high impact-resistance. Some typical properties are tabulated in table 7-6. Systems are available to meet specific application requirements as to flame resistance, electrical resistance, heat distortion temperature, stiffness and impact resistance.

Table 7-6 Structural foams: range of properties from systems using 'Suprasec' DNR, 'Suprasec' VM 50 or 'Suprasec' VM 30

Property	Test method	Range of values	Units
Shrinkage (from the mould)		0.4 to 0.6	%
Flexural strength. (Stress at break)	DIN 53423 ASTM D790-71	30 to 45 26 to 35	MPa MPa
Flexural modulus.	DIN 53423 ASTM D790-71	900 to 1250 800 to 1200	MPa MPa
Impact strength.	DIN 53453 BS 2782/3/306B	10 to 22 10 to 30	J J
Heat distortion temperature.	DIN 53424 ASTM D648-56 (66 psi OFS)	95 to 110 70 to 110	°C °C
Tensile strength.	DIN 53455 ASTM D638-72	14 to 20 12 to 20	MPa MPa
Surface hardness	DIN 53505 ASTM D2240-68	70 to 80 72 to 80	°Shore D °Shore D
Thermal conductivity	ISO 2581-1975	0.07 to 0.08	W/mK

(Measurements on flat moulded samples, 10-12 mm thick of average density 550 ± 60 kg/m^3)

The properties of rigid polyurethane foams

The properties of rigid polyurethane foams will vary with changes in the formulation and with changes in the ratio and the temperature of the chemical components. The foam properties may also change with changes in the process machinery and in its adjustment and operation. Changes in the ambient conditions, i.e. temperature and humidity, may also influence the chemical reaction and the resultant physical properties of the foam. Measurement of the properties of polyurethane foam enables the reproducibility of foam manufacture to be monitored. The measurement of physical properties may also indicate the suitability of the material for a given application. Tests of the solvent resistance or chemical resistance of rigid polyurethane foam are important for some applications.

Where possible, the foam properties should be measured by standard test methods such as those specified by the International Standards Organisation (ISO). The applicable ISO methods are tabulated in table 7-7. Results obtained by the ISO tests are more

159

Table 7-7 ISO test methods for rigid polyurethane foam.

Title of test method	ISO number
1. Rigid cellular plastics: Determination of linear dimensions.	ISO 1923-1972
2. Cellular rubbers and plastics: Determination of apparent density.	ISO 845-1984
3. Cellular plastics: Compression test of rigid materials.	ISO 844-1985
4. Rigid cellular plastics: Bending test.	ISO 1209-1976
5. Cellular plastics: Determination of tensile properties of rigid materials.	ISO 1926-1979
6. Plastics: Rigid cellular materials: Determination of 'apparent' thermal conductivity by means of heat flow-meter.	ISO 2581-1975
7. Rigid cellular plastics: Determination of shear strength.	ISO 1922-1981
8. Cellular plastics: Test for dimensional stability of rigid materials.	ISO 2796-1980
9. Cellular plastics: Determination of the temperature at which fixed permanent deformation occurs under compressive load.	ISO Test report 2799-1978
10. Rigid cellular plastics: Determination of water absorption by immersion method.[1]	ISO 2896-1974
11. Determination of water vapour transmission of rigid cellular materials.	ISO 1663-1970

[1] Revision in progress
Tests for friability (DP 4590) and closed cell content (DP 6187) of rigid cellular materials
are being prepared.

meaningful, for both practical and legal purposes, than those obtained from non-standard tests. There are also many National Standard test methods, many of which are technically similar to the ISO tests, which are also used to characterise rigid foams. The more important of these are listed in Chapter 11. Physical tests are also carried out on fabricated items and on composite materials such as foam-cored laminates and building panels.

The most important quality control tests are the determination of the apparent density of the foam, the measurement of compression strength, and monitoring the dimensional stability of the foam under conditions of temperature and humidity that are related to those expected in the intended application. Other tests, important in some applications of rigid foams, include the measurement of shear strength (laminates and sandwich panels), the measurement of water absorption (buoyancy and low-temperature insulation), and the determination of creep under load (load-bearing applications in building). The measurement of thermal conductivity is, of course, important for most insulation applications.

Standard tests are made using specified test specimens cut from the sample material according to a sample cutting plan. The sample cutting plan must define both the position and the orientation of each test specimen. This is essential to reproducible testing because polyurethane rigid foams are anisotropic and show variations in properties from place to place within the foam. These variations result from flow patterns and pressure variations during their formation. Some foams may also show variations caused by differences in cure temperature resulting from variations in the rate of cooling immediately after manufacture.

Measuring the density. Measurement of the apparent density (the weight of foam per unit volume) is the simplest test of product uniformity. The standard test calls for an accurately-cut standard specimen, but production control is often carried out by measuring the average density of a large section of a product. The apparent density of most rigid polyurethane foams with closed cells may be measured without significant error soon after manufacture. Foam of low density with a high content of open cells will give erroneous results before cooling because of the buoyancy of the relatively warm air within the foam.

The mechanical strength of rigid polyurethane foam. The strength of a rigid foam increases with the apparent density of the foam. Strength may be measured in compression, tension and shear. Typical values for a foam of density 32 kg/m^3 are listed in table 7-8.

Table 7-8 **The mechanical strength of PUR/PIR foam: typical values at a density of 32 kg/m^3**

Property	Typical value	
Compression strength	**Maximum**	**At 10% strain**
Parallel to the direction of foam rise (ISO 844 - 1985)	ca 200 kPa	140-180 kPa (1.4-1.8 kgf/cm^2) (20-25 lb/in^2)
Perpendicular to the direction of foam rise (ISO 844 - 1985)	ca 120 kPa[1]	130-180 kPa (1.3-1.8 kgf/cm^2) (18-25 lb/in^2)
Tensile strength (ISO 1926 - 1979)		
Parallel to the direction of foam rise	350 kPa	
Perpendicular to the direction of foam rise	250 kPa	
Shear strength (1SO1922 - 1981)	160 kPa	
Flexural modulus (BS 4370: 1968)	ca 3000 kPa	

[1] Depends on the process conditions and the resulting degree of anisotropy.

161

The compression strength of a rigid foam is important in most applications. A sample of standard size and uniform thickness is compressed at a constant standard rate up to 10% compression between two parallel plates of larger surface area than the sample (see Chapter 11, figure 11-5), and the maximum stress sustained by the specimen is calculated. If the maximum stress is reached at less than 10% deformation, it is reported as compressive strength:

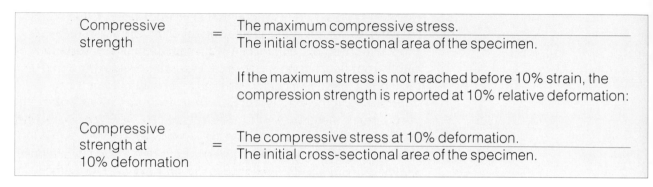

Compressive strength $=$ $\dfrac{\text{The maximum compressive stress.}}{\text{The initial cross-sectional area of the specimen.}}$

If the maximum stress is not reached before 10% strain, the compression strength is reported at 10% relative deformation:

Compressive strength at 10% deformation $=$ $\dfrac{\text{The compressive stress at 10\% deformation.}}{\text{The initial cross-sectional area of the specimen.}}$

As low density foam is usually anisotropic, the strength will vary with the direction of measurement. It also varies with the rate of stressing although the effect of the strain-rate decreases as the rigidity of the polymer increases. Highly cross-linked polymers show little change in compression stress with change in the strain-rate. This is a valuable property when rigid polyurethane foam is used in protective packaging and in crash padding. Most low-density foams show yield behaviour at high stress levels of the form illustrated in figure 7-37. Polyurethane and polyisocyanurate foams shows very low creep (the rate of change of strain with time under

Figure 7-37   Typical stress/strain curve of a 34 kg/m^3 rigid polyurethane foam.

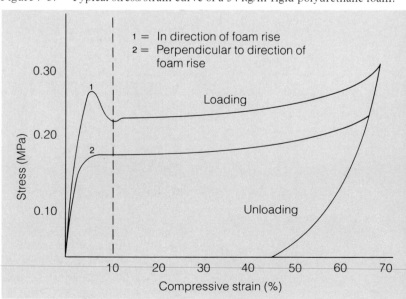

constant load) at low strain levels and at normal operating temperatures. It is usual to design load-bearing foam structures with a safety factor of 4 to 5 based on standard foam compression strength measurements.

For a perfectly elastic solid, the stress/strain curve is linear and is unaffected by the strain rate (Hooke's Law). The constant ratio of stress/strain is called Young's modulus (E). Polyurethanes, in common with all polymeric materials, only approach conformity with Hooke's Law at very low levels of loading, where the material recovers its original shape very quickly after the load is removed. The elastic modulus of rigid polyurethane foam is derived from the approximately linear initial portion of the stress/strain curves in the compression, tension and shear tests to give the compression modulus, tensile modulus and the shear modulus respectively.

The flexural modulus, or stiffness, of foam sandwich structures with skins that are higher in tensile strength than the foamed core is proportional to the third power of the thickness of the composite. When a stiff laminate supports a bending load, one face of the laminate is under compression and the opposite face is stretched; between the two facings is a neutral plane where the stress is, for pure bending, theoretically zero. In practice, there is always a shear force component and the low density foam core is subjected to shear stress. Where high load-bearing ability is required from foamed sandwich structures with skins of high tensile strength, the minimum foam density used must, therefore, be high enough to ensure acceptable shear strength. Shear strength is determined by bonding foam specimens of standard size to stiff steel plates and applying a longitudinal shear force (see Chapter 11, figure 11-7). The sides of the specimen are displaced at a standard rate and the applied force is measured. The shear strength is the maximum force applied per unit area and is expressed as kN/m² or kPa.

Thermal conductivity

As illustrated in table 7-1, rigid polyurethane foam is one of the best thermal insulation materials available in quantity today.

Thermal conductivity, λ-value, is defined as the rate of heat transfer through unit thickness, across unit area, for unit difference in temperature. In SI units, thermal conductivity, λ-value:

$$\lambda = \frac{J}{s} \cdot \frac{1}{m^2} \cdot \frac{m}{K} \text{ or } \lambda = \frac{W}{m.K}$$

where the symbols have the standard SI meanings: J = joules, s = second, m = metre, W = watt, K = temperature, degrees Kelvin (see table 7-9 for conversion to c.g.s. units or to the k-value, (Btu in/ft²h°F) often quoted in the USA). Thermal conductivity is measured under 'steady state' conditions (see Chapter 11).

The thermal conductivity varies with the temperature. It is usually measured at a mean temperature that is relevant to the application,

163

Note 1.
Conversion factors were computed for the three values of the calorie and Btu listed below. All three of the conversion factors are given where these differ significantly.

(a) Calories (g), mean value = 4.19002 joules = 0.00397403 Btu.

(b) Calories (g), at 15°C., = 4.18580 joules = 0.00397003 Btu.

(c) Calories (g), US Bureau of Stds., = 4.184 joules = 0.0039683207 Btu.

Note 2.

$$\frac{W}{m.K} \times 10^{-2} = \frac{W}{cm.K} = \frac{J}{s.cm.K}$$

$$\frac{W}{m.K} \times 10 = \frac{mW}{cm.°C}$$

Table 7-9 Thermal conductivity units: some conversion factors

	W $\frac{W}{m.K}$	Btu.in $\frac{Btu.in}{ft^2.h.°F}$	kcal.in $\frac{kcal.in}{ft^2.h.°C}$	kcal $\frac{kcal}{m.h.°C}$	Btu $\frac{Btu}{ft.h.°F}$	cal $\frac{cal}{cm.s.°C}$
$\frac{W}{m.K}$	1	6.93811	3.14255 (a) 3.14572 (b) 3.14707 (c)	0.859185 (a) 0.860050 (b) 0.860420 (c)	0.578176	2.38662×10^{-3} (a) 2.38903×10^{-3} (b) 2.39006×10^{-3} (c)
$\frac{Btu.in}{ft^2.h.°F}$	0.144131	1	0.452941 (a) 0.453397 (b) 0.453592 (c)	0.123836 (a) 0.123960 (b) 0.124014 (c)	8.33333×10^{-2}	3.43988×10^{-4} (a) 3.44334×10^{-4} (b) 3.44482×10^{-4} (c)
$\frac{kcal.in}{ft^2.h.°C}$	0.318213 (a) 0.317892 (b) 0.317755 (c)	2.207794 (a) 2.205572 (b) 2.204623 (c)	1	0.273403	0.183983 (a) 0.183798 (b) 0.183719 (c)	7.594538×10^{-3}
$\frac{kcal}{m.h.°C}$	1.16389 (a) 1.16272 (b) 1.16222 (c)	8.07523 (a) 8.06710 (b) 8.06363 (c)	3.657600	1	0.672936 (a) 0.672258 (b) 0.671968 (c)	2.777778×10^{-3}
$\frac{Btu}{ft.h.°F}$	1.72958	12	5.43529 (a) 5.44077 (b) 5.44311 (c)	1.486025 (a) 1.487523 (b) 1.488163 (c)	1	4.12785×10^{-3} (a) 4.13201×10^{-3} (b) 4.13379×10^{-3} (c)
$\frac{cal}{cm.s.°C}$	4.19002×10^{2} (a) 4.18580×10^{2} (b) 4.18400×10^{2} (c)	2.90708×10^{3} (a) 2.90416×10^{3} (b) 2.90291×10^{3} (c)	1.316736×10^{3}	3.60×10^{2}	2.42257×10^{2} (a) 2.42013×10^{2} (b) 2.41909×10^{2} (c)	1

i.e. a mean of 0°C for refrigeration and 10°C for building applications. The R-value of thermal insulants is also widely used, especially in the USA. The R-value is the thermal resistivity, that is, the resistance to the flow of heat across unit area, through unit thickness for unit difference in temperature.

$$R\text{-value} = \frac{1}{\lambda}$$

The low thermal conductivity of rigid polyurethane foam results from its low density and the fine closed-cell structure filled with CFMs. A typical rigid polyurethane foam with a density of 32 kg/m^3 consists of 3% polymer and 97% gas (by volume) trapped within the closed cells of the foam. Clearly, the thermal conductivity of the foam will depend to a large extent on the conductivity of the gas within the cells. The lowest conductivity is obtained by blowing the foam with CFM-11. The thermal conductivity of a foamed material may be expressed as the sum of four components:

$$\lambda_F = \lambda_G + \lambda_R + \lambda_S + \lambda_C$$

i.e. the heat transmission through a foam is the sum of that conducted through the gas contained in the cells, λ_G; that radiated across the cells of the foam, λ_R; that conducted through the solid polymer of the cell walls, λ_S; and that transferred by convection of the gas within the cells, λ_C. Convection is significant only in cells of diameter above 10 mm; λ_C is effectively zero in all commercially produced polyurethane rigid foams. Over the range of cell sizes usually manufactured, the radiant heat transfer component, λ_R is directly proportional to the cell diameter. Radiant heat transfer becomes proportionately greater as the foam density is reduced and is increasingly significant at lower densities.

The heat transmission through the solid polymer contributes about

Figure 7-38   Effect of cell size on thermal conductivity (at 0°C).

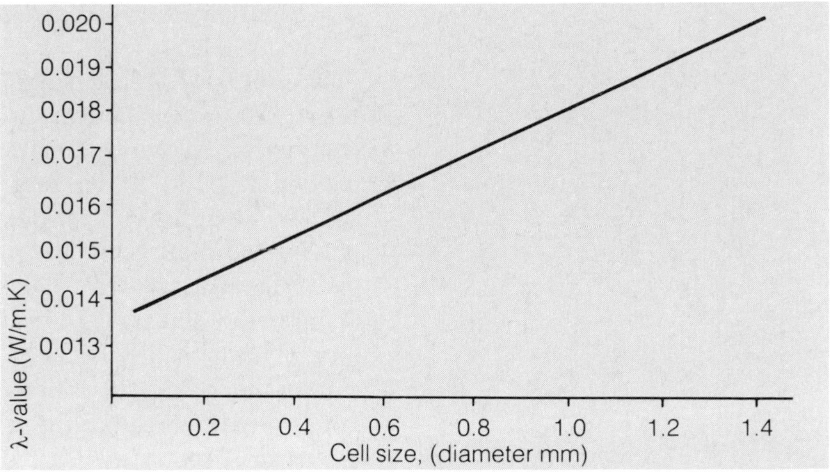

30% of the total heat transfer through typical commercially made foam. Obviously λ_S increases with increasing foam density as more polymer is present. However, the change in λ-value over the usual range of foam densities used in insulation applications – from about 28 to 50 kg/m^3 – is not significant.

The largest variations in thermal conductivity are attributable to variations in λ_G resulting from changes in the composition of the gas in the closed cells of the foam. Table 7-10 lists the thermal conductivities of the gases that may be present in rigid polyurethane foam.

Table 7-10 **The thermal conductivity of the gases in rigid polyurethane and polyisocyanurate foams**

Temperature °C	°F	CFM-11	CFM-12	Air	Carbon dioxide	Water vapour
−17.8	0	0.00571	0.00778	0.0227	0.0133	0.0145
−6.7	20	0.00623	0.00813	0.0235	0.0141	0.0154
0	32	0.00643	0.00844	0.0240	0.0146	0.0159
4.4	40	0.00657	0.00864	0.0244	0.0149	0.0163
15.6	60	0.00692	0.00899	0.0253	0.0157	0.0170
26.7	80	0.00743	0.00951	0.0260	0.0166	0.0179
37.8	100	0.00778	0.00986	0.0269	0.0175	0.0188

λ-values in SI units (W/m.K)

The rate of heat flow through a foam is also affected by orientation of the foam cells. As previously discussed, during the formation of the foam the expanding bubbles tend to elongate in the direction of foam rise, i.e. in the direction subjected to the lowest restraining pressure. The λ-value will be greatest in the direction of foam rise. As the degree of cell elongation and the orientation of the cells will depend upon the reactivity of the chemical system and the foam-making process and conditions, the initial λ-value, within a range from about 0.013 to 0.017 W/m.K, will vary with the process. The lowest values are obtained with products from the ICI vertical lamination process because it produces a foam with very small cells having their smallest diameter parallel to the temperature gradient (in service).

The effects of ageing on the thermal conductivity of rigid polyurethane and polyisocyanurate foams. The very low thermal conductivity of polyurethane rigid foams having small closed cells containing CFM-11 will be retained indefinitely if the foam is contained within impermeable skins. The rigid polyurethane and polyisocyanurate polymers that form the cell walls are more permeable to gases of low molecular weight. Polyurethane rigid foams of low density exposed to air show an increase in λ-value with time. This is mainly because of the inward diffusion of air into the cells of the foam. The bulky molecules of CFM-11 and CFM-12 do not pass easily through the cell walls at ordinary temperatures and are retained for many years. However, CFMs may be lost more rapidly at elevated temperatures. Figure 7-39 illustrates the theoretical stages in the change in the thermal conductivity of a foam exposed to the air. The changes shown are those resulting only from diffusion effects.

Figure 7-39   The theoretical change of λ-value with time, (foam completely exposed to the atmosphere).

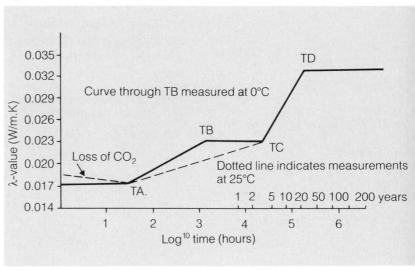

Note:
The time scale refers only to small samples about 35 mm thick. Larger speciments will change more slowly.

The time to T_A. At the end of foam rise, the hot foams have cells that contain gas above atmospheric pressure. When the foams cool to room temperature the pressure within the cells falls below that of the normal atmosphere. During this cooling period any carbon dioxide present in the cells will quickly diffuse out and the λ-value of the foam will fall to a minimum.

Time T_A to T_B and T_B to T_C. During this time interval the λ-value of the foam increases, primarily because of the inward diffusion of air. The plateau, T_B to T_C, occurs only at temperatures below 24°C,

166

the boiling point of CFM-11. It represents a constant partial pressure of CFM-11 resulting from the presence of liquid CFM-11.

Time T_C to T_D. This time interval represents the slow loss of CFM-11 vapour by outward diffusion to produce a foam with a λ-value corresponding to that of air-filled cells. The effect of cell diameter on λ_R remains unaffected by the composition of the gas.

The real world, however, differs from this theoretical picture. Measurements of the λ-value of foams installed up to twenty years ago show that the actual increase in conductivity is often much smaller than that forecast. The effect of ageing on the λ-value of a polyurethane foam depends largely on the size of the foam specimen, the ambient temperature(s), and the gas permeability of the facing material (in practice, uncovered low density foam is very rarely used). A mass of data is available, but it is important to separate the results of ageing in the laboratory from those obtained by the examination of specimens taken from foam actually in service.

In the laboratory, small uncovered foam specimens cut from larger pieces and aged under constant conditions at temperatures from 0 °C to 70°C, show an increase in λ-value to a constant value. This constant λ-value is often called the 'equilibrium' value because it is the value that corresponds to the air within the closed cells being in equilibrium with the air at the temperature of the experiment. The equilibrium λ-value of small specimens is reached in about one year at 60°C or about five years at 0°C. It corresponds to the level T_B - T_C in figure 7-39. Typical λ-values obtained in the laboratory on small uncovered specimens cut from commercially produced rigid poly-urethane and polyisocyanurate foams, are an initial value of about 0.017 W/m.K increasing on ageing to an equilibrium value of about 0.023 W/m.K. The latter figure is often used for design purposes but, as discussed below, a lower figure may be more realistic.

In service, the effect of ageing on λ-value depends largely on the facing material.

With impermeable facings such as steel, aluminium, GRP and many more thermoplastic and multilayer sheet materials, the foam core retains its low initial λ-value indefinitely. The initial value, as previously discussed, depends on the foam system and on the method of making the foam sandwich or laminate.

With permeable facings such as uncoated paper, plasterboard and uncoated scrims, the λ-value rises toward the equilibrium value as air diffuses into cells of the foam. The rate of increase in λ-value, however, is much slower in practice than found in laboratory ageing tests. A number of examinations of polyurethane foam insulation

---

Some typical initial λ-values are:

Continuous laminates
0.017 - 0.018 W/m.K

Vertical laminates and press-injection laminates
0.014 - 0.017 W/m.K

Adhesive-bonded laminate (sliced slabstock foam)
0.017 - 0.020mW/m.K

that was installed during the 1960s suggest that the time interval to the equilibrium value of thermal conductivity varies from about 50 years at a mean temperature of 10°C, to about 10 years at 50°C.

The effect of water vapour. The water vapour permeability of rigid polyurethane and polyisocyanurate foams is low. Typical values obtained by the ISO R1195 (1970) type of test (see Chapter 11), under tropical conditions of 90% R.H. at 38°C, is a permeability of 17 to 18 g/m^2/day for foam sheet 25mm thick. The effect on λ-value is often very small. It is always so when the foam is used to prevent heat losses from a warm surface and the cooler face of the foam is nearer to the moisture-laden air. The use of a water vapour barrier is recommended, however, for all insulation used in buildings and most foam-cored building laminates incorporate a moisture barrier of polyethylene film or aluminium foil. An effective moisture barrier, on the warm side, is particularly important in the insulation of cold surfaces in refrigerators, freezers and cold-stores. In the absence of a water vapour barrier, on the warm side, the water vapour will diffuse through the foam and condense on the cold face where at temperatures below 0°C, it will form ice in the cells of the foam. This will increase the thermal conductivity of the foam and may also rupture the cells.

The effect of liquid water. The amount of water absorbed by rigid polyurethane foam at normal pressure is small. Small samples of foam (100mm × 100mm × 25mm) immersed in water at 20 °C for one month, absorb about 3% to 5% of water by volume depending on the cell size of the foam. This uptake of water corresponds to the saturation of the surface layer of foam cells only. The water uptake per unit surface area is about 3 to 4 ml/cm^2 of cut foam surface. Foam samples taken from roof insulation which had been in place for over 15 years were usually found to contain less than 1% of water.

Measuring the thermal conductivity of rigid polyurethane foam.
There are two ISO methods of measuring the thermal conductivity of cellular plastics. ISO 2582 describes the determination of thermal conductivity using a guarded hot-plate. This is the absolute method using an apparatus (see Chapter 11), with heated edge guards. The energy input to the edge guards is adjusted until no heat is lost through the edges of the specimen of foam. The temperature drop across the faces of the specimen is then measured under steady state conditions when the power input to the hot plate is equal to the heat flow normal to the foam surface. The method is also fully described in various national standards (table 11-1). The alternative method of ISO 2581 uses a calibrated heat-flow meter. Calibration is done using a specimen of known heat conductivity. The rate of heat flow through the foam specimen is measured when the specimen is subjected to a known temperature gradient (see Chapter 11). The

168

method is quickly carried out and is suitable for quality control in the factory in contrast to the guarded hot-plate method which may require about 2 days for each determination. Heat-flow meters are also described in many national standards (table 10-11).

Quick, single-position tests on thermal conductivity may be made using a conductivity probe. This is a cylindrical instrument with a centrally-positioned heater of fine wire and one or more thermocouples to measure the temperature of the outer surface. For a known rate of heat generation per unit length of heater wire the temperature at the surface of the probe depends on the thermal conductivity of the surrounding foam. The conductivity probe is not very accurate. The main error arises from variable contact between the surface of the probe and that of the foam.

Other tests for rigid polyurethane foams

There are a number of other tests that are generally useful to characterise rigid foams.

Dimensional stability test. One of the most important, especially for low density foams, is the dimensional stability test, in which the foam is subjected to accelerated ageing at a standard temperature and its dimensional stability measured by observing any changes in the dimensions of the samples. Ageing at $-15°C$, for example, is an important quality control check in foams for freezer insulation.

The heat resistance of the foam may be determined by the method of ISO 2799. This gives the temperature at which a specimen of foam under compression stress becomes permanently deformed.

Closed-cell content. As several important properties of rigid foam vary with the proportion of closed cells in the foam, a check on the proportion of closed/open cells is useful. The proposed ISO method (ISO DIS 4590) is similar in principle to that given in BS 4370: Part 2: 1973, Method 10, and NF T 56-129. This test method measures the volume of air displaced by a foam specimen of known dimensions. The foam sample is placed in a closed chamber and the volume of the air contained in the chamber is increased by a predetermined amount. The pressure of the expanded air is measured and compared with the pressure obtained by a similar volume expansion without the foam sample. The air displacement of the sample is then calculated by the application of Boyle's Law. The standard method includes procedures to correct for the cut open surface cells of the specimen and to minimise errors attributable to adiabatic cooling (Chapter 11). The reproducibility of the method, using specimens of uniform size say, 100mm × 50mm × 50mm, is satisfactory for quality control purposes without the time-consuming correction for cut surface cells being necessary.

Flammability tests are of three types:

– Small-scale laboratory bench tests for quality control, specification or sorting purposes.
– Intermediate-scale tests designed to classify a material or construction for use in buildings.
– Full-scale fire testing under controlled conditions.

The uncorrected result may be referred to as the "apparent" closed cell content".

The bending test described in ISO 1209 is useful not only as a rigid quality control test but it also forms a basis for structural design using foam-cored laminates. It is a three-point loading test on a conventional tensile or compression test machine (Chapter 11).

Flammability testing. Because most rigid polyurethane foam is used in building construction, it is in that context that most flammability testing is carried out. The hazards to life in real fires in buildings are discussed in Chapter 10. It is concluded that polyurethane rigid foam, when used in suitable constructions, does not add significantly to severity of fires or to the hazards in a burning building. The objective of fire testing is both to avoid any increase in the hazards in a real fire and to guide progress toward reduction in the fire hazard whilst improving the cost and energy-efficiency of the building.

Small-scale tests include flash point and oxygen index tests which measure a property of the polymer rather than of the foam or of a particular foam composite structure. ASTM-D 1929-1977 and ISO 871-1968 are used to determine the minimum temperature resulting in flammable decomposition products. For most rigid polyurethane and polyisocyanurate foams this temperature is above 300°C. The apparatus specified in ASTM-D 1929 is also used to determine the auto-ignition temperature. This is about 500°C for most polyurethane foams. These temperatures are not lower than those found for most natural organic materials. Compared with 500°C for rigid polyurethane foam, typical self-ignition temperatures of other building materials are 375°C for red deal and 275°C for the bitumen commonly used in roofing.
ASTM-D 2863-1976 and BS 2782 (1978), are oxygen index tests. These tests which give good reproducibility, determine the minimum oxygen concentration required to support flaming combustion of a small test piece under standard conditions. Such tests cannot be used to predict the performance of a material in a real fire, but have often been used to rank the effect of novel additives on flammability.
The simplest standard ignition/flammability tests are those using a small horizontally-mounted strip of foam or foam composite which is ignited at one end by a standard flame under standard conditions. The best known of these, ASTM-D 1692, was used to rank materials in order of flammability. The classifications used, such as "self-extinguishing" and "burning", were discontinued in 1978 because they were often misleadingly assumed to indicate the behaviour in a real fire. ISO 3582-1978 is a similar test designed for the monitoring of manufacturing processes or to compare the horizontal burning

characteristics of different materials. It is not intended to be used to assess the potential fire hazards in use.

Small-scale tests that are more relevant to the testing of rigid polyurethane and polyisocyanurate foams for use in buildings are those using a vertical strip of foam, and also the alcohol cup test of BS 2782: Method 140E. The latter uses a small test sheet of foam held at 45° to the horizontal above the flame from a standard quantity of ethyl alcohol. Specimens are rated according to the amount of flame and the severity of charring. Possibly the most useful small-scale tests are vertical chimney tests such as the American Butler Chimney tests and that of DIN 4102. The latter is specifically intended for the classifications of building materials. It includes both a simple vertical burn test using a single strip of material and a chimney test using four long strips to form the sides of a square section chimney (see Chapter 11). DIN 4102 allows the classification of combustible building materials into classes, B1 – difficult to burn, B2 – normal, and B3 – readily flammable.

Surface spread of flame tests for building materials are carried out on small specimens in the laboratory. Tests are also carried out on larger specimens, up to full-scale assemblies. BS 476: Part 7: (1971); DIN 4102, Part 3; and ASTM E162 use radiant panels mounted at a specified angle to the small test piece so that the radiant intensity reaching the sample is within the limits specified in each standard. A flame is then applied to the hotter end of the test sample and the rate of flame spread is measured.

Smoke and toxic gas production may be evaluated on the small scale. Standard smoke tests include the ISO smoke box test, ISO DP 5659, the XP2 smoke chamber, ASTM-D 2843, and the smoke density apparatus of DIN 53436. The latter has also been used as the basis of work on the toxicity of the combustion products produced by burning or pyrolysing polyurethanes under controlled conditions. Small scale tests must not be assumed to indicate the behaviour of materials in real fires.

Intermediate and large-scale tests are not specific to rigid polyurethane and polyisocyanurate foams but are the tests used for building materials. In addition to the National Standards there are the important tests specified by fire insurer's organisations. There are no internationally agreed standards. Intermediate-scale tests include surface spread of flame tests such as the Steiner tunnel test, ASTM-E 84, which specifies that a tunnel 8 metres long shall be covered by a sheet of the material to be tested, 500mm wide. A 320 kW gas flame is applied at one end of the tunnel and the rate of flame spread is measured. BS 476: Part 7, specifies a large radiant panel 90 cm × 90 cm, mounted at an angle of 90° to a specimen of dimensions 90 cm × 23 cm, that simulates the surface of a wall.

Figure 7-40   Radiant panel for the surface spread of flame test to BS 476: Part 7: (1971).

Measurements of the rate and extent of spread of flame are used to classify the material or composite structure.

Tests specified for insurance classification include those by the US Factory Mutual and those by the UK insurance companies' Fire Offices' Committee (FOC) (see footnote). The FOC corner tests for non-load bearing walls is on a smaller scale than the Factory Mutual corner test. The full-scale corner Factory Mutual test requires the construction of a room with a ceiling height of 25 feet and uses 340 kg of wood for the fire. The FOC corner test requires only a room height of 8 feet, a test wall 8 foot long, two corners and, where this is to be tested, the ceiling constructed from the composite panels to be tested. The test is made by igniting a specified wooden crib (30 lb of white fir sticks) mounted in a standard manner. The test material is classified for use in different types of buildings depending on the resistance to flame spread, the rate of smoke evolution and the total smoke generated, the readings from thermocouples mounted in the test room and the structural integrity of the test panels after the fire.

Fire resistance tests are defined (BS 4422: Glossary of terms associated with fire) as "The ability of an element of a building construction to withstand the effects of a fire for a specified period of time without loss of its fire-separating or load-bearing function or both". The standard tests, ISO R 834; BS 476: Part 8: 1972; and ASTM-E 119 are designed to test full-sized structures by subjecting them to furnace temperatures that follow a standard time/ temperature curve. This standard curve, for example, corresponds to 556°C at 5 minutes, 821°C at 30 minutes and over 1,000°C after 2 hours. One face of the test wall or panel is heated in this manner. The temperatures of both sides of the composite panel are monitored. Failure of the panel may be due to the formation of cracks or is deemed to occur when the unheated face reaches a temperature more than 140°C above the initial temperature. It has been shown that panels filled with polyisocyanurate foam, especially polyisocyanurate foam containing glass-fibre or wire reinforcement, have substantially increased time to failure compared with panels filled with other insulants. Sandwich panels having non-combustible facings and polyisocyanurate foam cores are used, for example, to form insulated fire-break panels in multi-storied buildings.

Several of the test procedures discussed above are described in greater detail in Chapter 11.

Figure 7-41 FOC wall and ceiling lining test on glass-fibre reinforced polyisocyanurate foam boards with aluminium foil facings.

1.

2.

3.

4.

Fast burn: Time after crib ignition:
1. Four minutes; 2. Nine minutes
Slow burn: Time after crib ignition:
3. Twenty minutes; 4. Twenty-eight minutes

8   Polyurethane elastomers

Although nearly 80 percent of the commercial production of
polyfunctional isocyanate is consumed by the manufacture of the
foams and RIM materials described in earlier chapters, the
remainder is used to make an extremely wide range of polymers
including elastomers. These are used in a large number of diverse
areas such as construction, transport, aerospace, furniture,
domestic appliances and medicine. This chapter describes the wide
range of polyurethane elastomers, their manufacture and
application.

Types of polyurethane elastomers

Types of polyurethane elastomers commercially available include
thermoplastic elastomers in a form suitable for conventional
thermoplastic processing, cast elastomers made by mixing and
casting reactive liquid components, elastomeric fibres and various
one- and two-component systems for making elastomeric coatings
on textiles and other flexible substrates. There are also several types
of millable polyurethane rubbers which can be processed by the
traditional methods used in the rubber industry. Almost all of these
polyurethane elastomers are based upon segmented block
copolymers of the general molecular structure, $(AB)_n$, having
alternating soft and hard segments. The soft segments, so called
because the molecular segments are flexible or 'soft' at normal
ambient temperatures, are saturated aliphatic polyester or polyether
chains; the hard segments, (rigid or stiff at normal ambient
temperatures), are the reaction products of a polyfunctional –
usually difunctional – isocyanate and a chain extender of low
molecular weight, which is usually a diol or a diamine. Polyurethane
elastomers may be made in a wide range of hardnesses from about
10° Shore A to over 60° Shore D. In general, the fully-cured
elastomers are tough, abrasion-resistant materials of high
mechanical strength having good resistance to many solvents and
chemicals, with the exception of strong bases and strong acids,
oxidising agents and a few strongly polar solvents. Polyurethane
elastomers are usually based upon one of the aromatic diisocyanates,
TDI, MDI, or NDI, but the trend is toward the increasing use of
distilled MDI and derived products. There is also some small

175

consumption of HMDI and IPDI to make elastomeric coatings having a high resistance to weathering and discolouration by light.

Thermoplastic polyurethane elastomers (TPU)

Thermoplastic polyurethane elastomers are supplied as granules or pellets for processing by the well-established thermoplastic processing techniques such as injection moulding and extrusion. By these means elastomeric mouldings having a very valuable combination of high strength with high abrasion and environmental resistance, may be mass-produced to precise dimensions. TPU are generally classified by their degree of hardness and are available in a range from about 80° Shore A to over 60° Shore D. Softer mouldings with hardnesses as low as 60° Shore A, may be obtained by adding specially-formulated plasticising granules to the conventional grades of polyester-based TPU. The physical properties and the range of working temperature of TPU fill the gap between conventional moulded rubber goods and such engineering thermoplastics as nylons and polycarbonates. Some automotive applications are illustrated in figure 8-1.

Figure 8-1   Typical automotive components in thermoplastic polyurethane elastomer

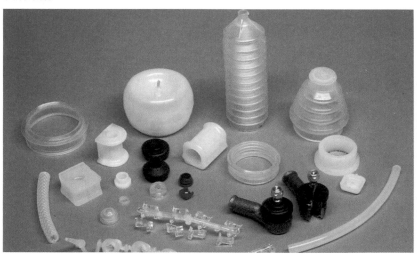

TPU are usually made from pure MDI which is reacted with a substantially linear polyether or polyester diol and with a chain-extending diol of a low molecular weight, such as 1,4-butanediol, in either a one-step or a two-step reaction process. The polyester diols are usually the condensation products of adipic acid and one or more simple aliphatic diols in the series from ethylene glycol to 1,6-hexanediol. To obtain improved low-temperature performance with relatively high resistance to hydrolysis, 6-hydroxycaproic acid polyesters, made by the polymerisation of ε-caprolactone are also used. Polyether TPU are usually based upon

176

poly(oxytetramethylene) diols, (polytetrahydrofurans).

The high modulus of TPU is attributable to a two-phase polymer microstructure resulting from the physical bonding and aggregation of the hard segments of adjacent polymer chains. The resulting hard block domains act as cross-linked, physically-bonded crystalline centres dispersed in the soft segment domain of flexible polyether or polyester chains. The hard block domains, therefore, act as molecular reinforcing-fillers.

The degree of hard segment aggregation or domain formation depends not only upon the weight ratio of the polyurethane hard segment to the polyether or polyester soft segment, but also on the choice of glycol, the type and the molecular weight of the polyester or polyether diol, and also upon the manufacturing process and the reaction conditions. Hard segments from MDI and linear aliphatic glycols have the general structure:

Figure 8-2

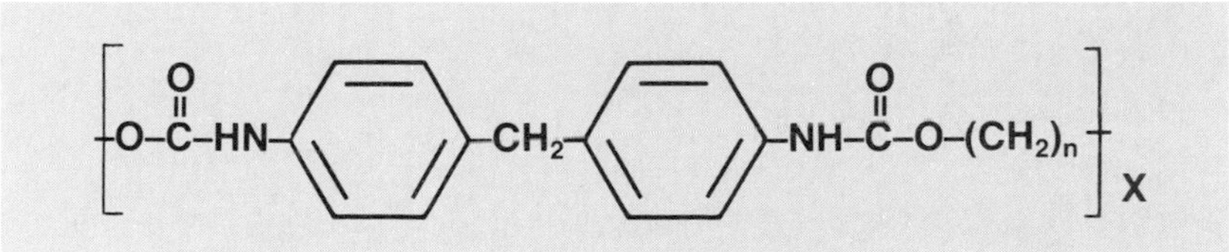

Hard segment domains of the highest packing density and the highest melting point are obtained when n is an even number. In these instances the hard segment domains consist of aligned chains which are staggered to form straight $C = O \cdots H-N$ hydrogen bonds in three dimensions. Hard segment domains of the odd-numbered series, however, adopt a higher energy conformation in which the chains are contracted in order to form the hydrogen-bonded network.

The soft segment structure also affects the degree and the type of hard segment aggregation. Polyethers, in general, are less compatible with MDI than polyesters and the TPU derived from them show a higher degree of phase separation. The polyurethane hard segment domains also tend to be larger and more complex than those found in polymers based on adipic acid polyesters. The latter usually contain many small, evenly-dispersed hard block domains although over 50% of the hard segments remain in the soft segment domains. Increasing the hydrocarbon chain length, in both polyesters and polyethers, will usually increase the tendency to phase separation. Increasing the molecular weight of polyester or polyether diol also favours domain separation.

The physical bonding of the hard block domain is unusually stable and the physical properties of TPU change only slowly over the normal range of operating temperatures. The lower useful operating temperature is limited by the T_g of the flexible polyester or polyether soft segment which is always well below normal ambient

177

temperatures. The T_g of most polyesters and polyethers tends to fall as the molecular weight increases and for good elastomer performance at low temperatures the molecular weight of the polyol should be at least 2,000. The upper operating temperature limit depends mainly upon the softening point of the hard block domain, i.e. the upper limit is influenced mostly by the breaking of interchain hydrogen bonds at temperatures above about 110°. As the concentration of hard segment is increased, producing elastomers of higher hardness, the plateau region of the modulus/temperature curve is less marked (figure 8-3).

Figure 8-3   Shear modulus/temperature relationship for some typical polyester-based thermoplastic polyurethane elastomers

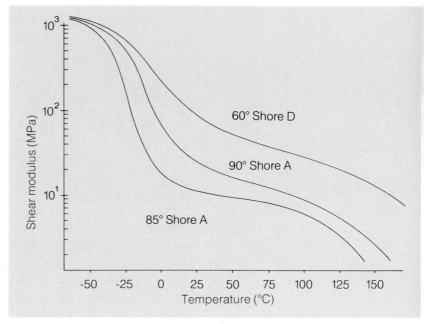

In general, the tensile and the other mechanical properties of thermoplastic polyurethane elastomers vary with the hardness. Some typical properties are tabulated in table 8-1.

Special application requirements, such as high resistance to heat and hydrolysis or to chemical attack, good flex performance at low temperatures, or good impact resistance at temperatures down to –50°C, are met by specially developed grades that are available from several suppliers.

TPU are electrical insulating materials with volume resistivities in the range from 10^9 to about 10^{13} ohm.cm. Antistatic and electrically-conducting TPU may be obtained by the incorporation of special grades of graphitic carbon. A filled grade of TPU with a volume resistivity of less than 0.1 ohm.cm. has been proposed for the electrical trace heating of pipes and tanks for space heating.

Table 8-1 Thermoplastic polyurethane elastomers: some typical physical properties

Property			
Hardness range (°Shore A/D)	97/60-68	92-93/40-44	83-89/30-34
Density (kg/m^3)	1.2-1.25	1.16-1.25	1.14-1.20
Tensile strength (MPa), (DIN 53504, Die S2)	40-55	35-45	25-35
Elongation at break (%, DIN 53504)	350-500	450-500	450-550
Graves tear strength (kN/m, DIN 53515)	130-190	80-100	55-80
Abrasion resistance (Volume loss, mm^3, DIN 53516)	40-65	35-45	25-40
Resistance to solvents (% volume change, 70 h at 23°C., BS 903: Part A 16)			
ASTM Fuel A	< 0.1%	< 3%	< 5%
ASTM Fuel B	< 10%	< 20%	< 25%
ASTM Oil No. 1	< 0.2%	< 1%	< 1%
ASTM Oil No. 3	< 1%	< 1%	< 2%
Girling brake fluid	< 5%	< 20%	< 40%
Methanol	< 20%	< 20%	< 30%
Petrol	< 5%	< 5%	< 10%
White spirit	< 2%	< 5%	< 10%
50/50 Ethylene glycol/water	< 1%	< 1%	< 1%
Typical processing temperature, (°C)			
Injection moulding	200-230	190-215	190-200
Extrusion	—	180-200	180-200

Processing

Drying. Thermoplastic polyurethane elastomers are usually supplied ready for use in sealed, moisture-proof bags. All TPU are to some extent hygroscopic and will absorb moisture if left exposed to the atmosphere. Material exposed at room temperature to air at 50% R.H. will reach equilibrium with atmospheric moisture at a water content of 0.3% to 0.5%, depending on the grade and the chemical structure of the TPU. Such material must be dried to a moisture content below 0.1% to ensure satisfactory mouldings free from streaks and bubbles and to avoid the risk of hydrolysis of the polymer at the processing temperature. Drying should be done

179

quickly, by circulating hot air at 60°C to 70°C, or by the use of reduced pressure.

Injection moulding. Most grades of thermoplastic polyurethane elastomer can be injection moulded. They are best processed with screw preplasticisation using standard screws of the polyethylene type with a compression ratio between 2:1 and 2.5:1. Accurate temperature control is essential. Typical barrel temperature profiles are indicated in figure 8-4, the lower limits of the range applying to the softest grades that have a hard segment content around 35 to 40%. Overheating should be avoided, as significant polymer degradation may occur above 230°C. Insufficient heating, on the other hand, may also cause degradation because of local overheating arising from high shearing forces. Control of the screw speed and the injection speed are also important. Polyether-based TPU are more sensitive to overheating than the polyester-based products. Moulds should be temperature controlled, usually within the range from 20°C to 30°C. Lower temperatures may aid the release of thick mouldings and higher temperatures are sometimes helpful in producing large thin-walled mouldings such as automotive steering rack gaiters. TPU mouldings will shrink by up to about 2.0% from the mould dimensions and the shrinkage may vary with the cross-section of the moulding.

Figure 8-4   Temperature profiles for injection moulding and extrusion.

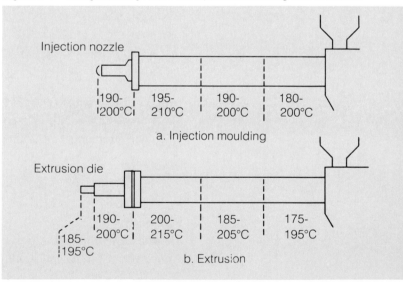

Extrusion profiles. Thermoplastic polyurethane elastomer profiles, tubes, hoses and cable sheathings can be formed by extrusion. Three-stage screws having a high l/d. ratio in the range from 25 to 35 and with a compression ratio about 3:1 are most suitable. High compression screws cause excessive shear heating and are unsuitable. The screw should be driven by an oversized high-torque

180

motor to ensure uniform rotation of the screw and uniform propulsion of the viscous TPU. A breaker plate and screen pack (80-120 mesh filters) should be included. Melt viscosity varies with the shear rate and the processing controls should include melt pressure and temperature control on the back of the die, in addition to accurate control of the barrel zone temperatures (figure 8-4 b).

Compounded thermoplastic polyurethane elastomers. In addition to varying the chemical structure of TPU, suppliers offer materials containing additives. These include hydrolytic stabilisers, U/V absorbers, fire-retardants, lubricants, pigments, bacteriocides, reinforcing fibres and plasticisers. Additives such as pigments may also be added by dry tumble-blending with the polymer granules. It is essential to use only dry powders. Colour masterbatches are often used for pigmentation.

Glass-fibre reinforcement. Glass-filled TPU granules containing chopped glass strands 2 mm to 3 mm in length are available for injection moulding, and non-woven glass-fibre mat impregnated with TPU for compression moulding. Compression mouldings made from a polyester-based TPU of 80° Shore A hardness and with 40% w/w of glass fibre, show an increase in modulus up to about 7,000 MPa and a tenfold reduction in the coefficient of thermal expansion. The heat sag resistance of such glass-fibre-reinforced materials is claimed to be high enough to allow paint stoving at temperatures up to 150°C.

Film and sheet. TPU are available in grades suitable for blow-moulding into thin film (from about 25 to 150 microns in thickness) using the bubble-process. Sheet material is made by extrusion to thicknesses of between 100 microns and 1,000 microns, and to a wider range of thickness by calendering. The latter requires specially formulated TPU in a restricted range of hardness.
TPU film has very high abrasion and puncture resistance but is permeable to water, CO_2 and air. It is unaffected by dry heat sterilisation and remains flexible down to about −40°C. TPU film is used for aircraft drop bags, emergency wound dressings, and as heat-sealable laminating foil.

Applications of thermoplastic polyurethane elastomers
Thermoplastic polyurethane elastomers are relatively expensive materials that find very wide use in applications requiring their toughness and high resistance to abrasion, lubricating and fuel oils, combined with a wide range of operating temperatures. Some of these applications are listed below.

Automotive. Bushes and bearings for oscillating movements made in polyester-based TPU of hardness in the range from 55° to

181

60° Shore D. Softer polyester-based TPU (40° to 50° Shore D) are used for flexible couplings, for ball-joint seatings, for timing belts and seals. Other items, made from 33° to 40° Shore D material, include fuel line connectors, bumper trim, grommets, suspension diaphragms for pneumatic and hydraulic systems, ball-joint and rack-and-pinion gaiters, and sheathing for exposed springing and for cables subject to abrasion.

Engineering. 60° Shore D materials are used to make injection moulded gears, bearings, diaphragms for hydraulic systems, hammer heads, ball cups and wear plates and, in textile machinery, for high speed loom pickers and bobbin centraliser bushes. Hard TPU are also used for abrasion-resistant rollers in mechanical handling systems and for heavy-duty separation screens in mining – as an alternative to the widely-used ore screens of cast polyurethane.

Hose and cable sheathing. TPU are widely used for hoses and for cable sheathing. Polyester-based TPU, with hardness in the range from 33° to 40° Shore D, give the best balance of properties for applications such as sheathing for hydraulic hoses and exposed cables. Cables for use in the earth, or in environments such as animal houses where they are exposed to micro-organisms, should be sheathed with a polyether-based TPU containing a bacteriocide. Polyester-based TPU buried in the ground may suffer microbiological attack resulting in the formation of acidic materials that will accelerate the hydrolysis of the polyester.

Other applications of TPU include animal ear-tags (polyether-based), moulded soles for football boots and other sports shoes, wear resistant moulded heel pieces for ladies' shoes and for ski boots. The latter application involves the moulding of the complete ski boot-shell, usually in a polyester-based TPU formulated to retain its flexibility and impact resistance at low temperatures. It is one of the biggest single applications for TPU.

Figure 8-5   Ski boots.

Figure 8-6   Castors, abrasive disc holders, hose sheathing and animal ear-tags.

Cast polyurethane elastomers (CPU)

These are elastomers made by mixing and pouring a degassed reactive liquid mixture into a mould. There are a large number and variety of CPU with a wide range of physical properties and costs. The high-quality CPU may be divided into those that are mixed, cast and cured at an elevated temperature and those processed at or near normal room temperature. The products include both linear and partially cross-linked materials, the former being chemically and physically similar to TPU. Many high quality CPU are still made by a two-stage process, although there is increasing use of one-shot systems based on MDI variants. One-pack liquid casting systems, employing blocked diisocyanates which are activated by heating, have also been used for many years on a small scale. However, the properties of the elastomers obtained from one-pack systems are limited by the presence of the blocking agent.

Hot casting processes for high quality elastomers are usually based on an isocyanate-terminated polyester or polyether prepolymer, with an available NCO content in the range from about 3% to about 6% by weight, although NCO contents over 10% are sometimes used. The prepolymer is mixed with a chain extender at a temperature in the range from about 50 to 130°C and is immediately poured into heated moulds or it is rotationally cast or sprayed. Successful systems are based upon TDI, MDI and NDI. Polyester and polyether polyols, usually linear diols, are both used. Polyesters in common use are poly(ethylene adipate), (PEA), and poly(ethylene/tetramethylene adipate), (PETMA), each typically having a molecular weight about 2000. Polycaprolactone diols with molecular weights from about 800 to 2000 are also used. Polyether CPU are based upon polyoxytetramethylene diols (polytetrahydrofurans) or the cheaper polyoxypropylene or

Table 8-2 **Typical physical properties of cast elastomers made by traditional hot polymerisation**

Type of material	Polyether/TDI	Polyester/NDI	Polyester/MDI
Typical chain-extending agents	1,4-BD,TMP, MOCA* *m*-phenylene diamine.	1,4-BD TMP	1,4-BD
Post cure (hours/°C)	1/100 to 4/100	24/110	1/110 to 3/110
Hardness (°Shore A)	60 to 95	65 to 96	65 to 95
Tensile strength (MPa)	20 to 35	20 to 40	22 to 31
Elongation at break (%)	500 to 400	700 to 350	670 to 470
Tear propagation resistance, Graves (kN/m)	36 to 55	20 to 96	40 to 128
Resilience at 25°C (% rebound)	85 to 25	70 to 25	65 to 40

Key to abbreviations
1,4-BD = 1,4-Butanediol
TMP = Trimethylolpropane
MOCA = 1,1′-Methylene bis (2-chloroaniline) or 3,3′-Dichloro -4,4′-diamino-diphenylmethane

*MOCA is a known animal carcinogen. It must be used only with extreme caution and with extreme handling precautions. For full safe-handling advice consult the supplier before use.

poly(oxypropyleneoxyethylene) diols or lightly-branched polyols. TDI-based CPU are often made from storage-stable prepolymers produced by chemical suppliers. Some of these are made with 2,4-TDI and are available with isocyanate values about 3% and a low level of free TDI. Storage-stable MDI prepolymers are also available from a number of suppliers and MDI is also widely used in two-stage semi-continuous CPU systems. All NDI-based CPU are made by continuous or semi-continuous two-stage processes, because NDI prepolymers are not storage-stable. Some well-known products illustrating the use of TDI, MDI and NDI, are listed in table 8-2. Hot cast elastomers may be removed from the mould or casting surface about 15 to 30 minutes after pouring, depending on the system. A post-cure at an elevated temperature (50° to 130°C) for several hours is often recommended to ensure the development of the optimum physical properties. The chain extension reaction, both with diols and diamines, is exothermic. High modulus CPU systems produce the most heat because they have a higher proportion of reacting groups. The actual temperature rise during the casting of a high modulus elastomer will be in the range from about 40 to about 90°C. It will depend largely on the molecular weight of the soft segment, the thickness of the casting and the rate of heat loss to the mould.

The aliphatic diisocyanates, 1,6-diisocyanatohexane (HDI), and 1-isocyanato - 3-isocyanatomethyl - 3,5,5,-trimethylcyclohexane (IPDI), are also used in casting systems where there are special requirements such as high light fastness.

TDI prepolymers. The best known polyether/TDI prepolymers are based on polyoxytetramethylene glycols, having equivalent weights in the range from 450 to 1,000. The user reacts the prepolymer with a chain extender at about 100°C. The chain extenders used include 1,4-butandiol (1,4-BD), trimethylolpropane (TMP), and hindered diamines, especially 3,3'-dichloro-4,4'-diaminodiphenylmethane (MOCA), and 'Caytur' 7, (a eutectic mixture of cumene diamine and *m*-phenylene diamine). In general, the hardest elastomers with the highest modulus are obtained from diamine chain extenders as these give hard block domains with the highest physical cross-link density. Cast elastomers based on polyoxytetramethylene soft segments and TDI polyurea hard segments give CPU of the highest resilience and 'snap-back', with good low-temperature performance.

NDI prepolymers. Cast elastomers from NDI must be made from recently prepared prepolymers. A prepolymer is made by adding a polyester polyol, e.g. a PEA or PETMA, to molten NDI at about 120°C. The resultant NDI prepolymers have a limited storage life and are mixed with the chain extender, at about 100 to 110°C and cast immediately. Chain extenders used in these systems include

184

1,4-BD, 1,3-BD, TMP and sometimes water, in addition to selected diamines.

MDI/polyester and MDI/polyether prepolymers. Prepolymers for the manufacture of CPU are available from many suppliers. However, the relatively low toxic hazard involved in handling MDI in closed vessels has led to the widespread use of semi-continuous two-stage processes.

Hot casting. A typical process to make elastomers with hardnesses in the range from just below 60° to about 95° Shore A uses polyester polyols. The polyester polyol, dehydrated by stirring at 110° to 120°C under reduced pressure for between 15 and 30 minutes, is cooled to around 85°C and added slowly to molten pure MDI, at 60°C. The reaction mixture is stirred for about 30 minutes, keeping the temperature between 80 and 90°C by controlled cooling. A degassing stage is then usual to remove any trapped air or dissolved carbon dioxide. The prepolymer is mixed with vacuum-dried chain extender (1,4-BD), at a temperature of 80 to 90°C and poured into moulds at about 110°C. Demould times vary from about 10 minutes to around 50 minutes depending on the formulation of the CPU and the thickness of the part.

The main advantage of polyether-based CPU lies in their superior resistance to hydrolysis. They are thus preferable for use in wet conditions especially at high ambient temperatures. Polyester-based systems have the better thermal stability and resistance to oils and solvents, UV light and oxidation. These differences are small when compared to those of natural rubber which has very poor resistance to heat, oxidation and solvents compared to polyurethanes. Polyester-based elastomers may contain added carbodiimide derivatives which greatly improve hydrolysis resistance at high humidities. Polyether-based CPU usually contain anti-oxidants.

Room temperature processing. A wide range of MDI prepolymer systems are available which are designed for hand or machine mixing and pouring at room temperature or at only moderately elevated temperatures up to about 50°C. Two types of easily handled 2-component systems are available. One type is based on full MDI prepolymers and the second type on quasi-prepolymers. A full MDI prepolymer is made by pre-reacting MDI with all the polyol to form component A. The second component of the system, component B, consists of the chain extending agent together with any minor additives. This type of system has a wide component ratio, usually about 10:1. The use of quasi-prepolymer systems allows the component mixing ratio to be adjusted closer to 1:1. This greatly simplifies the mechanical problems involved in mixing and dispensing the cast polyurethane elastomer system.

Most commercially available 2-component systems provide a degree of control over the hardness of the cast elastomer by fine adjustment of the mixing ratio of the two components, A:B. Polyester-based cast elastomer systems that are designed for handling and mixing at or near to room temperature are based upon special polyesters with a sufficiently low melting point. The cast polyurethane elastomers produced from these systems do not match the highest levels of tensile properties obtained by the best hot casting processes but the low temperature castings have excellent resistance to abrasion and flexing, (table 8-3).

Table 8-3 Some cast elastomers from 2-component systems using MDI prepolymers

Type of system	Polyester/MDI prepolymer			
	Full prepolymer	Quasi-prepolymer		
Component A (parts by weight)	100	100	100	100
Component B (parts by weight)	10	130	100	85
Mould temperature (°C)	70	80	80	80
Post-cure at 80°C (hours)	24	24	24	24
Hardness (° Shore A)	85	80	80	80
Tensile strength (MPa)	26	32	34	36
Stress (MPa) at 100% strain	7	6	9	12
Stress (MPa) at 200% strain	9	9	16	21
Stress (MPa) at 300% strain	12	15	22	24
Resilence (% ball rebound)	48	34	28	31
Type of system	**Polyether/MDI Quasi-prepolymer**			
Component A (parts by weight)	100	100		100
Component B (parts by weight)	125	100		70
Mould temperature (°C)	80	80		80
Post cure at 80°C (hours)	24	24		24
Hardness (° Shore A)	70	80		90
Tensile strength (MPa)	25	30		35
Stress (MPa) **at 100% strain**	2.5	4.8		8.5
Cold flexing temperature (minimum in °C)	−55	−50		−48

One-shot cast polyurethane elastomers have been made from liquid polyesters and diisocyanates for over 30 years. The best known applications are the manufacture of potting compounds and soft printers' rollers from lightly branched polyadipates and TDI. Printers' rollers with hardnesses from about 15° Shore A to about 35° Shore A are made from poly(diethylene adipate) and TDI by varying the ratio of polyester diol to TDI (table 8-4).

Table 8-4 Typical formulations for the production of soft printers' rollers

	Parts by weight			
Poly(diethylene adipate)	100	100	100	100
TDI (65:35)	9.5	10	10.5	11
Hardness (°Shore A)	15-20	20-25	25-30	30-35

Rollers with higher hardness may be obtained by the addition of a silica filler but, especially for use with inks containing polar solvents, it is preferable to use polyesters that are more crystalline. The hardness range is then extended up to about 55° Shore A (table 8-5).

Table 8-5 **Typical formulations and physical properties of one-shot polyester/TDI cast polyurethane elastomers**

Polyester polyol (parts by weight)	100	100	100	100
Catalyst (parts by weight) (Ferric acetylacetonate)	0.013	0.013	0.013	0.013
80:20-TDI (parts by weight)	9.2	9.6	10.0	10.5
Post cure at 110°C (hours)	3	3	3	3
Hardness (° Shore A)	30-32	33-35	43-45	53-55
Tensile strength (MPa)	7.3	7.6	15.3	23.5
Stress (MPa) at 100% strain	0.5	0.5	0.9	1.1
Stress (MPa) at 300% strain	0.9	1.0	1.8	2.3
Elongation at break (%)	850	750	700	650
Tear propagation resistance (Graves) (kN/m)	13.7	14.7	21.6	27.5
Resilience at 25°C, (% ball rebound)	46	46	63	68
Du Pont abrasion index, (NR = 100)	24	76	230	270

Processing conditions: Polyester at 70°C; TDI at 20-25°C; mould at 110°C.

Processing machines. Small items from cast polyurethane elastomers may be moulded by batch mixing, and reasonable results are obtainable even by manual stirring. However, commercial operation demands mechanical equipment for reproducible results. CPU are usually processed using specially-designed two-component metering and dispensing machines.

Materials are kept at the required temperature, in the range from 60 to 130°C, by jacketed tanks, lines, valves and pumps. Many machines incorporate automatic degassing systems and can be obtained with a vented, temperature-controlled vessel for the in-situ manufacture of prepolymers. Machines may be obtained as simple 'casting' machines or as moulding machines. The latter allow the automatic dispensing of predetermined amounts of reactive mixture into discrete moulds. The former are simpler manually-controlled machines for continuous pouring and manual manipulation in operations such as lining of pipes and vessels and continuous manufacture of cast sheet.

Polyurethanes for tyres
One widely publicised market with major potential is that of tyres.

Currently, polyurethane is mainly used for cast solid tyres, such as those used for castors on furniture. Other major methods of using polyurethanes are the filling of conventional rubber tyre casings and the liquid injection moulding (LIM) or RIM of polyurethane casings for pneumatic tyres. In addition, microcellular and integral skin tyres are manufactured for articles such as trolleys and children's bicycles.

Solid tyres. Solid tyre manufacture accounts for a major proportion of all cast elastomer production. These tyres are used on vehicles such as fork-lift trucks and other industrial off-road machines. Cast polyurethane elastomers in the hardness range from 90° to 96° Shore A are used. Increasing the hardness increases the load-bearing capacity, reduces the rolling resistance and minimises the weight of tyre required. Other factors such as traction and wet-grip are improved by reducing the hardness. Solid cast elastomer tyres are usually cast directly onto the metal hub. The metal hub is first shot-blasted, degreased and primed with a bonding agent. The hub is then placed in a metal mould and heated to about 100°C when the tyre is cast from a hot-curing elastomer. New systems are being tried, especially in the market for replacement tyres. Press-on tyres, for example, enable tyres to be changed using a hydraulic or screw press. They consist of a cast elastomer tyre which has been bonded during manufacture to a steel bush that is an interference fit with the wheel hub.

Tyre filling. Soft cast elastomers are used to replace the air in standard pneumatic tyres. Such elastomer-filled tyres have a prolonged and puncture-free life and show substantial savings in arduous off-the-road operating, especially with heavy vehicles having large expensive tyres. Cast elastomer-filled tyres are restricted to maximum sustained speeds of less than 25 to 35 mph (40 to 56 km/h) depending on the size of the tyre.

Tyre filling is usually carried out by injecting a reactive, two-component mixture into a tubeless tyre/wheel assembly by means of the valve-stem in the wheel. As the reactive liquid fills the tyre, air is displaced through a small hole drilled or punched in the diametrically-opposite tread. This hole is sealed after filling with a plug or screw. Filled tyres are left overnight at room temperature to cure.

Both prepolymer and one-shot systems are used to make soft elastomers (usually slightly blown to give a high density microcellular structure), in the hardness range from about 30° to about 40° Shore A, depending on the application. The main requirements are high resilience (low hysteresis) and a low compression set.

One of the most popular systems uses two components with a 1:1 ratio for ease of application using simple machines. The isocyanate

Table 8-6 **Typical properties of tyre filling cast polyurethane elastomers**

Specific gravity	1.04
Hardness (° Shore A)	30 to 40
Tensile strength (MPa)	1 to 2
Elongation at break (%)	150 to 300
Resilience (% ball rebound)	55 to 65
Compression set (25% compression for 22 hours at 70°C.)	1 to 4

component is a TDI/polyether polyol prepolymer, with an available isocyanate value of about 5%. The second component is a mixture of polyether diol and triol with catalysts. Alternative one-shot systems are based upon MDI variants such as Suprasec VM30, reacted with polyether polyols.

RIM or cast elastomer pneumatic tyres. The use of a cast elastomer or RIM process to make a pneumatic tyre is potentially a most attractive production method compared with the traditional process for making rubber tyres. A RIM process is easily automated, consumes little process energy and is potentially a cleaner process than those required to handle solid rubber raw materials. It is probably a large future market for polyurethanes, but one inhibited mainly by the cost and availability of the chemical raw materials, the current over-capacity of conventional tyre factories and the continuing development of the high speed, radial tyre. Cordless cast polyurethane elastomer tractor tyres are, however, being produced in restricted markets. Cast polyurethane elastomer tyres appear to become economically competitive for tyres of large size, where the number required is insufficient to justify the high capital cost of conventional rubber-processing machinery. Tractor tyres are made from cast elastomers based on pure MDI and high molecular weight polyether diols by a single injection into a closed mould. Bead inserts are encapsulated in the moulding to reinforce the rim of the tyre but otherwise the tyre is unreinforced. Tractor tyres, with a diameter of 190 cm, made from cast elastomers, are claimed to last twice as long, and to have better traction and lower rolling resistance than radial-reinforced conventional tyres.

Experimental CPU or RIM polyurethane tyres for passenger cars have been made by a number of people including several established tyre makers. Cordless and conventional radial constructions have both been used. The most widely-publicised work has been done by Polyair Maschinebau GmbH in Austria. Their development tyre for passenger cars is produced by a two-stage RIM process using MDI and high molecular weight polyether polyols. The body of the tyre is first moulded in a closed mould containing the wire bead-reinforcement, using a stiff high modulus elastomer (about 90° Shore A). The mould is then opened and a circumferential glass-fibre restrictor belt is wound over the carcass which then passes to a second mould where the tread compound of lower modulus (about 65° Shore A) is injected. Except for the circumferential windings, which are necessary to prevent tyre growth at high speeds, the tyre is cordless. Performance is claimed to equal that of a standard radial tyre but with a longer life and a better run-flat capability. Heat build-up during running is said to be much lower for the cordless RIM PU tyre than for the conventional radial tyre.

Figure 8-7 Polyurethane tractor tyres.

Integral-skin foam bicycle tyres. Rotationally cast, integral skin, flexible polyurethane foam tyres are puncture-proof. They are used on children's bicycles, pedal cars and trolleys. Integral-skin foam systems are available that are technically satisfactory for use on bicycles for adults, but they are not competitive in weight or cost with conventional pneumatic tyres.

Polyurethane fibres (spandex)

The first commercial polyurethane fibre was developed in Germany during the early 1940s. It was made by reacting HDI with a slight excess of 1,4-butane diol. The polymer, having a high level of intermolecular physical bonding, gave a stiff fibre that found some use as a monofilament and for synthetic bristles. It was also used as a high-performance thermoplastic. The only polyurethane fibres of commercial importance today, however, are the elastic fibres known as spandex fibres. (Spandex is a generic term approved by the U.S. Federal Trade Commission. They define spandex yarns as: "a manufactured fibre in which the fibre-forming substance is a long chain synthetic polymer comprised of at least 85% of a segmented polyurethane".) Spandex fibres are chemically similar to thermoplastic polyurethane elastomers, in which the soft segments are polyether or polyester chains in the molecular weight range from 1000 to 3000, and the hard segments usually consist of substituted polyurethanes and polyureas. They are usually made by a continuous two-stage process using an isocyanate-terminated prepolymer that is chain-extended with a low molecular weight diol, diamine or hydrazine. Several methods of spinning have been used; dry spinning from solution, reaction spinning, wet spinning and hot-melt extrusion. Wet spinning uses a solution of a linear polyurethane elastomer in a polar solvent, usually dimethylformamide (DMF), which is extruded and drawn into a bath of liquid in which the dimethylformamide is soluble but the elastomer is not. In reaction spinning a prepolymer is extruded into a solution of a diamine to make a partially cured fibre that is finished by moisture curing. The most important process today is dry spinning from the DMF solution in which polymerisation is carried out. This type of process is used to make Dorlastan, the polyester-polyurea filament yarn made by Bayer AG, and Lycra, the polyether-polyurea filament yarn made by Du Pont Co. Lycra, the first and most successful spandex fibre, is made from a poly(oxytetramethylene) glycol reacted with excess diisocyanate (HMDI) and then chain-extended in solution with aliphatic diamines. The solution of elastomer in DMF is pigmented with titanium dioxide and spun in a current of hot air to remove the solvent. The resulting fibre is a dull white multifilament yarn in which the filaments are bonded together. Dorlastan is a similar dull white bonded multifilament yarn based on MDI, but the soft segment consists of hydrophobic polyester chains based on adipic acid and aliphatic glycols such as 1,6-hexane diol. Stabilisers are included before spinning to impart resistance to light and hydrolysis.

There are many other types of spandex fibres. A great deal of work has been done on melt-spinning to avoid the use of expensive solvents and solvent recovery systems but the process is less important commercially than dry spinning from solution because of the inherent limitations on elastomer melting point and the sensitivity of the fibre properties to variations in the heating cycle. Melt spinning has been used in the manufacture of heterofibres with thermoplastics such as nylon.

Spandex fibres quickly established a substantial and competitive market in clothing, replacing natural rubber threads in some applications and creating new opportunities for the designer because of their versatility. Natural rubber threads are made either by slitting cured, calendered sheet or, more usually, by extrusion from latex. They cannot be made as fine as spandex threads, and are usually covered by winding with fine yarns to improve their "handle" in clothing manufacture, although uncovered rubber tapes are becoming increasingly used. Spandex fibres, on the other hand, can be produced as fine yarns and as staple fibre with satisfactory colour and handle, and are used uncovered in many applications. Compared with natural rubber threads, spandex fibres have better resistance to oxidation, sunlight, perspiration, cosmetics and sun-tan oils and to dry-cleaning fluids. They may be dyed using acid, basic or disperse dyes to match both natural and synthetic fibres. However, they are inferior to natural rubber threads in stress retention and hysteresis losses, and some grades are less resistant to hot detergent washing treatments. Spandex fibres are degraded by solutions of bleach (sodium hypochlorite) and chlorine, but are generally satisfactory in the low chlorine levels used in swimming pools (< 0.5 ppm).

Table 8-7 Typical properties of spandex fibres

Property	Spandex	Natural rubber	Other synthetic fibres
Elongation at break (%)	500-800	500-900	10-40
Tensile strength (break tenacity, g/tex.)	6-12	2-4	30-70
Thickness available (tex.)	2-60	17 upwards	2 upwards
Stress at 200% elongation (g/tex.)	0.4-1	0.15-0.3	N.A.
Melting point (°C)	ca 250	> 150	140-250
Moisture regain (% at 21°C/65% R.H.)	ca 1.0	0.2	0.2-5.00

Applications for spandex fibres. The traditional applications for spandex fibres are in supportive clothing such as belts, girdles, corsets, brassieres, garters, surgical stockings, sock tops and swimwear, but there is an increasing market in stretch fabrics for both clothing and furnishing. Many of these utilise 4 to 10% of spandex staple fibres but in other applications such as self-supporting sock tops, as little as 3 to 4% of spandex yarn is sufficient to give the required support without discomfort.

Elastomeric polyurethane coatings

Elastomeric polyurethane coatings have very high abrasion and soil resistance. They are used as fabric coatings to make simulated leather and other decorative and wear-resistant materials for clothing and furnishing. Thin polyurethane coatings are also widely used to improve the surface finish and the resistance to scuffing and soiling of leather, PVC-coated fabrics and furnishing fabrics. Poromeric materials are microporous polyurethane elastomers that are coated onto textiles, often non-woven materials, to make leather-like sheet materials for use in footwear. Poromeric foils are also available for adhesive lamination to leather splits to make a material for composite shoe-uppers. Such poromeric/leather laminates combine the breathability, abrasion- and water-resistance of poromeric polyurethanes with the moisture absorption of leather and its ability to conform to the shape of the foot.
Polyurethane coatings may be applied as one- and two-component systems that are diluted by solvents or as one-component water-based polyurethane latices. Heavyweight coatings for industrial applications are made by spraying two-component solventless or low solvent systems, by the adhesive lamination of calendered TPU sheet and by extrusion melt coating systems.

Two-component coating systems. The first polyurethane coated fabrics were made in the 1950s. They were made by coating a fabric with a thin layer of an isocyanate-terminated prepolymer (from a poly(ethylene glycol adipate) and TDI) and curing the coating with ethylene diamine vapour. There were many production problems caused by the side reactions of the isocyanate-terminated coating, and the cured products had a poor and variable resistance to hydrolysis. Satisfactory two-component polyester-based polyurethane coatings are now based upon hydroxyl-terminated, urethane-extended poly(glycol adipates) and/or polycaprolactones that are mixed with a polyisocyanate curing agent before coating. The curing agents are usually trifunctional TDI derivatives or MDI variants. Although coatings based upon aromatic isocyanates tend to yellow in sunlight, the addition of suitable fillers, pigments and stabilisers produces a degree of lightfastness that is adequate for many applications. Non-discolouring coatings of high lightfastness are obtained by the use of aliphatic diisocyanates such as IPDI and HMDI. The latter is sometimes called hydrogenated MDI. Both of these diisocyanates are easier to handle than HDI, which has also been widely used to make coatings which do not discolour in daylight. The wide range of one- and two-component systems is based upon either polyester or polyether polyols reacted with one or more polyisocyanates. The choice of system will depend upon the performance required of the coating. In general, polyester-based coatings have the highest toughness and resistance to abrasion and may be used over the widest temperature range (–50 to about 120°C). Polyether-based coatings have the highest resistance to hydrolysis and microbiological attack.

One-component coating systems are of two kinds, those consisting of elastomers dissolved in polar solvents and aqueous dispersions of cross-linkable elastomers. The former are chemically similar to thermoplastic polyurethane elastomers. They are of a lower molecular weight (about 40,000) than the elastomers obtained from isocyanate-cured two-component systems. One-component systems cure only by solvent evaporation and have a lower solvent resistance than the elastomers obtainable from two-component systems or from aqueous polyurethane lattices. The latter have advantages over solvent coating systems. In addition to avoiding the difficulty and expense of solvent disposal or recovery, aqueous polyurethane dispersions contain reactive groups that may be cross-linked with melamine/formaldehyde or epoxy resin dispersions.

Coating methods. Textiles may be coated directly by knife or roller, or by a transfer process. The choice depends on the product requirements, the type of textile substrate, the thickness of the coating and the flexibility and handle required. Transfer coating is widely used to coat knitted fabrics and raised fabrics. It avoids the risk inherent in direct coating of impregnating the fabric.

Figure 8-8   Methods of coating thin layers.

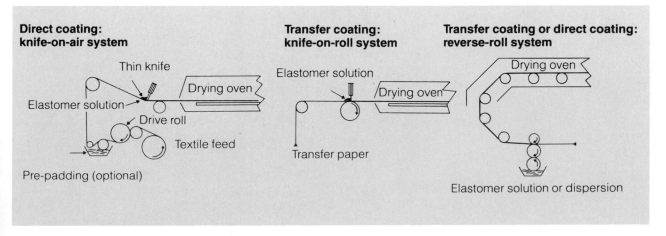

Impregnation by the polymer will bind together the textile fibres giving a stiff fabric with greatly reduced tear strength. Transfer coating uses a temporary carrier, usually an embossed silicone-coated release paper. This is coated, usually by knife or reverse roller coating (figure 8-8), with one or more layers of polyurethane elastomer. The solvent or water is removed from the coating by heating before it is contacted with the textile substrate in a heated nip-roller and bonded. The embossed release paper is then removed and rewound for further use. Coating machines are usually equipped with two or three coating heads to allow the application of two or three layers in one pass. It is common practice to use a one-component top coat followed by a softer two-component tie-coat. A typical machine arrangement is illustrated diagrammatically in figure 8-9 and a modern coating line in figure 8-10.

Figure 8-9 A typical transfer-coating line.

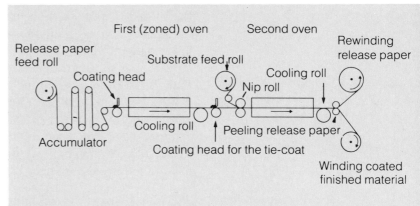

Figure 8-10 A modern textile coating plant.

Foamed coatings are usually produced by the direct spreading of aerated froths made from polyurethane latex. An anionic polyurethane latex with a high content of solids is frothed by mechanically mixing with air, spread by a knife and dried in a conventional zoned oven. The resulting foamed polyurethane elastomer coating may be used as a resilient inter-layer but is often abraded by an abrasive roller to make artificial suede materials for light-weight clothing.

Soon after manufacture, coatings may be embossed by heated rolls. In transfer coating processes the surface finish and the embossed pattern on the release paper determine the surface appearance of the coated fabric. Coatings may be printed using flexible inks based on polyurethanes or acrylics.

One- and two-component systems are both capable of giving coatings of either hard or soft elastomers with hardnesses from about 65° to over 80° Shore A with elongation at break up to 700%. Typical properties of coatings made with aqueous dispersions are indicated in table 8-8.

Table 8-8 **Typical polyurethane latex coatings**

Type of material	Anionic aqueous dispersion of 35% solids – aromatic isocyanate-based polyurethane	Anionic aqueous dispersion of 65% solids – aliphatic isocyanate-based polyurethane
Cross-linking agent	1% aqueous methylol resin (melamine/formaldehyde). Top coat of high light fastness	3% aqueous epoxy resin dispersion
Application	Top coat	Foamed coatings, suede effect
Tensile strength (MPa)	31	30
Elongation at break (%)	400	700
Stress at 100% elongation (MPa)	6	2
Stress at 300% elongation (MPa)	–	4

There are many other applications for aqueous dispersions of polyurethane elastomers. These include coatings for patent leather, anti-shrink and anti-felting finishes for wool, and sealing and finishing coatings for paper and board.

Thick coatings

Thick coatings are used for such industrial applications as conveyor belting for mining use, pipe linings for handling abrasive slurries and as protective membranes in building. Thick coatings are usually applied by spray using two-component reactive systems sometimes called "RIM-spray". The combination of toughness and high elasticity with the capability of forming large seamless membranes is particularly useful in building where sprayed elastomers are used to seal insulated flat roofs (figure 8-11).

Figure 8-11   Typical application of polyurethane elastomeric coating over rigid polyurethane roof insulation.

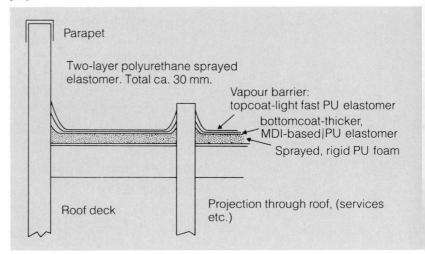

The optimum coating thickness for roof coatings is 28 to 30 mm using two layers. The first coating, about 20 mm thick, is based on an aromatic diisocyanate and a polyether polyol. It is covered by a thinner coating of a light-stable polyurethane that is pigmented for both high heat reflection and decoration.

9 Polyurethane adhesives, binders, paints, lacquers, cellular coatings and other materials

The low energy consumption of isocyanate-based, low temperature polymerisation processes, and the ease with which both the process conditions and the product properties may be adjusted, has led to many specialised applications for polyurethanes. This chapter briefly describes some of the more important of these. Unlike those described in earlier chapters, several are applications where the relatively expensive polyurethane materials occupy only a small part of the field and where their special properties satisfy a requirement for particularly high performance. The chapter also lists some potentially useful polyurethane processes that have, so far, found little commercial application.

Adhesives, binders and sealants

Polyurethanes can be used to join together most other materials. They make tough, vibration-resistant adhesives with a wide range of operating temperatures. Used as binders for both organic and inorganic materials, the ease with which they form gap-filling foams has found many uses.

Adhesives

During the past 20 years the use of adhesives has greatly increased as a result of the availability of improved adhesives based on synthetic polymers. Polyurethanes form a particularly versatile class of adhesive polymers, and it seems certain that their usage will continue to increase. In addition to the well-known, high-volume, applications in footwear, packaging and lamination, adhesives are not only replacing screws, rivets and nails but are becoming a preferred alternative to spot welding, soldering and brazing. Use of adhesives can produce joints of higher strength, allow the assembly of components without the need to drill or perforate them, and avoids the heat distortion resulting from high temperature jointing processes.

A satisfactory adhesive must:
- Wet the surface of the adherends
- Set or polymerise to form a strong solid that attracts the surfaces of the adherends and remains stable under the environmental conditions applied to the joint in service.

197

In general, liquids will only wet solid surfaces having a higher surface energy than their own, i.e. the surface tension of the adhesive liquid should be less than the free surface energy of the solid. Inert materials having low surface energy such as 'Fluon' (polytetrafluoroethylene), polysiloxanes, polyethylenes and the like, need surface modification by flaming, etching, UV light, corona-arc discharge or abrasion before they can be wetted even by adhesives containing free isocyanates. Most other surfaces can be wetted by polyurethane adhesives.

Polyurethane adhesives polymerise to form strong bonds without the need to use high temperatures. The high interfacial bond strengths obtained are derived not only from the physical forces resulting from intimate contact, but also from the ability of the polymerising polyurethane to form both hydrogen bonds and covalent chemisorption bonds with many surfaces. Such bonds can have excellent long-term durability, especially when the polyurethane adhesive is based upon hydrophobic polyether or hydrocarbon polyols. The ability of isocyanates to react with the unimolecular layer of water present on the surface of metals, such as steel and aluminium that have been cleaned by standard processes (such as defined in ASTM D-2651) that include washing, is probably the main reason for the high resistance to humidity of metal-to-metal urethane bonds. Polyurethane adhesives include both one- and two-component systems, with and without solvents to assist wetting, heat-activated thermoplastic adhesives, cross-linkable liquid polyurethanes and various aqueous dispersions. Some adhesives are mixtures of polyurethanes or isocyanates with other polymers.

Solvent-free adhesives are usually polyether or polyester polyols that are mixed with a liquid polymeric MDI or an MDI variant to form a liquid adhesive with a pot life of about two hours. Solvent solutions of contact adhesives are commonly urethane-extended polyadipates which can be chain-extended and cross-linked with polyisocyanates. Polyisocyanate solutions available for curing include the reaction product of TDI and TMP, TDI trimer, MDI variants and 4,4',4"-triisocyanato-triphenylmethane. Solvent solutions of polyisocyanates are also useful as primers and adhesion promoters, e.g. the use of a phenol-blocked MDI to promote the adhesion of polyester – poly (ethylene terephalate) – tyre-cords to rubber, and the use of polymeric MDI, in a volatile, water-free solvent, to prime metal.

The established uses of polyurethane adhesives include:
– Laminating adhesives for flexible materials – textiles, plastic films, aluminium foil, paper and board.
– For bonding rubbers, textiles, leather, plasticised PVC and polyurethanes.
– For bonding metal to metal, and metal to rubber and plastics.

Table 9-1 Typical peel strengths of plastic laminates (using about 2.5 g/m² of a polyester-based polyurethane adhesive.)

Laminate	Peel strength (g/25 mm)
Cellulose film/Polyethylene	600
Polypropylene/Polyethylene ('Propafilm'C)	425
Polyester/Polyethylene ('Melinex'800)	550
Nylon/Polyethylene	630
Aluminium foil/Polyethylene	630
Aluminium foil/'Propafilm'O	350
Aluminium foil/'Melinex'800	525
'Melinex'800/ Cast polypropylene	>870

Adhesive lamination. The traditional materials are solid, hydroxyl-group terminated, urethane-extended polyadipates that are applied to the substrate as a solution in ethyl acetate or methyl ethyl ketone. They are available in a range of molecular weights for use either as heat-activated thermoplastic adhesives, or in two-component systems chain-extended and cross-linked by the addition of a polyisocyanate. Most lamination, however, is carried out using solution-polymerised polyurethane elastomers that are supplied in solvent containing up to 75% solids or as dispersions in water. A typical range includes general-purpose laminating adhesives and high-performance systems for heat-sterilisable laminates and for food packaging.

Adhesives supplied in solution are usually diluted before use with an appropriate solvent to the viscosity required by the method of coating and the coating weight. Application to the substrate is usually by roller. Solvent is then removed in a vented oven and lamination completed using a heated nip roller. The amount of adhesive required varies from about 2 to about 8 g/m² depending on the type of laminate. The lower amount is usually sufficient for the simple lamination of plain (imprinted) plastics films (table 9-1). Printed films require from 3 to 5 gm/m². Higher levels may be required for textiles, unevenly printed boards and for sterilisable laminates.

Print lamination. This is the lamination of a transparent plastics film to printed paper or board to produce the glossy protective finish required for articles such as book covers, record sleeves, display cards, table mats, etc.. In Europe several thousand tonnes/year of adhesives are used for print lamination much of which is polyurethane adhesive. The use of polyurethane adhesives resulted

Figure 9-1 Plastic film laminates made with polyurethane adhesives.

199

from the replacement of 17-micron cellulose acetate film with the cheaper 12-micron oriented polypropylene film (OPP), which has superior flex-crack resistance. The polyacrylate adhesives used to laminate cellulose film cannot be used to laminate polypropylene satisfactorily. Print lamination is a high-speed continuous process. The adhesive solution, mixed with an isocyanate curing agent, is applied to the OPP film by a doctor blade or by a roller. The coated film is dried in a tunnel oven at not more than 100°C and then laminated to the printed paper or board in a nip roller at between 50 and 60°C. The adhesive layer is fully cured in 48 hours at room temperature. Water-based polyurethane dispersions are becoming available for print lamination although their adhesive properties do not fully match the best solvent-based systems. The absence of organic solvents simplifies the operation and venting of the laminating machine, although more energy is required to dry the aqueous dispersion coating.

Figure 9-2   Production of polyurethane fabric laminates

Textile laminates. Polyurethane adhesives are used to make fabric-to-fabric laminates and foam-to-fabric laminates for use in clothing and in the upholstery of vehicle seats and furniture. Fabrics are combined to increase the stability of knitted and open-weave fabrics and to obtain special effects for fashion clothing. Foam-backed upholstery fabrics have high crease-resistance and do not slip easily on the supporting cushion. They also have the high permeability to air and moisture essential for comfort in vehicle seating.

Two methods are used to combine fabrics, each of them a continuous process: by applying a liquid solution or dispersion of an adhesive; or by 'flame-bonding'. Polyurethane adhesive solutions or aqueous dispersions may be applied by knife or roller coating to give a thin continuous layer, or by means of off-set printing rollers or by spraying (aqueous dispersions only), to obtain a discontinuous highly-permeable layer of adhesive. Flame-bonding is a melt-bonding process using a continuous thin sheet of a special grade of flexible polyurethane foam. The foam is passed through a gas flame to melt the surface and then, while still tacky, is contacted with the textile material in a nip-roller. The process, originally developed to make fabric-to-foam laminates, is also used to laminate two fabrics by using very thin foam foils (about 0.5 mm), so that substantially all of the foam is melted to form the adhesive. Flame-bonding gives highly permeable laminates and the process machinery is capable of operation at rates up to about 50 metres/minute or several times the rate obtainable with processes that rely on adhesive spreading. The bond strength of flame-bonded fabrics increases with the amount of foam used to form the bond. A typical peel strength of about 0.05 kg/cm is obtained by flame-bonding with 0.5 mm of a 22 kg/m^3 foam. The bond strength is unchanged by dry-cleaning with perchloroethylene or light petroleum. Flame-bonded textiles, and those bonded with polyurethane adhesive, will both withstand washing at 95°C (ISO Wash Test No. 4). A compact flame-bonding machine is shown in figure 9-3. The actual bonding process is done immediately beneath the hood which is positively vented to remove fumes from the heated polymer.

Figure 9-3   A compact flame-bonding machine.

Shoe adhesives. The shoe industry is a major user of polyurethane adhesives. Single-component and reactive two-component systems are both used. The main applications are in lasting (the stretching of the shoe upper over a last and bonding the upper to the in-sole), and for attaching moulded soles. Polyurethane adhesives are particularly effective for bonding soles made from PVC, thermoplastic rubber and SBR compound soles. Effective priming treatments include a wipe with a polyfunctional isocyanate solution (about 3% polymeric MDI in MEK or a chlorinated solvent).

Engineering adhesives. Adhesives are increasingly used to replace traditional fixing methods using bolts, screws, etc. Metal-to-metal bonding is often achieved with epoxy-based adhesives but there are many applications where the resilience and toughness obtainable from polyurethanes are advantageous. Polyurethanes are used in the direct glazing of windscreens, rear and other fixed windows, to the metal bodies of automobiles. A popular system uses a moisture-curing polyurethane which is applied, as a high-viscosity liquid bead, by a robot. The uniform bead is then sandwiched between the screen and the metal body where it cures rapidly to give a strong, vibration-damping, direct glazing material. A similar process is used to attach vehicle windows which have been edge-encapsulated with a RIM polyurethane elastomer. The use of RIM edge-encapsulation together with direct glazing provides a simple method of making aerodynamic window-seals by robotic assembly. Polyurethanes may also be used to apply obscuration bands to windscreens. These bands protect the direct glazing material from the effects of direct sunlight and – on highly raked screens – are also used to avoid over-heating of the interior trim. Obscuration bands based upon HDI are applied by spraying and are then quickly cured, at room temperature, by passage through a chamber containing controlled amounts of a mixture of water vapour and tertiary amine vapour. Another application is the assembly of moulded GRP truck and tractor bodies. Screws and rivets are unsuitable for joining GRP as they give too high a stress concentration from the vibrations and shocks transmitted in service. A two-component polyurethane adhesive spreads the stresses and damps the vibrations. Two-component diamine-cured systems with a short gel time have proved to be successful. They are applied by piston dispensing units which supply and mix the two components in the required ratio. The items to be bonded are held together under a slight pressure for 3 to 30 minutes depending on the gel time of the adhesive system and the temperature of the items. Typical bond strengths are indicated in table 9-2.

Table 9-2 Typical bond strengths of polyurethane adhesive joints ('Pliogrip' series)

Adhesive system	System 1		System 2[+]	
Adherends	('Pliogrip' 6000) SMC/ SMC	SMC/ Metal	('Pliogrip' 6040) SMC/ SMC	SMC/ Metal
Shear strength (MPa)				
at − 40°C	4.4 D	6.6 D	5.3 D	6.5 D
at 20°C	5.2 D	7.0 D	4.6 D	5.5 D
at 82°C	3.3 D	2.8 D	4.4 D	3.0 D
at 120°C	1.9 C	—	2.5 D	—
at 180°C	—	—	1.5 C	—
After 14 days at 90°C	5.9 D	7.8 D	4.4 D	5.8 D
After 14 days at 38°C/100% R.H	4.3 D	5.4 D	4.9 D	5.7 D
After 240 hours salt spray	5.3 D	4.2 D	—	4.7 D
Flexural peel strength (ASTM 970)	2.8 D	—	1.8 D	—

Key to abbreviations
D = Delamination of the SMC substrate.
C = Cohesive break of the adhesive.

[+] System with high heat resistance to meet the paint oven of the FORD and MACK truck specifications. (30 minutes at 190° and 10 minutes at 205°C respectively)

Binders

Binders are adhesive systems used to bond together particulate materials. Examples of the use of isocyanate-based binders include the reconstitution of scrap flexible foam, the manufacture of wood-chip and particle boards, the bonding of foundry sand, and the manufacture of moulded articles and sheet materials from recovered rubber granules and natural products such as straw.

Reconstituted flexible polyurethane foam. There are four stages in the manufacture of rebonded foam (figure 9-4):
– Chopping the foam into suitable pieces.
– Coating the pieces of foam with an adhesive or a foaming adhesive mixture.
– Compressing the coated pieces in a mould.
– Curing the rebonded foam.

Figure 9-4   Stages in the batch reconstituted block-foam process.

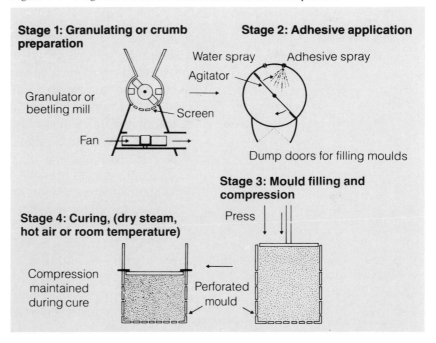

The conversion of slabstock flexible foam into cushioning leaves up to about 12% of waste off-cut material – depending on the shape of both the foam blocks and the cut parts. This waste off-cut material is passed through a suitable cutting or tearing granulator, fitted with a screen having a mesh size of between 5mm and 15mm depending on the product required, to produce what is called 'crumb'. Larger pieces, having a smaller surface to volume ratio, require less adhesive but need higher compression in the mould to ensure a product free from voids. They also tend to expand slightly on cutting the reconstituted foam giving a product with an uneven cut surface. The chopped and graded foam crumb is coated with a reactive

203

polyurethane adhesive by spraying and tumbling. The adhesive is based upon a low-cost polyether triol, together with TDI or polymeric MDI. TDI is chosen for making reconstituted flexible foam slabstock, but small mouldings are best moulded and cured using MDI or MDI variants. The adhesive may be an isocyanate-terminated prepolymer containing some free isocyanate. These adhesives are often made by simply mixing the required ratio of isocyanate and polyol at room temperature a few hours before use. However, most reconstituting processes use one-shot systems. The foam crumb is first moistened with water and then coated with a TDI/polyether mixture.

Pigments and fillers may also be added at this stage. Cold-curing systems require the addition of catalysts; these are conventional, low-odour, foam catalysts. The coated foam crumb is dropped into a mould and compressed.

The degree of compression, to between 50% and 10% of the uncompressed height, depends upon the density and load-bearing properties required of the finished product. The material is maintained in the compressed state until sufficiently cured to allow demoulding without significant expansion. Most block-making systems are cured at room temperature or in warm, vented rooms. Surface cure may be accelerated by infra-red heating. Small moulded parts may be cured in conventional ovens, or more efficiently by using microwave heating. Cold-curing systems based on MDI variants are used in the manufacture of sound and energy absorbing items.

Reconstituted foam may also be made by semi-continuous and continuous processes. Continuous processes on a large scale have been used for many years for the manufacture of carpet underlay. The foam crumb is coated with reactive adhesive and distributed on to a polyethylene film carried on a conveyor. It is then levelled, covered with a second film, compressed, perforated and cured by heat and/or steam. The laminate is then slit horizontally into two polyethylene-faced foam sheets and reeled up.

Semi-continuous processes are used in the manufacture of sound-absorbing automotive trim. The foam crumb, coated in a continuous extruder, is dispensed into moulds and cured, often in contact with a decorative facing material. Reconstituted foam, with a density in the range from about 40 to 100 kg/m^3, is used for carpet underlays and, in furniture, for padding the front edges of chair frames, arm-rests etc. beneath a soft foam cushion. Materials of higher density have many applications in seating where a thin cushion is required over a solid base, e.g. bar stools, lecture hall and theatre seating, and for matting in sports halls and gymnasia. All grades are used in packaging and in sound-absorbing padding.

MDI-based binders for wood-chip and wood-fibre boards. The past ten years has seen a steady growth in the use of polymeric MDI

Figure 9-5   Roofing board of MDI-bonded chipboard and MDI-based rigid polyurethane foam.

Figure 9-6　Box made from a board consisting of waste vegetable fibre bound together with polymeric MDI.

compositions to replace methylol resin solutions (urea-formaldehyde, phenol-formaldehyde and melamine-formaldehyde syrups) in the manufacture of reconstituted wood products. The use of MDI compositions as a binder provides significant advantages in both the manufacturing process and in the quality of the product. MDI compositions can often be used on existing machines for making particle board with little or no modification. Special grades of polymeric MDI are available which can be distributed directly onto the wood chips by spraying. Two methods are used. The first uses low-viscosity polymeric MDI of 1 to 2 d Pa s. These polyisocyanates have a low vapour pressure at room temperature and are applied to the wood chips by spraying into the blender. Water emulsions of sizing agents or of paraffin water-repellants must be sprayed separately. The alternative method, which gives the most effective distribution of the binder, uses specially-developed MDI compositions that are easily emulsified in water without the use of emulsifying agents. The emulsion may be used in equipment designed to handle aqueous methylol resins. There are two products available from ICI, 'Suprasec' 1042 and 'Suprasec' 1249, both of which form stable emulsions of polyisocyanate in water. One of them, 'Suprasec' 1249, has self-releasing properties when used as the binder for the outer layers of 3-layer particle board. These stabilised polymeric MDIs react only slowly with water. Over 90% of the available isocyanate groups remain unreacted 2 hours after stirring the emulsifiable MDI into water (figure 9-7). The low viscosity emulsions (figure 9-7) are easily handled on conventional equipment.

Figure 9-7　Properties of MDI/water emulsions from 'Suprasec' 1042.

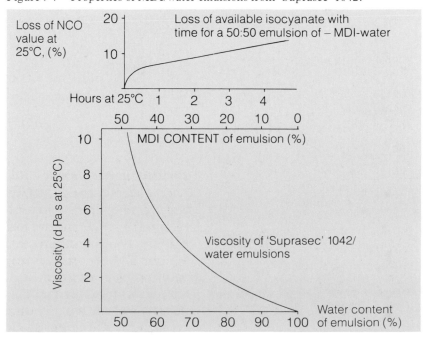

205

The reaction of MDI with water is catalysed by alkaline materials. The presence of alkaline phenolic resin residues or of alkaline melamine compounds will cause premature reaction of the MDI in water emulsion. Urea-formaldehyde resins, which usually have a pH near to 7, and aqueous wax emulsions are compatible with 'Suprasec' 1042/water and with 'Suprasec' 1249/water emulsions. The most economic method of using MDI is by continuous emulsification. Emulsifiable MDI is injected, at a metered rate, into a water jet, using a high pressure impingement mixer. This facilitates adjustment of the MDI/water ratio with variations in the particle size and the water content of the wood particles. Particle board bound with polymeric MDI has a high strength (figure 9-7), good weather resistance and high dimensional stability. MDI-bonded particle boards are also free from objectionable or toxic odours and, unlike some boards bonded with phenolic resins, they are neutral and non-corrosive. MDI-bonded boards have been approved for both interior and exterior building applications and also for use in lining dry food containers.

Figure 9-8   The internal bond strength of three-layer board bonded with polymeric MDI compared with that bonded with emulsifiable MDI.

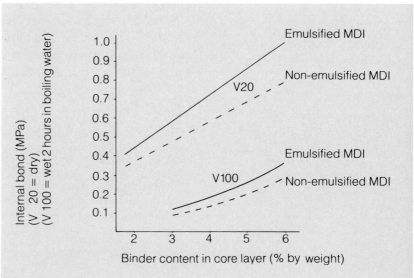

Bonded rubber granules. Reclaimed rubber is obtained during the retreading of car and truck tyres but scrap tyres may also be milled and separated from the steel reinforcement. New processes for reclaiming scrap tyres are making large quantities of rubber granules available in particle sizes from about 1cm down to less than 0.5mm. Rubber granules may be obtained by grinding at normal ambient temperatures or at cryogenic temperatures. Grinding or tearing scrap rubber at ambient temperatures tends to give rubber particles that are rough surfaced and elongated, whereas grinding at cryogenic temperatures produces smooth, almost cubic granules.

Rubber granules may be used as fillers, or may be reconstituted and revulcanised using conventional rubber processes. Rebonding rubber granules with polyurethane binders offers an alternative method which requires less capital investment and a lower consumption of energy than reconstitution by rubber milling. The graded rubber granules are coated with adhesive, compacted in a mould or press, and cured. Both one-shot and prepolymer adhesive systems are used. The properties of the rebonded composite polyurethane/rubber material depend upon the type of rubber and the distribution of particle sizes of the granules, the constitution and the amount of the polyurethane adhesive and the degree of compaction during curing. A wide range of physical properties is thus possible. The products range from dense moulded articles to large porous sheeting and flooring. A major use of polyurethane-bonded reclaimed rubber is in flooring for gymnasia and sports halls and for the surfacing of outdoor athletic tracks and ball courts.

Sports surfaces may be laid directly onto an asphalt surface, using graded rubber granules coated with an isocyanate-terminated moisture-curing prepolymer. The porous, bonded rubber granule layer is usually coated with a pigmented flexible polyurethane lacquer or with a coating, 3 mm to 4 mm thick, of sprayed, pigmented polyurethane elastomer, the thickness depending on the degree of wear resistance required.

Figure 9-9   A typical outdoor running track construction.

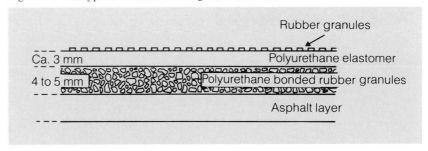

Isocyanate-cured foundry resins. Foundry resins are the binders used with foundry sand to make moulds and mould cores. So called 'cold' processes produce accurate sand castings without heating thus saving heating energy, time and capital investment compared with hot-curing processes using urea-formaldehyde and phenol-formaldehyde resins. Productivity per core-making machine can be doubled.

In the Ashland process, a phenolic resin and polymeric MDI are mixed with sand. The mixture is poured into a mould in an enclosed box. A rapid-curing reaction is then initiated by the introduction of a volatile amine catalyst carried in a stream of air or carbon dioxide. The process was introduced over ten years ago and it has become widely used over the past five years especially in the manufacture of

Figure 9-10   A foundry sand core bound with polymeric MDI and a phenolic resin.

complex aluminium castings for the automotive industry. The process uses thousands of tonnes of phenolic/MDI binder each year in a ratio of 1 to 2% by weight of binder to sand.

Consolidation of loose rock. Rigid polyurethane foam finds a specialised use in the consolidation of loose rock. Tough, high density (200 to 500 kg/m^3) foam, based upon polymeric MDI, is injected through holes bored in the weak areas of tunnel walls or in roof support layers. Two methods of injection are used. Two-component rigid polyurethane foam systems are metered and injected using special machines driven by compressed-air pumps delivering up to about 6 litres/minute of foam reaction mixture or, more commonly, a cartridge system is used. The cartridges consist of two concentric tubes of plastic foil. The outer tube contains the polyol blend and the inner tube contains the amount of polymeric MDI required to form the foam. The PU cartridge is inserted into a 50mm hole bored into the face. A pointed wooden rod of square section is mounted in a drill and inserted into the hole. It is then rotated and moved around for about 20 seconds to break the foil tubes and to mix the foam components. The hole is then plugged. The reacting foam mix expands to fill the cracks and fissures around the borehole and so consolidates the rock-face behind the roof-supports.

Polyurethanes in the construction industry

As discussed earlier, this is the industry that uses most rigid polyurethane foam usually for insulation. There are many other applications of polyurethanes as well as interesting applications of established rigid polyurethane foam products. These include polymer concretes and flooring compounds made from aggregates coated with polyurethane binders; elastomeric polyurethane membranes; moulded structural foam items such as gully pots and sanitary ware; and the use of polyurethane structural adhesives and sealants.

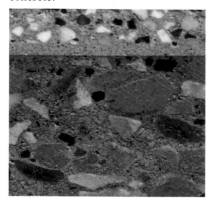

Figure 9-11 Section through an industrial floor of polyurethane-bound concrete.

Paints and lacquers

The toughness, abrasion- and mar-resistance of polyurethane extended polymers are valuable in protective and decorative surface coatings. In addition to the use of diisocyanates as intermediates in the manufacture of urethane-modified resins for conventional air-drying paints, there are several types of surface coatings containing isocyanate groups which react during the final polymerisation and cure of the coating. Both aromatic and aliphatic isocyanates are used. Coatings based on aromatic isocyanates tend to yellow in daylight, but they are satisfactory for durable interior finishes on wood and for applications such as coatings for wire and cable. Non-yellowing coatings are made using aliphatic diisocyanates such as

HDI and IPDI. HDI is widely used, usually as the biuret which is often sold as a 75% solution in a solvent. Polyurethane coatings are available as both one- and two-component systems.

Two-component coatings consist of a polyol blend and an isocyanate component. The two parts are blended together immediately before application to the substrate. The mixture has a limited pot life as the coating cures at room temperature by the reaction of the hydroxyl group of the polymer with the polyisocyanate to give a urethane-linked polymer of high molecular weight.

There are several types of one-component coatings. Liquid moisture-curing prepolymers consist of polymers with terminal isocyanate groups, that cure at room temperature by the reaction of the isocyanate end-groups with moisture. Solvent-diluted and solvent-free coating systems are both available. Baking enamels are formulated from mixtures of polymers having reactive hydroxyl groups polymers and 'blocked isocyanates'. These do not react at room temperature. The blocking agents are released at elevated temperatures (from 140°C depending on the blocking agent) to generate reactive isocyanate groups which react quickly with the hydroxyl groups of the resin at the baking temperature. Powder coatings based on diisocyanates are extremely durable. These are homogeneous blends of solid polyester polyols and solid 'blocked' diisocyanates which are powdered and applied by electrostatic powder spray or by hot dip-coating and then baked to complete the polymerisation.

Two-component coatings

Isocyanates for use in surface coatings must have a low vapour pressure if they are to be used without special handling precautions. TDI-based coatings employ the reaction products of TDI and low molecular weight polyols. The most widely used is the trifunctional adduct from trimethylol propane and three moles of TDI (figure 9-13), but the use of polymeric MDI and of special MDI compositions is significant for heavy duty coatings.

Figure 9-13 TDI/trimethylol propane adduct for cross-linked coatings.

In addition to both polyester and polyether polyols, polyols based upon natural products such as castor oil are used. Some covalent

209

cross-linking of the polyol assists in obtaining a satisfactory cure and good resistance to hydrolysis. Once the polyol and the isocyanate components are mixed, the viscosity of the mixture begins to increase and the pot life is limited. Theoretically, two-component coatings of this type are cured by the urethane reaction, but in practice especially at room temperature, there is always some reaction with atmospheric moisture. At oven-curing temperatures above 100°C, cross-links may also be formed during the curing reaction.

Moisture-curing prepolymers are also sometimes used as two-component systems, the second component being a catalyst, such as a tertiary amine or metal naphthenate, to ensure a fast cure. These systems also have a limited pot-life as they 'skin' very quickly on exposure to the atmosphere.

One-component coatings

Moisture-curing prepolymers. These are isocyanate-terminated polymers that are liquid or soluble in a suitable carrier solvent. They are usually made by reacting polyester or polyether polyols having molecular weights of about 1000 with diisocyanates. For satisfactory stability in storage of moisture-curing prepolymers, water must be excluded during their preparation and at all times. Solvents, pigments and fillers must be free from water before incorporation into the prepolymer.

Moisture-curing prepolymers are made using TDI, but require special treatment to reduce the level of free volatile diisocyanate to the very low level allowed. MDI is widely used because of its low volatility. Alternatively, adducts of TDI may be used or a combination of TDI and a non-volatile isocyanate in a two-stage process. Non-yellowing coatings require the use of aliphatic diisocyanates.

Moisture-curing coatings cure by the reaction of the isocyanate end-groups with atmospheric or other moisture to form substituted urea linkages. The carbon dioxide produced in the curing reaction diffuses from the coatings without bubble formation provided that the coatings are not too thick. There are no significant side reactions and moisture-cured coatings give very reproducible properties.

One-component baked enamels. These include mixtures of branched polyester and polyether polyols with 'blocked' isocyanates. The latter are simple isocyanate adducts, stable at ordinary temperatures, which decompose to produce free isocyanate on heating. One-component baking systems are cured by stoving at 140 to 200°C, depending on the thermal stability of the adduct used. That used most frequently is the urethane formed by the reaction of an isocyanate and phenol. The urethane releases free isocyanate at 140°C. The phenol volatilises in the curing oven and may be recovered from the exhaust system.

Powder coatings. Powder coatings are used to give very durable high-gloss coatings, especially on metal. The powder is usually a mixture of a solid, branched polyester polyol and a solid "blocked" diisocyanate. A typical polyester polyol is derived from terephthalic acid and a diol such as bisphenol A, with a branching agent such as glycerol or trimethylol propane. The blocked diisocyanate is an adduct such as that from caprolactam and IPDI that releases the isocyanate groups when the powder coating has melted to form a glossy film. Powder coatings are usually prepared by mixing together the solid polyol and the blocked diisocyanate, homogenising the mixture by melting in an extruder at a temperature below the decomposition temperature of the blocked isocyanate, and passing the chilled product directly into a pulveriser.

Urethane alkyds. Urethane alkyds or "Urethane oils" are resins with hydroxyl end-groups. They are made by reacting hydroxyl-terminated esters or partial esters of unsaturated fatty acids with a diisocyanate, usually TDI. These urethane alkyds do not contain unreacted isocyanate groups. They are dissolved in hydrocarbon solvents to make conventional paints that cure by air oxidation of the unsaturated groups in the presence of metallic driers, to give tough coatings with high abrasion resistance.

Cellular coatings

There are several methods of applying thin layers of high-density foam to a substrate. These are used mostly to make flexible products with a foam layer from 2mm or 3mm thick up to about 7mm or 8mm, although foam layers up to 100mm thick have been made experimentally. The biggest established application is for backing carpets but the process is also being applied to the production of resilient foam layers for many other uses. These include the manufacture of sheet material for sports shoe insoles and for exercise mats for aerobics. Frothed foam layers are used to make a playing surface for tennis courts. Many other products are being developed using this coating technology.

Carpet-backing with polyurethane foam
Woven carpets were often stiffened with gums and starch-based products but polymeric secondary carpet backings with good tensile properties became important with the development of high speed tufting machinery. Tufted carpet depends upon a secondary backing of adhesive material to anchor the tufts. The most popular tuft anchorage coatings are SBR lattices. They give good tuft-lock at low cost. The next step in reducing the cost of tufted carpet was the application of a resilient foam backing over the anchor coat. This increased walking comfort and enabled the weight of the expensive

211

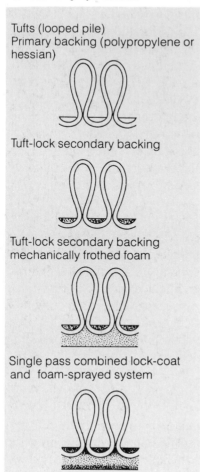

Figure 9-14   Tufted carpet backed with cellular polyurethane.

Tufts (looped pile)
Primary backing (polypropylene or hessian)

Tuft-lock secondary backing

Tuft-lock secondary backing
mechanically frothed foam

Single pass combined lock-coat and foam-sprayed system

pile fabric to be reduced.

There are, therefore, two sequential steps in the established process for backing a low cost tufted carpet. These are, first, the application and partial drying of a tuft-locking latex having a high content of solids and a low content of filler. This is followed by the application of a mechanically-frothed SBR latex foam. The latter is often heavily filled with ground chalk or limestone to give the required load-bearing properties at low cost.

In addition to adhesive lamination processes, in which preformed sheet foam or rebonded flexible foam sheet is brought into contact with a tacky tuft-lock coating, a reactive polyurethane foam mixture may be applied directly to the carpet backing. Two methods are in use:

– The most widely used method employs a mechanically-frothed, slow-reacting, flexible foam mixture which is spread onto the primary carpet backing, at a selected thickness and then cured by heating to a temperature between 80 and 130°C.
– The second well-established method uses a "water-blown" foam mixture which is distributed over the primary carpet backing by a traversing spray. Infra-red heaters are used to prevent heat loss from the reacting material and to control the rate of surface cure. The partially-cured foam back is then embossed by a heated roller.

Both methods of application give products of high quality with large savings in process energy and high rates of production compared with the application of aqueous lattices. Both polyurethane systems allow tuft-locking and foam-backing to be carried out in a single spreading and curing operation.

Mechanically frothed polyurethane foam backing. The main advantage of this process is its use of established latex foam backing equipment with the addition only of metering units for the polyurethane components. Three components:

– a mixture of polyether polyols with catalyst(s), a silicone surfactant and a filler,
– a polyisocyanate (now usually polymeric MDI) and,
– air or inert gas

are metered separately to an 'Oakes'-type mixer. This frothing mixer must have efficient cooling to remove the heat resulting from shear action. The resulting stable creamy froth is deposited by a traversing pipe onto the carpet back, smoothed to the required thickness by a conventional knife spreader, and passed into a heating tunnel where the froth expands by about 30% and reacts to form a polymeric foam (figure 9-15).

Figure 9-15   Mechanically frothed polyurethane foam backing.

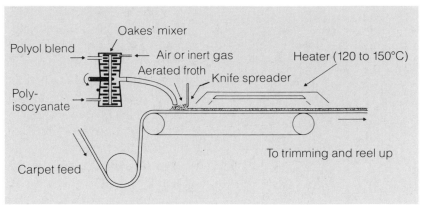

Figure 9-15   Mechanically frothed polyurethane foam backing.

Figure 9-16   Tufted carpet with polyurethane tuft-anchorage and foam-backing applied in one step by the ICI process.

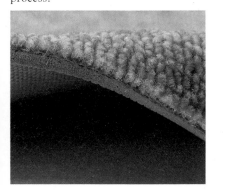

Oven cure temperatures range from about 120°C to 150°C, depending on the formulation of the foam. Either hot-air or infra-red heating may be used to cure the foam in about 3 minutes without overheating the thermoplastic primary backing or tufted pile. The energy requirement is about 600 Kcal/m^2 or about 25% of that required to cure a similar latex foam backing and tuft-locking coat. The process uses delayed-action catalysts, such as nickel acetylacetonate, which give a long pot life at room temperature. Froth systems based upon selected polymeric MDI compositions and reactive polyether polyols have also been used with low levels of such more usual polyurethane foam catalysts as dibutyltindilaurate and acid salts of tertiary amines. Both TDI and MDI based systems are used. Both give satisfactory properties (table 9-3). As the froth reacts and cures without significant chemical blowing from the reaction of water and isocyanate, the polyurea hard segments of the conventional flexible foam structure are absent. The foam is stiffened by increasing the covalent cross-link density, by the use of 'polymer polyols' with grafted polyvinyl filler, and by the incorporation of inorganic particulate fillers. Foam densities are relatively high as it is difficult to make stable mechanical froths with densities much below about 200 kg/m^3.

Sprayed foam backing. This system, originally patented by ICI in the 1950s, has been in commercial use for some years. The process is characterised by the use of a high-speed traversing spray to distribute a highly-reactive foam mixture directly onto the back of untreated tufted carpet. The carpet is carried through the foam application and curing tunnel on stenter pins that are shielded from the spray application. The reactivity of the foam-forming mixture is adjusted to allow from about 15 to 25% of the applied mixture to penetrate the base of the tufts and the primary backing of the carpet. This gives stable carpet structure with a good tuft-lock. The foam thickness is controlled by adjusting the amount of foam mixture deposited and by the rate of travel of the carpet through the

213

application zone. An additional minor degree of thickness control is given by the embossing roll.

The process gives a unique product with a tough abrasion-resistant integral skin, the thickness of which is controlled mainly by adjusting the initial level of infra-red surface heating in the curing tunnel. The foam is blown and the polymer is stiffened by the reaction between water and polyisocyanate. The foam has the required load bearing and walking comfort characteristics at a lower minimum density than the alternative systems. A disadvantage of the direct spray process is the tendency of the foam surface to reproduce the texture and the faults of the primary backing and tufting. The foam back is therefore embossed, soon after the foam surface becomes tack-free, by an engraved roller at 130 to 150°C (figure 9-17).

Figure 9-17   Sprayed polyurethane foam-backing process.

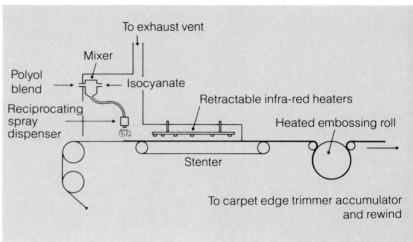

The basic process has been used to apply foam to other substrates such as paper, needlefelts, foils, PVC flooring and non-woven textiles. The process may be adapted for the manufacture of flexible laminates. The properties of some polyurethane foam-backs are tabulated in table 9-3.

Table 9-3 Typical properties of polyurethane foam-backs

System/type	Mechanical froth (MDI-polyether)	Direct spray (TDI-polyether)
Density (kg/m^3)	475[1]	130[2]
Thickness (mm)	3 to 4	3 to 4
Tensile strength (KN/m^2)	700 to 880	600 to 700
Tear strength (N/m)	2500	1300
Elongation at break (%)	150	170
Compressibility (%)	24	18 to 24
Tuft-lock (kg)[3]	4 to 10	3.5 to 9
Resilience (% DIN 54316 rebound)	31	35 to 45
Fibre bundle penetration (%)	High	100
Pilling/fuzzing resistance	Excellent.	Excellent.

Key
[1] 90% chalk filled.
[2] Excluding integral skin.
[3] Tuft-lock depends on the tuft fibre and its construction.

214

Compared with SBR foam backings, those made from polyurethane foam have higher abrasion resistance, a higher fibre-lock, better resistance to heat ageing and to compression from chair castors, and better thermal and sound insulation. The polyurethane carpet-backing process requires less heat energy and less labour and gives improved working conditions compared with rubber latex processes.

Some other applications of isocyanates and polyurethanes

The vulcanisation of natural rubber with diisocyanates
Aromatic diisocyanates are effective cross-linking or vulcanising agents for rubber. The isocyanate derivatives used are stable powders, unreactive at room temperature, which dissociate to yield free isocyanate at normal vulcanising temperatures. The most popular urethane vulcanising agents are diisocyanates blocked by reaction with two moles of nitrosophenol. An important product for the vulcanisation of natural rubber is a blocked MDI (see figure 9-18). TDI dimer, which is used to cross-link millable polyurethane elastomers, is also a useful synergistic reinforcing additive in some rubber vulcanisation systems.

The reaction of the nitrosophenol (from the dissociation of the MDI-nitrosophenol adduct), with natural rubber requires catalysis by a metal dithiocarbamate. The same catalyst also serves to catalyse

Figure 9-18   Vulcanisation of natural rubber with MDI.

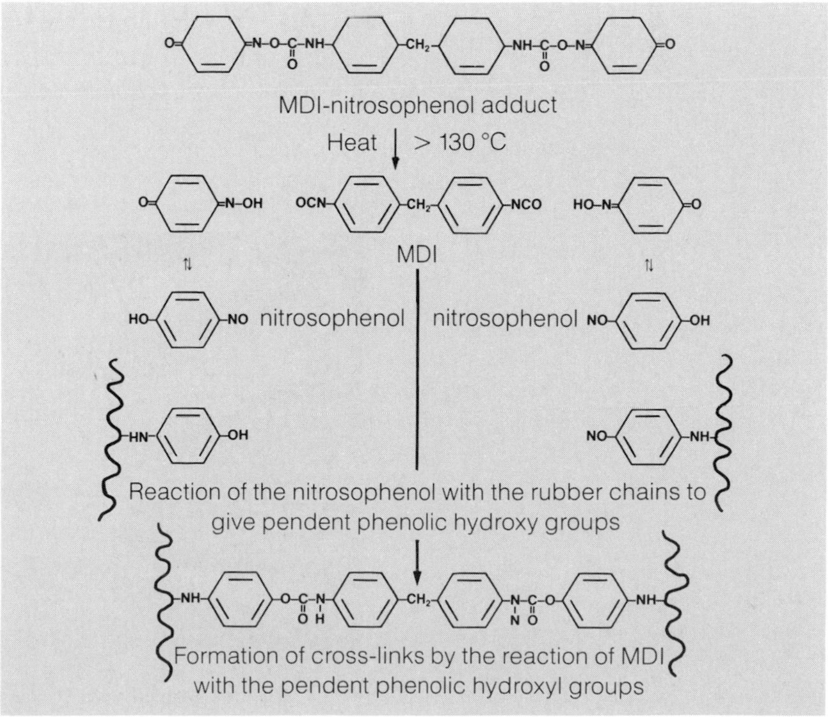

215

sulphur vulcanisation in mixed isocyanate/sulphur synergistic systems. The MDI subsequently reacts with the phenolic groups to produce covalent cross-links between rubber chains.

MDI-based vulcanising agents are making an increasing contribution to rubber processing that is fast and of uniform quality. They give good heat stability and greater reversion resistance than sulphur vulcanizates. Diisocyanate-based vulcanising agents are synergistic with sulphur-based vulcanizates and the addition of the diisocyanate-based agent gives improved ageing resistance of the rubber. The Malaysian Rubber Producers' Research Association particularly recommends the use of an MDI-nitrosophenol adduct as a vulcanising agent for high temperature, fast vulcanising processes employing curing temperatures of 180°C and above. Mixtures of the adduct and TDI dimer are suggested for the fast injection moulding of items such as bushes, flexible couplings and engine mountings.

Sealants

Include gap-filling foam systems provided in convenient aerosol containers. Moisture-curing, one-component systems and two-component systems are both available in this form. The two-component systems are packed in special containers divided into two compartments. Before use, the division is punctured and the contents mixed by shaking. Once mixed, of course, the whole of the contents must be used immediately. One-component froth systems, which were invented by ICI, may be dispensed intermittently from the can. They are widely used to fill gaps in rigid polyurethane foam and other insulating materials because they act as both a flexible sealant and as an insulant. One-component froth systems based upon polymeric MDI are widely available in small aerosol containers (figure 7-32).

Figure 9-19   A 27,000 tonne grain warehouse which is sealed by rigid polyurethane foam to minimise the loss of fumigant gas.

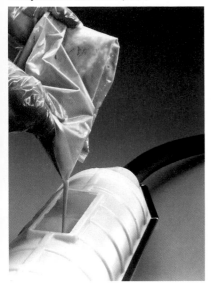

Figure 9-20   Polyurethane resin is poured into the polystyrene shell and encapsulates the cable joint.

Potting compounds

Pour-in-place 2-component and 3-component polyurethane systems are widely used throughout the world to encapsulate and protect the joints in power cables. Until the early 1960s, joining power cables involved soldering the joints of the conducting cables, insulating with hot bitumen compounds and the use of cast-iron boxes for protection and earth continuity. Modern multiple cable earthing systems and the development of mechanical cable connectors that can be applied safely to live cables were factors which encouraged the development of the more cost effective polyurethane systems. The polyurethane cable jointing systems used today are two-component or three-component packs which are mixed on site to form a reactive, liquid polyurethane resin which is poured into a plastics shell surrounding the cable joint, (figure 9-21).

Figure 9-21   A diagram of an 11 kV joint insulated with polyurethane resin.

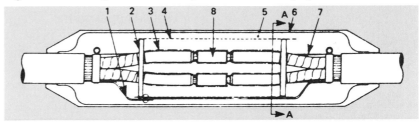

1. Armour continuity bond
2. Spacer
3. Interface/stress control tape
4. Expanded metal safety screen
5. HV polyurethane resin compound
6. Polystyrene box
7. Armour connection to core screen
8. Conductor ferrule

The polyurethane reaction mixture encapsulates the cable joint, gels quickly and cures with a relatively low exothermic temperature rise of about 40°C. The result is a water-tight seal with sufficient resilience to absorb and dissipate both the mechanical stresses and the heat arising from the internal joints in the power conductor. Polyurethane systems are available containing fillers which increase the thermal conductivity of the encapsulating resin for low voltage, high current carrying cables.

PUR structural foams are also used to make electrical junction or joint boxes and distribution boxes as well as their well-known use to mould loudspeaker cabinets, TV cabinets, and housings for computer terminals and other office equipment.

Self-skinning flexible foams and cast elastomers are also used to make seals and gaskets for electrical housings. Many of these are now made by pouring a liquid reaction mixture into a groove in the casing or the lid of the housing, so that the sealing strip is formed and bonded to the housing in one operation. Thixotropic reaction mixes are available which allow the manufacture of a half-round seal on a flat surface. Typical applications using pour-in-place seals are electrical junction boxes and housings for dust- and moisture-proof fluorescent lights.

217

Figure 9-22 MDI-based, custom cold-moulded, seat cushions for invalide chairs.

Electrical and electronic equipment

Polyurethanes are electrical insulators* and their use in this area has grown substantially during the last few years.

Polyurethane wire enamels and powder coatings have been used for many years. They are tough and flexible with a wide operating temperature range and the coated wires may be soldered directly, without cleaning the wire or removing the polyurethane enamel. Many types of polyurethane elastomers are used in the electrical supply and machinery industries. Thermoplastic polyurethane elastomers are mainly found as tough protective wiring sleeves on heavy machinery, both indoors and outside, including wiring installations in the oil and gas industry. Cast polyurethane elastomers are used for moulded insulators, transformer housings and junction boxes, and for the encapsulation of sensitive equipment.

Medical applications

All types of polyurethanes, elastomers – both cast and thermoplastic – rigid foams and flexible foams are used in medicine. Polyether-based cast and thermoplastic elastomers are valuable because of their flexibility combined with good resistance to hydrolysis and to the formation of local blood clots. Polyurethane elastomers are being used without surface treatment for devices such as artificial hearts, heart valves, connector tubing for heart pacemakers, haemodialysis tubes, blood bags and other components.

In addition to the long-established use of flexible and semi-rigid foams in sterilisable pillows and orthopeadic mattresses, individually moulded cold-cure foam is used to make cushions and mattresses for special treatments where the patient needs to be uniformly supported without pressure points. Examples of individually-matched mouldings include chair cushions for severely disabled people (figure 9-22), and the moulded turning beds used in treating spina bifida patients. Flexible polyurethane foams are also used as wound dressings. Sterile, soft foam sheet is used as a ventilated, self-draining burn dressing and specially developed, highly hydrophilic flexible foams meet the need for light absorbent padding to replace textile fibre dressings. Flexible foam sheet is also used for soft elastic support bindings.

Rigid and flexible foams may both be used to replace heavy plaster-of-Paris splints for broken limbs. One ingenious method uses a flexible foam sheet bandage that has been impregnated with an MDI-based, branched prepolymer. The impregnated bandage is moistened with water before use as the outer binding layers around the limb to be supported. The bandage hardens within minutes to form a tough but light and permeable support.

10 Health and safety :
making and using polyurethanes safely

All man's activities, at work, at play, and even asleep, involve a risk of injury or disease. The objective of safe working practices is to minimise or avoid any increase in the risk at work over the normal hazards of everyday life. The hazards involved in making and using polyurethanes – or any other product – are divided for both convenience and legislative purposes into three sections:

– Hazards affecting the worker in the factory
– Hazards affecting the safety of people outside the factory, e.g. hazards resulting from the transport of materials and from the disposal of effluents and other waste.
– Hazards affecting the customer using the final product.

In the factory making polyurethanes there are the normal hazards associated with the use of machinery. We will consider here only the particular hazards arising from the handling of polyurethane chemicals and from their use to make polyurethanes.
The safe handling of any chemical material requires an assessment of the toxicity of the chemical, i.e. its effect upon the human body. The hazard involved in handling any chemical material is quite different from its toxicity. The hazard is a function of its toxicity and the possible level and route of exposure in the workplace. Chemical materials may enter the body by four routes; by inhalation, by ingestion and by absorption or implantation through the skin.
Inhalation: dusts, aerosols and vapours or gases may enter the body with the breath.
Ingestion: materials which are swallowed, alone or with contaminated food, are carried into the gut where they may be absorbed into the body, damage the gut or pass through without effect.
Absorption: the skin protects the body against many harmful materials but some chemicals can pass through intact skin into the body. Others, of course, may cause visible changes or reactions in the skin itself.
Implantation: the passage of substances into the body through punctured skin.
In general, the intrinsic toxicity of the main chemicals used to make polyurethanes, isocyanates and polyols, by ingestion, absorption or

implantation, is low although some isocyanates can cause skin irritation, which can be severe for some aliphatic isocyanates. The hazard to health from skin contact with these materials may be avoided by safe handling procedures combined with satisfactory protective clothing which must include eye protection and impermeable gloves. The main hazard in making polyurethanes arises from the possible inhalation of isocyanate vapour, dust or aerosol droplets. All isocyanates are respiratory irritants and potential sensitisers, that is, they may cause allergic symptoms in susceptible people. Additionally, from more than 30 years of industrial experience handling isocyanates there is nothing to suggest that they are human carcinogens. Inhalation studies of TDI in animals has shown that TDI is not a carcinogen at exposure levels up to 7.5 times the recommended maximum exposure level in the workplace. Massive doses of TDI in corn oil, given via tubes directly into the stomachs of animals, have been shown to be carcinogenic.

Handling isocyanates

The commercially available diisocyanates and polyisocyanates may be safely handled by using the appropriate protective equipment and procedures. The objectives of these handling procedures are to avoid the inhalation of isocyanate vapour or aerosols and to protect the eyes and skin from contact with isocyanates. Detailed recommendations for handling particular isocyanates are obtainable from the suppliers of the materials.

Recommendations are also made by many government and trade organisations. Some of these sources are listed in table 10-2. The recommendations of the isocyanate suppliers for safe handling should be made available to all personnel involved with handling isocyanates. In many countries the approval of a government agency, such as the Health and Safety Executive in the UK, is required before isocyanate handling plant is installed and used.

As previously indicated, the main hazard to people making polyurethanes arises from the possible inhalation of isocyanate

Table 10-1 **The vapour pressures of some isocyanates at their usual processing temperatures**

Isocyanate	Melting point	Vapour pressure (at stated temperature) (Pa)
Polymeric MDI	Liquid at 25°C	6×10^{-4} (25°C)
Monomeric MDI (pure MDI)	42°	2.5×10^{-3} (42°C)
IPDI, (Isophorone diisocyanate)	Liquid at 25°C	4×10^{-2} (25°C)
TDI	Liquid at 25°C	3.3 (25°C)
HDI (Hexamethylene diisocyanate)	Liquid at 25°C	6.7 (25°C)
CHDI, (1,4-*Cyclo*hexane diisocyanate)	62°C	40 (62°C)
NDI, (1,5-Naphthalene diisocyanate)	130°C	150 (130°C)
PPDI, (*p*-Phenylene diisocyanate)	95°C	300 (95°C)

Table 10-2 **Some publications about handling isocyanates**

Publisher/source	Title
ICI Polyurethanes Literature Distribution Centre Central Way, Feltham Middlesex, TW14 0TG England	**MDI-based compositions: Hazards and Safe Handling Procedures** (Ref: PU 193-1E)
International Isocyanate Institute Inc. 119, Cherry Hill Road, Parsippany, New Jersey 07054, USA or to: III Safety Officer c/o ICI Organics Division Hexagon House P.O. Box 42, Blackley Manchester M9 3DA England	**Recommendations for Handling of Toluene Diisocyanate (TDI)** (Technical Information 1) **Recommendations for the Handling of 4,4′-diisocyanato-diphenylmethane, (MDI), Monomeric and Polymeric** (Technical Information 4)
British Rubber Manufacturers Association Ltd. 90-91, Tottenham Court Road London W1P 0BR England	**Isocyanates in Industry: Operating and Medical Codes of Practice**
United Kingdom Government, Health and Safety Executive Baynard House 1, Chepstow Place London W2 4TF England	**Isocyanates: Toxic Hazards and Precautions** (Guidance Note EH 16) Occupational Exposure Limits for 1987 (Guidance Note EH 40)
Manufacturing Chemists Association 1225, Connecticut Avenue NW Washington D.C. 20009 USA	**Properties and Essential Information for Safe Handling and use of Toluene Diisocyanate** (Chemical safety data sheet SD-73)
Chemical Industries Association Ltd. Alembic House 93, Albert Embankment London SE1 7TU England	**Information from Isocyanates Product Group: Isocyanates Incidents**
United States Dept. of Health, Education & Welfare Public Health Service National Institute for Occupational Safety and Health (NIOSH) Cincinatti, Ohio 45226 USA	**Urethane foams: Good practice for Employees Health and Safety** (1976)

vapour or aerosol. It is important to avoid breathing the vapour of any isocyanate. Most of the isocyanates in commercial use have, very approximately, similar inhalation toxicities. The relative hazard from an isocyanate vapour, however, varies widely with the volatility of a particular isocyanate and the temperature at which it is used. The hazard is approximately proportional to the vapour pressure of the isocyanate at the temperature at which it is handled.

The hazard, therefore, increases as the temperature of the isocyanate is increased. Table 10-1 lists the vapour pressures of some isocyanates at the temperature at which they are normally used.

Control Limits and Threshold Limit Values (TLVs)

TLVs are recommendations issued by the American Conference of Governmental Industrial Hygienists (ACGIH), as guidelines for good practices in places of work. TLVs refer to airborne concentrations of substances and represent conditions under which it is believed that nearly all workers may be repeatedly exposed day after day without adverse effect. The TLVs for some diisocyanates are defined as ceiling levels, (TLV-C) which must not be exceeded even instantaneously. These are known as Maximum Allowable Concentrations (MAC) in some countries. The 'Control Limits', which apply to the whole range of work activities in the UK, are in two parts and are expressed in terms of isocyanate groups (-NCO), in air, as follows:

8-hour, Time-Weighted Average (TWA), ... 0.02 mg/m^3.
10-minute TWA .... 0.07 mg/m^3.

These figures equate to 0.0058 ppm and 0.02 ppm respectively for diisocyanates. The TLV values and Control Limits for a number of diisocyanates are listed in table 10-3. Users should check on such limits and specifications as are recommended in their locality since the levels quoted do not apply in all countries.

Table 10-3 **Examples of TLV values and Control Limits for isocyanates**

Isocyanate	Mole wt.	10-minute TWA or TLV-C			8-hour TWA		
		UK (mg/m^3)	**USA** (ppm)	(mg/m^3)	**UK** (mg/m^3)	**USA** (ppm)	(mg/m^3)
Toluene diisocyanate, (TDI)	174.16	0.145	0.02*	0.15	0.041	0.005*	0.04
4,4'-Diisocyanatodiphenylmethane, (MDI)	250.26	0.208	0.02*	0.2	0.059	0.005	0.055
4,4'-Diisocyanatodi*cyclo*hexylmethane, (HMDI)	262.35	0.218	0.01*	0.11	0.062	0.005	0.055
1,5-Naphthalene diisocyanate (NDI)	210.19	0.175	0.02**	0.17	0.05	0.005**	0.04
1,6-Hexamethylene diisocyanate, (HDI)	168.2	0.14	0.02**	0.14	0.04	0.005**	0.035
Isophorone diisocyanate, (IPDI)	222.29	0.185	0.02**	0.18	0.052	0.005*	0.045*
p-Phenylene diisocyanate, (PPDI)	160.13	0.133	0.02**	0.13	0.038	0.005**	0.032
Trans-1,4-*cyclo*hexane diisocyanate, (CHDI)	166.18	0.138	0.02**	0.13	0.039	0.005**	0.033

Note: Most countries have legislation or recommendations governing the maximum level of isocyanate vapours in the air of the workplace. Those tabulated above are the limits used in the USA and the UK; some countries have different limits.

Note: UK values correspond to the *Control Limits* of 0.02 mg (–NCO)/m^3 – 8-hour TWA, and 0.07 mg (–NCO)/m^3 – 10-minute TWA.

USA values *ACGIH, 1986/1987, **NIOSH recommendations, 1980.

Today's limits to the allowable concentration of isocyanate vapour in the atmosphere of the workplace are derived from some 30 years' experience of handling isocyanates, supported by tests on animals, medical tests on people working with isocyanates and the case histories of people who, in the past, have developed respiratory problems after exposure to isocyanate vapour. The present TLVs applied in a well-controlled factory environment are believed to represent safe conditions for nearly all workers in normal health. People suffering from chronic lung conditions, such as asthma and bronchitis, or from allergic reactions, should not work with, or near, isocyanates. Detailed medical screening procedures are recommended by the British Rubber Manufacturers' Association (table 10-2). Prolonged or repeated exposure of susceptible people to isocyanate vapour may lead to increased sensitivity or an allergic, asthmatic reaction to traces of isocyanate vapour. Individuals who have developed sensitivity to isocyanates may experience wheezing, tightness of the chest and shortness of breath. Symptoms of both irritation and sensitisation may be delayed for some hours after exposure. A cough at night may be a symptom of sensitisation.

A hyper-reactive response to even minimal concentrations of isocyanate may develop in susceptible persons. Persons who have become sensitised to isocyanates should not resume work in, or near to, areas where isocyanates are handled.

As previously discussed, the hazard involved in handling a particular isocyanate will depend largely upon its vapour pressure at the temperature at which it is used (table 10-1). Polymeric MDIs which are liquid at ordinary temperatures, are usually handled at room temperature when the vapour pressure is so low that the TLV-TWA is unlikely to be exceeded – unless the isocyanate is sprayed or airborne aerosol particles are created in some other way. All the other diisocyanates listed in table 10-1 have a relatively high vapour pressure in the liquid state at their normal processing temperatures, such that the equilibrium vapour concentration above the liquid will exceed 0.02 ppm. PPDI, NDI and CHDI have the highest vapour pressure above the liquid and are usually processed in closed, but vented vessels. NDI, CHDI and PPDI are, however, all available as flaked solids in a form which may be handled and weighed by simple, remote control devices. Details of suitable equipment are available from the isocyanate suppliers. Although they are solids, the vapour pressures of NDI, CHDI, and PPDI are high enough to easily exceed 0.02 ppm at room temperature. Pure MDI is usually handled as a liquid above about 42°C. The precautions required to avoid breathing the vapour produced at this temperature are similar to those applied when handling TDI – although it should be noted that the vapour pressure of MDI at 42°C is much lower than that of TDI at 21°C.

Although the inhalation of isocyanate vapour is the main hazard in handling isocyanates, it is important to bear in mind that isocyanates are reactive chemicals. They have a tanning action on the skin and prolonged or repeated contact with isocyanates may cause skin irritation. Splashes into the eye will cause mild to severe irritation (in order of increasing effect, MDI < TDI < NDI < HDI), and should be treated by *immediate* flushing of the eye with copious amounts of clean, cold water or preferably using the contents of several sterile eye-wash bottles, which should be stored in a readily accessible position in the work-place. The eye should then be examined by a physician. Contamination of the skin should be treated by immediately washing with copious amounts of clean water followed by washing with ordinary soap and water. The soap not only helps to emulsify and remove traces of isocyanate but also decontaminates by catalysing the reaction between the isocyanate and water.

Summary of safe practices for handling isocyanates
1 People should be medically screened and thoroughly trained.
- Medical examination and screening (Ref. BRMA Code of Practice 10-2).
- Training in safe handling. This must include on-the-job training designed to ensure a full appreciation of the hazards to be avoided.
- Wear protective clothing:
  a) Safety spectacles, full goggles or face shields of an approved type.
  b) Impervious gloves – rubber or PVC, preferably reinforced. Gloves must be decontaminated after use. Gloves should be inspected frequently because contact with isocyanate may cause embrittlement and cracking of the impervious coating. A recommended alternative is to use gloves made from polyethylene which are used only once.
  c) Overalls, preferably of heavy cotton.
- Regular medical checks including lung function tests are recommended for all those who work regularly with isocyanates.
- Emergency equipment should be immediately available on the job and all concerned should be trained and practised in its use. Safety equipment should include a breathing air supply free from oil, water and dirt, with full-face masks, decontaminants, eye-washing equipment and safety showers.
- Plant operators should be trained to check the venting air-flow rates.

2 Venting. The processing and handling of all the relatively volatile isocyanates should always be done in positively vented enclosures which create an inward flow of fresh air, especially at the operators position. An air-flow rate of not less than 33 m/min is satisfactory. If this is not possible, e.g. in spillage incidents or when handling isocyanates outside the factory, operators must be protected by fresh air masks fed by an oil-free compressor or by portable air cylinders. Warning notices and barriers should be used to exclude those not wearing protective clothing and air hoods. Barriers and warning signs must always be used in major spillage incidents.

224

3 Bulk storage and processing plant for isocyanates. Volatile isocyanates should not be handled in open vessels. Storage should be in fully enclosed systems with transfer by pump. Isocyanate storage tanks should be blanketed with dry air or nitrogen (<100 ppm v/v water vapour, a dew point of -40°C). If there are several tanks, they should have interconnected vent lines to minimise the escape of isocyanate vapour into the atmosphere. The common vent should be at a high level, via a dryer, into an exhaust stack or scrubber intake. All isocyanate storage tanks must be fitted with suitable safety valves for both above and below atmospheric pressure. To guard against safety valve malfunction due to the formation of polyureas in a valve, bursting discs must also be fitted. Isocyanate bulk storage tanks should be surrounded by an impervious bund wall of sufficient height to contain the contents of the tank in case of leakage. Isocyanate tanks should be separated by a bund wall from polyol tanks. Although isocyanates are not particularly inflammable it is recommended that isocyanate bulk storages are separated from the workplace, and from any inflammable materials, by a fire-wall. It is suggested that any enclosure containing bulk stocks of isocyanates should be fitted with an isocyanate vapour monitor which is arranged to trigger an alarm outside the enclosure in case of a major leak of isocyanate. This ensures that no one will accidentally enter an isocyanate contaminated area without full emergency protective clothing.

4 Handling drums of isocyanates. Handling drums of isocyanates should only be done by trained personnel. A potential hazard arises when it is necessary to heat drums of isocyanate in order to melt the contents. Drums should be carefully inspected prior to heating to ensure that they are in good condition with no evident damage. Heating and melting operations should be carried out with extreme caution and under the supervision of a responsible person. Drums containing frozen TDI should be placed in a ventilated, heated store. The rate of melting can be increased by the use of drum rollers and fan-assisted, hot air heating. Solid MDI is often melted by placing the sealed, unopened drum in a steam chest or hot water bath. The use of steam heating instead of hot air reduces the formation of dimer by reducing the time required to heat the MDI before use. The operation requires great care, and should not be attempted without seeking detailed advice from the MDI supplier. On no account should isocyanates be heated by the use of naked flames, hot plates or by any other method of direct heating.

Decontaminants for isocyanates
Any spillage of isocyanate should be decontaminated immediately. Both solid and liquid decontaminants should be kept readily available in clearly labelled containers. Aqueous liquid decontaminants are the best for minor spillages and for treating

a) **General purpose liquid decontaminants.**

1) Water 90%
Conc. ammonia solution 8%
Liquid detergent 2%

(Note that the TLV for ammonia in the air is 25 ppm.)

2) Water 90- 95%
Sodium carbonate 5- 10%
Liquid detergent 0.2-0.5%

b) **Solvent decontaminant: Flammable**
Water 45%
Industrial alcohol 50%
Conc. ammonia solution 5%

(*Note:* The Fire Point of this mixture is 46°C)

c) **Solid decontaminants**
These consist of non-flammable absorbent carriers such as sand, kieselguhr, Fuller's earth, etc., moistened with the aqueous liquid decontaminant, a)1 above. A typical mixture is:
Sand 50%
Kieselguhr 45%
Liquid decontaminant 5%

contaminated clothing or equipment. Absorbent solid decontaminants are essential for larger spillages of liquid isocyanates. Solidified isocyanate spillages should first be removed by scraping and the residue treated with a suitable decontaminant. Sometimes the use of a decontaminating solvent is useful for cleaning traces of isocyanates which are solid at room temperature.

An absorbent mixture is sometimes used for small spillages. This contains sawdust instead of sand. Solid decontaminants should always be used to contain, absorb and neutralise isocyanates in major spillages.

Decontamination of isocyanate drums

Drums which have contained isocyanate should be decontaminated immediately after emptying. This is easily done by filling the drum with one of the general purpose liquid decontaminants, a) 1 or 2 above, and then allowing it to stand, open, with the screw closure removed, for at least 24 hours in a vented area or in an outside enclosure. The decontaminant solution may then be transferred to other drums and used several times. Disposal of the decontaminant solution should be carried out according to the local and national requirements. The re-use of isocyanate drums is not recommended. They must never be used as containers for food or food additives, even after decontamination and cleaning. If there is any doubt about the future use of such drums they should be punctured before disposal.

Monitoring isocyanate vapour in the air of the workplace

Regular analysis of the air inside factories using isocyanates is essential. It is the only certain method of confirming the safe operation of the process plant and the venting system.

The important consideration when monitoring is that it should simulate as closely as possible the exposure of personnel whilst carrying out their normal work operations. Concentrations of isocyanates in a factory can vary considerably, especially if there are air currents or if spraying is in progress.

When a new process involving the use of isocyanates is being introduced or an existing process modified, extensive monitoring of the atmosphere around the plant must be carried out to confirm that as low a level of isocyanates in the atmosphere as is reasonably practicable, and certainly not in excess of any exposure limit, is being maintained. It is strongly recommended that the atmosphere should be subsequently checked at regular intervals with the emphasis on monitoring personal exposure levels. This should include both time-weighted average and short term exposure samples, the duration of the sampling period to comply with local authority legislation.

Where isocyanate leakage (either as a liquid or vapour) is suspected or is known to have occurred more frequent sampling is necessary.

The analytical methods available to measure the concentration of isocyanate vapour in air are reviewed in a number of publications:

1) Technical Information 5, *Analysis of Isocyanates in Air* (1982), from the International Isocyanate Institute (see table 10-4).

2) *NIOSH Criteria for a recommended Standard for Occupational Exposure to Diisocyanates*, US Dept. of Health Evaluation and Welfare publication number 78 (1978) National Institute for Occupational Safety and Health. (Address in table 10-2).

3) *Environmental, Toxicological and Legislative Aspects of Polyurethanes* Hurd,R., *Cellular Polymers*, 1, (1) 53-71 (1982).

4) *Flexible Polyurethane Foams* Woods, G., 241-249, see bibliography, Appendix C.

5) *Methods for the Determination of Hazardous Substances:* MDHS 25 (1983) a U.K. Government Health and Safety Executive publication. (Address see table 10-4).

Localised monitoring can be carried out by taking individual samples for analysis and this is particularly relevant to discontinuous operations. Continuous monitoring equipment is available and is widely used, especially on continuous manufacturing plant using TDI. Lightweight personal monitors can also be worn. From these can be obtained a continuous record of exposure with time.

Analytical methods

In addition to chromatographic separation methods, both wet and dry colorimetric methods are available to measure low concentrations of isocyanate vapour in air. Colorimetric methods, in general, require less operator training and skill than chromatographic methods.

Colorimetric methods. Colorimetric aqueous solution methods were first developed for measuring TDI vapour in air. A known volume of the contaminated air is drawn at a constant rate through an acidic aqueous solution where the isocyanate vapour is absorbed and hydrolysed to the corresponding diamine. The latter is then diazotised and coupled with N-2-aminoethyl-1-naphthylamine, (N-naphthyl-ethylene diamine), (NNED). The violet-blue colour produced is measured spectrophotometrically at 550 nm, (2,4-TDI), or 598 nm (2,6-TDI), or compared with liquid or glass colour standards, to estimate the amount of TDI captured from the sample of air. The many modifications of the method, sometimes called the 'Marcali' or the 'Pilz' method after those who, separately, first developed it, have been designed to improve the accuracy and speed of the determination and to extend its applicability to other isocyanates.

A version of the Marcali/Pilz method which has been widely used employs the absorbing mixture of aqueous hydrochloric acid and dimethylformamide which was first used by Pilz. After diazotisation and coupling the resulting coloured solutions are compared with coloured glass standards in a colour comparator (ref. 1, table 10-4). The method is applied to TDI, pure and polymeric MDI and NDI, using glass standards. The glass standards are specific for each isocyanate because the colour of the derivative is specific. The glass standards correspond to 0.01, 0.02 and 0.04 ppm of isocyanate using a 10-litre air sample. The sensitivity may be increased by increasing the size of the air sample taken, preferably by increasing the sampling time and maintaining the recommended air flow rate. Other modifications of the method are recommended by the British Health and Safety Executive (ref. 2, table 10-4) and by NIOSH for TDI (ref. 3, table 10-4) and for MDI (ref. 4, table 10-4). The limitations of the Marcali/Pilz method, which is only applicable to aromatic isocyanates, are:

– It is non-specific in that other aromatic isocyanates present may interfere. The colour produced is different for different isocyanates and therefore different wave-length maxima are used for different isocyanates. Mixtures of isocyanates cannot be measured with accuracy unless the mixture is of known composition and the colour measurement has first been calibrated with that mixture.

– Aromatic amines interfere.
– The long sampling and colour development time of 10 to 20 minutes, depending on the particular modification of the method, the volume of absorber used and the accuracy required.
– Substituted ureas, formed by the reaction of diisocyanates with atmospheric moisture, do not interfere, i.e. they are not recorded.

Table 10-4 Analysis of isocyanates in the air: some references

Method	Reference
Colorimetric methods	
1) Review, **Analysis of Isocyanates in Air**	International Isocyanates Institute, Inc., Information Document No.5., 119, Cherry Hill Road, Parsippany, New Jersey 07054 USA
2) **Methods for the Determination of Hazardous Substances MDHS 49, 'Aromatic Isocyanates'** an in-field method using acid hydrolysis, diazotisation and coupling with N-2-amino ethyl-1-naphthylamine	Health and Safety Executive Sales Point, Baynards House, 1, Chepstow Place, London, W2 4TF, England, and also from H.M. Stationery Office book shops
3) NIOSH method for TDI in air	**NIOSH Manual of Analytical Methods, 2nd. Edition, Vol.1, P & CAM No. 141, 2,4-Toluene diisocyanate, (TDI), in Air,** US Dept.of Health, Education and Welfare, Public Health Service, NIOSH, Cincinatti, Ohio, 45226, USA
4) NIOSH method for MDI in air	**NIOSH Manual of Analytical Methods, 2nd Edition, Vol.1, P & CAM No. 142, *p,p'*-diphenylmethanediisocyanate, MDI,** Address as 2) above
5) 'Pilz' Technique for Aliphatic and Alicyclic Diisocyanates	Walker, R. F., and Pinches, M. A., **Analyst,** vol. 104,928-936, (1979)
6) Continuous paper-tape monitors and recorders	MDA Scientific Inc., 808, Busse Highway, Park Ridge, Illinois 60068, USA GMD Systems Inc., Old Route 519, Hendersonville, PA 15339, USA
7) 'Draeger Tube' estimation of TDI vapour level	Draegerwerk AG., Postfach 1339, Moislinger Allee 53-55, D-2400 Lubeck, West Germany
Chromatographic methods	
8) Thin layer chromatography using 'Nitroreagent'	Keller, J., and Sandridge, R. L., **Anal. Chem.,** 51, 1868-70, (1979)
9) TLC using 2-PP reagent	Elwood, P. A., Hardy, H. L., and Walker, R. F., **Analyst,** 106, 85-93, (1981)
10) HPLC using the ethyl urethanes of TDI	Bagon, D. A., and Purnell, C. J., **J. Chromatogr.,** 190, 175-182, (1980)
11) HPLC using 'Nitroreagent'	Graham, J. D., **J. Chromatogr. Sci.,** 18, 384, 1980
12) HPLC using 2-PP reagent	Hardy, H.L., and Walker, R. F., **Analyst,** 104, 890-891, 1979
13) **Methods for the Determination of Hazardous Substances MDHS 25, 'Organic Isocyanates'**	Health and Safety Executive Sales Point, Baynards House, 1, Chepstow Place, London, W2 4TF. England

Aliphatic diisocyanates may be estimated colorimetrically by a technique, also due to Pilz, using 2,4-dinitrofluorobenzene. The isocyanate vapour is absorbed and hydrolysed in a mixture of dimethylsulphoxide and hydrochloric acid, and then reacted with 2,4-dinitrofluorobenzene buffered with sodium bicarbonate. The resulting N,N'-2,4-dinitrofluorobenzene derivative is measured in a spectrophotometer. The method has been used for aliphatic and alicyclic isocyanates (ref. 5, table 10-4). Primary amines and aromatic isocyanates will interfere.

Dry, paper-tape methods. A dry test-paper method of analysis was developed in the laboratories of ICI. In the original ICI method a 5-litre air sample was drawn through a filter paper previously impregnated with sodium nitrite and Brenthol GB★, at a rate of one litre/minute. After 15 minutes the yellowish-brown stain produced on the paper was compared with a set of printed colour standards corresponding to TDI vapour levels in the range from 0.01 to 0.1 ppm. This method formed the basis of the first continuous monitoring device for TDI vapour in air. This monitor was developed during the late 1960s in the engineering laboratories of the Dunlop Company. The monitor employs an impregnated paper tape which is moved at a constant speed over a sampling port through which air is drawn at a fixed rate by a pump. The paper then passes under a photometric unit which converts the signal to TDI concentration in ppm of air. The method has now been further developed, using paper tape chemistry which responds equally to both TDI isomers and is claimed to allow readings to be obtained in less than 7 minutes. The manufacturers, MDA and GMD Systems Inc. (ref. 6, table 10-4), provide reference samples for both continuous monitoring and spot testing for TDI vapour. Measurements from 0.001 ppm are claimed to be possible. The continuous monitor will provide a continuous record of isocyanate vapour levels at a selected location and can be used to trigger a visible and audible alarm when vapour levels approach the TLV or other preset level. A chest-mounted miniature version of the monitor, the MCM version, is available for monitoring the exposure of individual people to isocyanate vapour.

The standard MCM personal monitor has been modified for use in a long term study, organised by the BRMA and supported by the British Medical Research Council and III, of the effects of TDI on people involved in foam manufacture. For use in this work the sensitivity of the MCM monitor has been doubled by increasing the sample air-flow through the paper tape from 100 to 200 ml/min. The modified MCM monitor will detect 0.002 ppm of TDI in air.

Limitations of paper-tape methods. Water vapour is essential to the reproducible operation of paper tape monitors and they are not recommended for use at less than 20% relative humidity. SO_2, NO,

★ Trade mark

229

NO$_2$ and halogens interfere. The method is, of course, only suitable for use with single isocyanates or with known reproducible mixtures. Abnormally low temperatures will slow the rate of colour development and give a low response. Paper tape is not recommended for the measurement of airborne isocyanate when the air contains a mixture of isocyanate vapour and liquid droplets, such as the aerosols sometimes formed in spraying operations. These may be estimated by absorption in a solvent mixture such as the Pilz mixture.

Paper tape monitors are mechanically limited. The MCM 4000 personal monitor, for example, moves the sensitised paper tape across the 3 mm wide sampling slot at a rate of 20 mm/hour. A spot on the tape is therefore exposed to the sample atmosphere for 9 minutes and the resulting stain, at any point, represents a 9 minute TWA. This factor, combined with the finite size of the area measured by the spectrophotometer, means that any reading obtained from an MCM personal monitor represents approximately a 10 minute TWA. The monitor is satisfactory for steady state continuous processes but cannot reveal the transient isocyanate emissions above the TLV which might occur in discontinuous production processes. Peak emissions, such as those from a polyurethane foam moulding process, have been estimated by 'Grab sampling', using a large gas pipette, followed by the rapid conversion of the diisocyanate vapour to a stable derivative. The high reactivity of diisocyanate vapours and their strong tendency to become absorbed onto the surface of the containing vessel, where they may react with surface moisture, makes essential the frequent standardisation of the equipment, timing and procedures of grab sampling and analysis. Recent work by the UK Health and Safety Executive has resulted in accurate 30-second TWA vapour measurements using 2-PP reagent (figure 10-1) and thin layer chromatography.

Chromatographic methods. High Performance Liquid Chromatography (HPLC), Thin Layer Chromatography (TLC), or High Performance Thin Layer Chromatography (HPTLC), may be used both to separate isocyanates in a mixture and to estimate isocyanate vapour concentrations. The air contaminated with isocyanate is passed at a known rate through an absorbing solution where the isocyanate is converted to a stable derivative. HPLC and TLC are applicable to aromatic, aliphatic and alicyclic diisocyanates, isocyanate mixtures, isocyanate adducts and prepolymers.

A widely used method is based upon "Nitroreagent" (figure 10-1). The reagent, N-*n*-propyl-4-nitrobenzylamine, reacts rapidly even with aliphatic isocyanates, to form stable urea derivatives containing an aromatic nitro group. These urea derivatives may be separated by HPLC, using an ultra-violet light detector to sense and measure each component. The simplest procedure is that described by Graham (ref. 11, table 10-4).

"Nitroreagent" can also be used as the basis of a visual TLC separation method. The aromatic nitro group of the stable urea derivative is reduced to the amine which is diazotised and coupled, on the chromatogram, to form a brightly coloured dyestuff. The amount of dyestuff, and hence the concentration of isocyanate, may be estimated by comparison with standards which are spotted alongside the test spot.

Another possible method for general use is that based upon the reaction of diisocyanates with 1-(2-pyridyl)piperazine, which was introduced by workers in the UK Health and Safety Executive laboratories in 1979 (HPLC method, (ref. 9, 12, table 10-4). The reagent, 1-(2-pyridyl)piperazine (2-PP reagent), is a stable high boiling point liquid which reacts rapidly with isocyanates to form stable urea derivatives with high molar absorptions of UV light, hence allowing sensitive analyses to be carried out. The reagent is less susceptible to air oxidation than nitroreagent.

Figure 10-1   Derivatisation agents for isocyanate analysis.

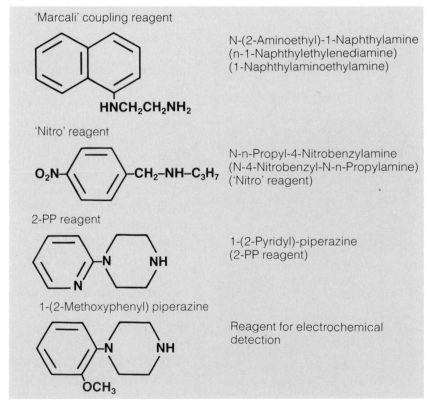

HPLC using electrochemical detection, in addition to ultra-violet light methods, gives increased sensitivity. A methoxyphenyl piperazine (figure 10-1), may be reacted with the isocyanate to form a derivative suitable for electrochemical detection. This reagent is the basis of the standard UK method for the analysis of isocyanates in the atmosphere, first published by the UK Health and Safety Executive in 1983. The method is applicable to both aromatic and aliphatic isocyanates and to mixtures of isocyanates. The concentration of individual isocyanates, alone or in a mixture, may be measured accurately at levels well below the Control Limit. The high sensitivity of the method permits the use of small air samples or shorter sampling times compared with those required for simple colorimetric methods. The method is also applicable to isocyanate-terminated prepolymers as well as monomeric and polymeric isocyanates.

There are thus several chromatographic methods which are capable of measuring isocyanate vapour levels with great accuracy and which will allow the separation and measurement of the constituents of mixtures of isocyanates. All, however, have so far required skilled operation in order to obtain reproducible results. The established paper tape monitors and the standardised Marcali/Pilz wet colorimetric systems, on the other hand, meet the requirements of operator protection in plants using known, common diisocyanates without the use of skilled analysts. Current development of the paper tape monitors is aimed at increasing both their sensitivity and their speed of reaction.

Isocyanate vapour emissions – waste disposal

In some countries there are statutory controls on the emission of isocyanate vapour into the atmosphere. In the UK these limits correspond to the Control Limit of 0.07 mg(-NCO groups)/m^3 in the exhaust air at the exit from the chimney but only 1/60th. of that concentration at ground level. In Germany, there is a total maximum permissible emission of 20 mg/m^3 of isocyanate molecules in the exhaust air at the stack exit. Some legislation in the USA defines the maximum allowable concentration of isocyanate vapour at the boundaries of the plant rather than the stack emission level. In some other countries there is no absolute limit but a degree of local discretion is used depending on the local environment and an assessment of the hazard.

The air exhausted from enclosed polyurethane manufacture is usually discharged directly into the atmosphere above the factory. The emission should be remote from the ventilating air intakes, so that contaminated air cannot be recycled. The exhaust air should be analysed for its isocyanate content by one of the standard methods. The concentration of isocyanate vapour in the air-flow from polyurethane moulding plants is usually found to be well below the TLV. This is because of the dilution of the discontinuous vapour emission with fresh air. High output continuous processes, such as TDI-based flexible foam manufacture and continuous spraying systems, may sometimes yield unacceptably high levels of airborne isocyanate. Airborne aerosols can be removed by impingement plates or filtration devices, and levels of isocyanate vapour above the permitted concentration may often be reduced simply by dilution with fresh air before discharge to the atmosphere. The alternative, where dilution is unacceptable, is to reduce the concentration of isocyanate vapour by chemical scrubbing or by absorption. Scrubbers using aqueous alkaline solutions with fluidised bed scrubbing plates, have operated satisfactorily for several years in removing up to 98% of the TDI vapour in the air exhausted from large flexible foam plants. Scrubbing with sodium hydroxide solution is expensive, however, and the cost of disposing of the spent alkaline solution is increasing. An alternative system, using activated carbon to absorb both the TDI vapour and the other gaseous emissions from the foam process, is being operated satisfactorily in several full scale trials. TDI vapour is also being removed by scrubbing with water, using high pressure water jets arranged so that the exhaust air is drawn through the venturi. This method is unlikely to be widely used as it is expensive in both power and water consumption. Installations using activated carbon to recover the CFM-11 used in blowing flexible foam, also absorb most of the isocyanate vapour present in the discharge from the foam making plant.

It should be borne in mind that isocyanate vapour emissions, especially those of aromatic diisocyanates, are believed to be transient. Isocyanates are reactive species and are known to absorb readily onto solid surfaces where they probably react with atmospheric moisture. Other airborne reactions may also occur but there is no evidence of a significant conversion of TDI vapour to toluene diamine in the atmosphere[*].

Experimental work to date has suggested the conversion of both 2,4- and 2,6-TDI to substituted ureas.

Waste disposal. The treatment of waste isocyanates is comprehensively discussed in Brochure 3, published by the III, which is also obtainable from III members. Three methods of disposal are recommended:
- Reaction with waste polyol to make a low grade foam
- Reaction with excess aqueous liquid decontaminant to form polyureas
- Incineration in specially designed and properly supervised equipment.

It is further recommended that, before disposing of large quantities of isocyanate, the advice of the isocyanate suppliers on the operation should be sought.

[*] Further reference should be made to the following information:
– Holdren,M.W., Spicer,C.W., Riggin,R.M., Battele, Columbus, Ohio.
Gas phase reaction of Toluene diisocyanate with water vapor, Am. Ind. Hyg. Assoc. J. 45, 9, 626-633, (1984).
– Duff,P.B., Olin Corporation, Stamford, Connecticut.
The fate of TDI in the environment, Proceedings of the SPI 6th International Technical/Marketing Conference (1983).

Table 10-5 TLV, STEL and IDLH levels of some chemicals
(ref. ACGIH (1986 - 1987) and NIOSH/OSHA (1980))

Chemical substance	TLV-TWA		STEL		IDLH
	ppm	mg/m³	ppm	mg/m³	ppm
Ammonia	25	18	35	27	500
Antimony trioxide	–	0.5	–	–	80
Carbon black	–	3.5	–		–
Carbon dioxide	5,000	9,000			27,000
Carbon monoxide	50	55	400	440	1,500
Dichlorodifluoromethane (CFM-12)	1,000	4,950			50,000
Diethylaminoethanol – *skin*	10	50	–	–	500
Dimethylformamide – *skin*	10	30			3,500
Diethylamine	10	30	25	75	2,000
Ethyl acetate	400	1,400	–	–	–
Ethanol (Ethyl alcohol)	1,000	1,900	–	–	–
Ethylene diamine	10	25	–	–	2,000
Ethyleneglycol	**C** 50	**C** 125	–	–	–
N-Ethylmorpholine – *skin*	5	23			2,000
Formic acid	5	9	–	–	100
Hydrogen cyanide – *skin*	10	**C** 10	–	–	50
Isopropanol (Isopropyl alcohol)	400	980	500	1,225	20,000
Methyl ethyl ketone (MEK)	200	590	300	855	–
*4,4'-Methylene bis (2-chloroaniline) (MOCA) – *skin*	*0.02 **A**2	*0.22 **A**2	–	–	–
Methylene chloride	*50 **A**2	*175 **A**2			–
Organo (alkyl) mercury compounds	–	0.01	–	0.03	10
Organotin compounds – *skin*	–	0.01	–	(0.2)	200
p-Phenylenediamine – *skin*	–	0.1	–	–	–
n-Propanol (Propyl alcohol) – *skin*	200	500	250	625	4,000
Tetrachloroethylene (Perchloroethylene)	50	335	200	1,340	500
Trichlorofluoromethane (CFM-11)	**C** 1,000	**C** 5,600	–	–	–
Triethylamine	10	40	15	60	1,000

Notes
* **A**2 Industrial Substance Suspect of Carcinogenic Potential for Man.

C = Ceiling value

STEL = Short term exposure limit, (maximum of 15 minutes).

IDHL = Immediately dangerous to life or health; the maximum level from which one could escape within 30 minutes without any escape-impairing symptoms or irreversible health effects.

TLV = 'Threshold limit values refer to airborne concentrations of substances and represent conditions under which it is believed that nearly all workers may be repeatedly exposed day after day without adverse effect.' This is a direct quotation from the ACGIH publication on TLVs. The guidance given by the ACGIH should be read in full by all who have interest in the work environment.
Readers are advised to check local values of the relevant exposure limits since those as TLVs quoted may not apply to all countries.

skin = Substances absorbed through the skin, mucous membranes and eye by direct contact with the substance or by airborne substance.

Handling polyols, catalysts, blowing agents, chain-extending and curing agents

Handling polyols

Polyether and polyester polyols are normally of very low toxicity and normal standards of industrial hygiene with the use of gloves, suitable eye protection and overalls is recommended. Always refer to the polyol manufacturer's product data sheet to check for specific hazards, especially when handling preblended polyol components which contain additives. The hazards of preblended polyols are usually low because of the low level of catalysts and other additives. Waste polyols are best collected together in clearly labelled containers and passed to a waste disposal contractor for disposal. Waste polyols containing chlorinated flame retardants or blowing agents should be collected separately because they require special treatment. Waste polyol may also be reacted, under controlled conditions, with waste isocyanate to form a scrap or low grade foam.

Catalyst handling

Both tertiary amine catalysts and organo-metallic catalysts should be handled with care to avoid contact with the skin and eyes. Many of these materials can be absorbed through the skin and most tertiary amines are strong irritants.

The toxicity of organo-metallic catalysts varies from very low to high. Stannous octoate has very low toxicity but dibutyl tin salts in general are of much higher toxicity by absorption, ingestion or inhalation. The guide TLV for tin compounds is 0.1 mg/m^3 but the supplier's handling instructions must always be consulted because of the wide variation in the properties of different tin catalysts. Some of the organo-mercurial catalysts, fungicides and bactericides are highly toxic and should be handled not only with protective clothing but in positively vented, designated enclosures.

Tertiary amine catalysts are volatile alkaline materials which are strong irritants of the skin and eyes. The more volatile materials are also strong respiratory irritants. Skin and eye contact should be avoided by the use of protective clothing, eye protection and gloves, together with positive venting to carry the tertiary amine vapour away from the operator. Most of the tertiary amines used to catalyse polyurethane production have not been allocated a TLV. A guide limit of 10 ppm has been suggested but should be checked with the catalyst supplier. The very strong smell of some tertiary amines will often, however, dictate a lower limit. Those tertiary amines for which TLVs are recommended are listed in table 10-5.

Blowing agents

The important blowing agents used to make polyurethane foams are carbon dioxide from the reaction of isocyanate and water, the chlorofluoromethanes, CFM-11 and CFM-12, and dichloromethane (methylene chloride). The CFMs are non-toxic and have been allocated TLVs appropriate to asphyxiants such as carbon dioxide which is of low toxicity (table 10-5).

The vapours of the CFMs are much heavier than air and in the absence of positive ventilation may collect in low areas and displace the air. Tanks and vessels which have contained CFMs should be entered only with breathing apparatus. Impervious gloves should be worn as CFMs are excellent solvents and skin contact can result in dry chapped skin.

The TLV of dichloromethane is only 5% of that allocated to CFM-11, and this must be borne in mind when substituting dichloromethane for CFM-11 on established manufacturing plant. The efficacy of the venting system should be checked.

Chain-extending agents and curing agents

These are monomeric diols, diamines and amino-alcohols. Diol chain-extending agents are not highly toxic and they are easily handled at room temperature. Skin and eye contact should be avoided and the vapours, especially from hot diols, must not be inhaled.

Diamine chain-extending agents or curing agents should only be handled in well-ventilated areas with care to avoid contact with the skin or eyes and to avoid breathing the vapour. All the diamines used in polyurethanes can be absorbed through the skin. Aromatic diamines should be handled with especial care, preferably in positively vented enclosures. Always refer to the manufacturers' or suppliers' handling instructions when in doubt.

4,4'-Methylene bis (2-chloroaniline), (MOCA), is a very effective hindered diamine curing agent which has been widely used in elastomer formulations. It is strongly recommended that this material is never handled except in purpose designed closed systems. It is classed by the ACGIH as an "Industrial substance suspect of carcinogenic potential for man", and its use is best avoided. There are alternative polyurethane systems for most applications – many based on MDI variants – which do not require MOCA curing to obtain a satisfactory level of physical properties.

Fillers, flame retardants and other additives

People handling particulate and fibrous fillers should wear normal protective clothing; overalls, gloves, eye and face protection. This applies even with the fillers such as ground limestone, chalk, barytes, china clay, etc., which are familiar substances of low toxicity. The hazards involved in handling a solid filler depend not only on its chemical constitution but also on its physical form, for example, on the particle size and density of the material and the presence of water, powder lubricants or other surface additives. The supplier's handling instructions should always be consulted. There is usually a dust hazard in handling particulate or fibrous fillers in the dry state. They are best handled in positively vented enclosures which prevent dust reaching the operators. Where this is not possible suitable dust masks should be worn. The classification of airborne dusts varies from country to country. ACGIH classify mineral fillers with a low quartz content such as ground limestone, chalk and barytes, as "Nuisance particulates", with a TLV of 10 mg/m^3 or 5 mg/m^3 of "Respirable dust". There are many definitions of respirable dust. They are usually linked to the use of a particular sampling device which partitions aerosol dust into particular particle size ranges. The selected size range is related to the size of the dust particles which are retained in the respiratory pathways of those breathing in the aerosol dust. The effect of particle size variation on the hazard from nuisance particulates has been expressed as follows:

Table 10-6

Particle diameter, (EUDS* in μm).	Effect on the Respiratory System
> 50	Settles rapidly and is not easily drawn into the nose or mouth.
10 – 50	Mostly filtered out in the nose.
7 – 10	Mostly deposited in the upper respiratory tract.
0.5 – 7	Deposited in the respiratory bronchia and in the lungs.
<0.5	These very small particles remain airborne and are mostly exhaled with the breath

*EUDS = Aerodynamic diameter expressed as that of an Equivalent Unit Density Sphere, i.e. the uniform sphere of unit density which has the same terminal settling velocity in air as an irregular particle.

Fibrous particles, because of their much greater surface/volume ratio, settle more slowly and are more easily drawn into the respiratory system.

Fibrogenic** dusts, of which asbestos fibres and silica dusts are the most well known, are subject to statutory control in most countries. Glass fibre dust, although irritant, is not fibrogenic and may be handled with a dust mask and normal protective clothing.

Flame retardant additives vary from particulate solids to reactive liquids. The supplier's handling instructions should always be consulted. The commonly used halogenated phosphate ester additives, tris (2-chloroethyl) phosphate, TCEP, and tris (2-chloropropyl) phosphate, TCPP, are of low oral and dermal toxicity, and may be handled at room temperature using normal protective clothing. At high temperatures these materials give off toxic vapours. If halogenated phosphate esters are involved in a fire, breathing apparatus must be worn.

**Fibrogenic dust is defined by the International Labour Organisation as dust capable of causing collagenous pneumoconiosis.

Handling polyurethanes in the factory

Flexible foam

Flexible polyurethane foam is a non-irritant material which does not present a toxic hazard. It is non toxic by both ingestion and skin contact. The only significant hazard in handling flexible foam results from its low density and high surface-to-volume ratio combined with high air permeability. These factors result in a high rate of burning, once the foam is ignited. Flexible foam stores should be separated from working areas by a fire wall and the storage area should be protected by an automatic sprinkler system. This is because, once ignited, the rate of burning of large amounts of flexible foam is too high to be tackled with manual equipment. Smoking should be prohibited in all factory areas handling flexible polyurethane foam. Good housekeeping in factories converting flexible foam into cushions, bedding and upholstery is very important. The amount of foam in the working area should be kept to a minimum and all scrap and other off cuts should be removed regularly to a designated storage area.

Scorch leading to auto-ignition of flexible foam during manufacture has caused several serious fires in factories making blocks of foam. It should be stressed that auto-ignition can only occur during the curing period immediately following the manufacture of large blocks of foam. Once the foam has fully cured and cooled, auto-ignition cannot occur. Auto-ignition arises from the exothermic reaction of excess isocyanate in freshly made large blocks of low density, polyether-based foam, which may already have reached a temperature over 165°C from the exothermic heat of the foaming reaction. The excess isocyanate groups may react with atmospheric moisture and evolve reaction heat at a rate greater than the rate of heat loss through the large block of foam. The interior temperature will then rise until the polyether chains begin to oxidise and produce more heat. Scorch, a yellow-brown discolouration in the centre of large blocks of foam, results from oxidation during the curing stage. If a sufficient excess of isocyanate groups is present the rate of oxidation may increase rapidly and result in smouldering followed by flaming combustion.

The precautions taken to avoid scorch and auto-ignition in flexible foam slabstock manufacture may be summarised as follows:
- Limit the chemical blowing by carbon dioxide from the water/isocyanate reaction to a level less than 4.6 parts of water/100 parts of polyether triol of mol. wt. 3000 – 4000.
- Limit the maximum TDI index to 108, especially with low-density chemically-blown foams.
- Avoid any imbalance in the ratios of the chemical components used on the slabstock machine by suitable metering and monitoring arrangements. A system that automatically gives a

warning of flow imbalance is desirable. The highest risk of scorch arises from deficiency of polyol which is easily detected from the visible increase in the reaction rate of the foam mix.
- Avoid high chemical component temperatures which yield correspondingly high final reaction temperatures.

A number of other factors which affect scorch have been suggested. These include the presence of acidic impurities, which tend to inhibit the cross-linking and chain extension reactions and so leave more isocyanate groups available to react with atmospheric moisture during the foam-curing stage, and the choice and concentration of the catalysts used. Foam formulations having a tendency to scorch will show an increase in discolouration with increasing cell count or with the presence of collapsed foam layers.

Freshly manufactured blocks of flexible foam are cut to predetermined lengths, weighed and transferred into a block curing and cooling room. This room is separated from the main factory and from the cool foam storage by fire-resistant walls or, more usually, it is a separate single storey building. The hot blocks are stored horizontally. It is usual to monitor the interior temperature of representative blocks of low density foam during the initial curing time. Needle thermocouples inserted into the interior of the foam allow the recording of the temperature and its change with time. It is suggested that blocks should show a temperature fall of at least 20°C below the maximum recorded before they are allowed to leave the curing room. A continuing increase in the interior temperature of the foam block during the first few hours after manufacture should sound an alarm. Blocks showing evidence of smouldering should not be moved. The safest treatment is dousing with water by using a water lance inserted into the interior of the foam.

Rigid foam

Most rigid polyurethane foam is moulded in place as a reinforcing insulant so that the heat produced by the foam is slowly dissipated through the facings. Overheating of temperature sensitive facings such as polyethylene-coated foils and papers is possible if the freshly made laminate is stacked horizontally without first allowing a short cooling period. Rigid foam which is cast into large blocks, continuously or discontinuously, may scorch due to an excessively high core temperature. Foam systems using relatively high molecular weight polyfunctional polyols are available for slabstock production. These give a lower reaction temperature than those foam systems designed to give a fast reaction and rapid cure when injected into a composite structure. This lower temperature, combined with the use of selected catalysts and other additives, avoids the risk of scorch.

Rigid polyurethane foam is often cut with a reciprocating or circular saw having fine cutting teeth. The sawdust produced should be removed as it is formed. This is usually done by suction nozzles and baffles attached to the saw mountings so that the dust is deflected and removed as it is generated. As indicated on page 237, all dusts, including those that are chemically and physiologically inert, create nuisance and discomfort. The generally recommended maximum concentration of a nuisance dust corresponds to the TLV (a total of 10 mg/m^3 of dust with a maximum of 5 mg/m^3 of respirable dust). However, in view of results from animal experiments, the International Isocyanates Institute Safety Committee have recommended that the concentration of polyurethane foam dust in the air of the workplace should not exceed a maximum level of 5 mg/m^3 of air. When, in spite of adequate ventilation and dust extraction, there is a possibility that the concentration of airborne polyurethane dust will be greater than 5 mg/m^3 of air, approved dust filter respirators should be used. Overalls and gloves should also be worn because rigid foam dust may irritate the skin.

Rigid polyurethane foam sawdust which is not removed at source will settle on nearby horizontal surfaces. It should not be allowed to accumulate in the factory. It presents a hazard in case of fire which may spread across a layer of polyurethane dust. Any finely divided organic material may also present a hazard of explosion. The explosion of airborne clouds of flour and coal-dust are well known examples. Rigid polyurethane foam sawdust can be induced to form an explosive mixture with air, but usually most of the particles are too large to easily form dust clouds of sufficient density to create a primary explosion hazard. However, as with most organic dusts, dust which has settled out of the air and is allowed to remain creates the possibility of a secondary dust explosion. This is the explosion which may occur when settled, organic dust is lifted into the air by a primary, often relatively minor explosion, to form a dense dust cloud. Polyurethane dust should be removed regularly. This is best done by vacuum cleaning or by hosing or mopping with water.

The flammability of rigid polyurethane foam varies widely depending on the chemistry of the polymer. Foams having high levels of isocyanurate linkages derived from the polymerisation of MDI, are not easily ignited and burn slowly even when exposed to a major conflagration, whereas polyurethanes based upon polyether triols, for example, are easily ignited. The use of uncovered rigid polyurethane foam is not recommended and large areas of foam, e.g. in spray coating operations, should not be left uncovered but should be coated with a flame retardant paint or with a cladding that does not sustain fire.

Polyurethanes in fires

Polyurethanes, like all organic materials and some inorganic substances, will burn, The ignitability and the rate of burning of a material depends upon its chemical constitution and upon its physical form. The effect of physical form is well known and no-one would attempt to light a log fire without starting material such as twigs, dry leaves, paper or wood shavings. Flaming combustion is a gas phase reaction between volatile combustible materials and oxygen. Flames from solid materials are the result of thermal decomposition of the solid to produce flammable vapours which then burn. The flames become self-sustaining when the heat produced is sufficient to keep the surface of the solid above its decomposition temperature and so maintain the supply of combustible vapour. The maintenance of a high surface temperature and the generation of combustible gas is aided by a high surface to volume ratio.

Articles made from solid polyurethanes, elastomers, microcellular and self-skinning foam, are in general no more flammable than similar articles made from natural and synthetic rubber. The ignitability, rate of burning and smoke production from polyurethane elastomers is often lower than that from natural rubber and the common synthetic rubbers. High density, rigid, self-skinning foams, especially those having a high level of polyisocyanurate, are non-melting and difficult to ignite. When involved in a fire they tend to char rather than to contribute to flaming combustion, although they will burn in a sufficiently hot fire.

Low density foams in fires

The most widely used polyurethanes are low density foams – flexible foams for cushioning and upholstery, and rigid foams as heat insulators. Both types of foam are very effective in their particular applications; they are also non-toxic and have good resistance to ageing, moulds and vermin. Polyurethane foams, partly as a result of the sensational reporting of some fires involving flexible foams, are often considered to be a major fire hazard. This view does not appear to be supported by the statistical evidence; table 10-7 shows the incidence of fires in occupied buildings over the years from 1969 – 1983. This period saw a great increase in the use of plastics in furnishing but the annual number of fires shows little change. There is a similar variation, possibly related to the severity of the winter weather, in the number of people dying in fires (table 10-8).

Real fires are extremely complex reactions, usually under changing conditions which cannot be simulated in the laboratory. Nevertheless, tests have recently been developed which are designed to assess the degree of the fire hazard resulting from the use of foams in furnishings and in building structures. As polyurethane foam is rarely used alone and uncovered, this requires the testing of composite articles such as foam-filled upholstered furniture and foam-cored building panels rather than testing the foam alone. These tests are described in Chapter 7 (*Rigid polyurethane foam*) and Chapter 11 (*The physical testing of*

Table 10-7 Fires in occupied buildings (thousands), UK (1969-1983)

Year	Dwellings	All industrial	Places of public entertainment, clubs, etc.
1969	45·9	17·6	4·4
1970	45·3	18·2	4·4
1971	46·0	16·8	4·3
1972	52·9	17·4	4·7
1973	55·5	18·6	4·7
1974	55·1	17·7	4·5
1975	52·0	14·9	4·3
1976	51·1	14·7	3·9
1977	51·0	14·0	3·8
1978	53·3	13·4	3·9
1979	58·6	14·0	4·0
1980	56.9	12.8	3.7
1981	56.0	11.0	3.4
1982	56.4	11.3	3.3
1983	57.4	10.8	3.6

Table 10-8 UK fire statistics; fatal casualties (1969-1983)

Year	Dwellings	Other buildings and outdoors
1969	688	173
1970	627	212
1971	574	248
1972	785	293
1973	765	276
1974	787	259
1975	720	200
1976	690	205
1977	652	197
1978	733	213
1979	865	231
1980	822	213
1981	780	195
1982	728	191
1983	710	193

polyurethanes). In general, fire performance is assessed by tests related to the application risks at each stage of a potential fire. Small scale tests are used to evaluate and compare the ignitability, rate of burning, the amount and the rate of heat and smoke evolution and also the production of lachrymators and toxic combustion products.

Ignitability

Flexible polyurethane foam upholstery. The carelessly dropped smouldering cigarette is the commonest source of ignition in all fires which start in upholstered furniture. Upholstery and bedding should therefore resist ignition by a smouldering cigarette. Statutory tests, in the UK (see Chapter 11), and in many states in the USA, ensure that upholstered furniture sold to the public is resistant to ignition by a smouldering cigarette. More severe tests, requiring resistance to various flaming ignition sources, are commonly specified for public buildings. The highest flame resistance is required for prison bedding. The test set by the UK Property

Services Agency, for example, requires that the foam mattress is removed from its cover and any barrier layer is removed. A third of the uncovered mattress is cut off and serves as the base of an igloo which is built from the other two thirds cut into ten roughly equal size pieces. Four number 7 size wood cribs – each comprising 4×20 sticks of wood, about 4×126 g of wood + 24 g of wood wool (BS 5852: Part 2) – are placed inside the foam igloo and ignited. Such high resistance to arson is met at high cost and with some loss in comfort because of the high hardness of the fire-resistant construction. The 4-inch thick, fire retardant, mattress is usually made up from four 1-inch thick sheets of foam, each impregnated with a dispersion of aluminium hydroxide in a binder, and laminated together to make the mattress.

The performance of foam-filled upholstered furniture in the cigarette test depends largely upon the nature of the covering material although the construction of the furniture is also important. Cotton and other cellulosic fibres, including rayon, tend to smoulder. This tendency is reduced as the proportion of thermoplastic fibre in the covering textile is increased, but too high a content of easily melted fibre will reduce the resistance to open flame ignition and may also significantly increase the rate of burning of the upholstery once it is ignited. One solution to the problem is the use of an interliner between the outer textile cover and the foam cushion. Interliners that have been found to inhibit the smouldering initiated by a cigarette include flame retardant polyester fabrics, polyesters coated with PVC or with chlorinated elastomers, polyester scrims with a thin PVC foam backing, high resilience polyether polyurethane foams containing a high level of chlorinated phosphate ester flame-retardant and conventional flexible polyurethane foam layers which have been impregnated with about 300% by weight of aluminium hydroxide carried in a suitable binder.

Once ignited, either directly or from other common combustible materials – electrical equipment, newspapers and draperies – polyurethane foam filled upholstery may burn rapidly compared to traditional upholstery. It has been shown that fires in living rooms, with an open door and containing modern upholstered furniture, become fully involved within 5 to 7 minutes. With all the doors and windows closed the initially intense fire may be extinguished by lack of oxygen. A great deal of work has been done to investigate the effect of many different foam and fabric cover combinations, using full scale burning tests of upholstered furniture.

This work has shown that by a suitable choice of the foam filling and especially of the cover material, together with the use of an interliner, as necessary, the ignitability and the rate of burning of foam-filled upholstered furniture can be reduced to give burning rates similar to those obtained with traditional furniture with steel springs and flock upholstery.

Rigid polyurethane foam insulation. In most countries the materials used in buildings are classified as combustible or non-combustible materials. Non-combustible materials are inorganic building materials such as brick, stone and concrete, or composites containing only a minor amount of organic material.

Polyurethane and polyisocyanurate foams, in common with wood, polystyrene and phenolic foams, all plastics, glass-fibre insulation containing more than about 6% of organic binder, and virtually all common furnishing materials, are classified combustible by ISO 1182, BS 476 and many similar national standard tests.

Rigid polyurethane and polyisocyanurate foams are mostly used in buildings as faced panels. The ignitability and the rate of burning – usually measured as the rate of flame spread across the surface (see Chapter 7, figure 7-40) – depend mainly upon the properties of the panel facing material. Fire resistance, or the resistance to heat penetration through the panel, is improved by an insulating foam core so long as it remains intact during a fire. The increased resistance may be small with a low density rigid polyurethane foam core but a polyisocyanurate foam core, because it can form an insulating carbonaceous char, significantly increases the fire resistance of a panel. The highest resistance is obtained using polyisocyanurate foam reinforced with glass-fibres or other high melting, net-like structures (see Chapter 7, figure 7-41).

The potential contribution of any combustible material to the intensity of a fire is expressed as its "Calorific Potential". This is the maximum amount of heat produced by completely burning a unit weight of the material. Both the total heat produced and the rate of liberation of the heat are very important in fires. Because rigid polyurethane and polyisocyanurate insulating foams usually have a very low density and give good insulation at a lower thickness than other insulants, their heat evolution in a fire is comparatively low (table 10-9). Low density rigid polyurethane foams which become completely exposed to a high temperature fire will, because of their low density, burn very quickly giving a brief but rapid evolution of heat. Because of their tendency to char, polyisocyanurate foams do

Table 10-9 **The maximum heat produced on burning some building materials**

Material	Maximum heat produced on burning: Equal weights (MJ/kg)	Equal volumes (MJ/m³)	Density (kg/m³)
Rigid polyurethane and polyisocyanurate foams.	26 to 27	800	30
		1,100	40
		1,300	50
Wood, pine.	18	9,000	500
oak.	17	12,000	700
Polystyrene,	42	45,000	1006
foam.	42	700	16
Asphalt.	40	48,000	1200
Phenolic foams	25	500 to 900	20 to 35

not behave in this way, except when uncovered foam is suddenly exposed in intense fires.

Smoke and toxic combustion products

Combustion products that are toxic to humans are produced by burning both natural and synthetic materials. The commonest combustion products from all organic materials, both natural and synthetic, are carbon monoxide (CO), and carbon dioxide (CO_2). The relative proportions of each produced in any fire depend mainly on the availability of oxygen and the temperature of the fire. Other toxic gases produced in fires include nitrogen oxides, acrolein and hydrogen cyanide (HCN); the latter is produced on burning all materials containing nitrogen, both natural and synthetic. The hazards in real fires, in diminishing order of importance, are:

- The reduction of the oxygen content of the air and the generation of carbon monoxide gas.
- The development of high air temperatures.
- The presence of smoke and lachrymators.
- Direct consumption by the fire.
- The presence of toxic gases other than carbon monoxide.

Carbon monoxide (CO) is highly toxic. It is colourless, tasteless and odourless. Above 8,000 ppm it causes immediate death. The effect is due to the high affinity of CO for the haemoglobin of the blood with which it combines and prevents the uptake of oxygen from the air. Death results from oxygen starvation. Most deaths in fires results from the inhalation of CO. Under similar conditions a burning polyurethane foam, flexible or rigid, will produce about the same amount of CO as an equal weight of natural products such as cotton and somewhat less than wood, but it may burn much faster thus producing the CO more quickly.

Smoke from a fire is a significant toxic hazard but an important effect of smoke is the loss of vision which hinders escape and rescue. Smoke density is not on its own a reliable measure of loss of visibility. The presence of lachrymators, especially acrolein from smouldering cellulosic materials (wood, paper, rayon), may prevent vision at low concentrations even when there is little visible smoke. Smoke evolution is measured in the laboratory by a number of methods (Chapter 11). Laboratory scale smoke measurements may, however, be misleading. This is especially the case with materials which melt easily and drip away from the flame. Such materials tend to give low smoke figures compared with non-melting materials in small scale tests.

Carbon dioxide (CO_2) is produced on burning any organic material, natural or synthetic. It is an asphyxiant, i.e. it is not itself toxic but

245

in high concentrations it dilutes and reduces the available oxygen level in the air. It also stimulates the rate of breathing (18,000 ppm of CO_2 increases the rate of respiration by 50%) thus increasing the hazard from more toxic combustion products which are also present.

Hydrogen cyanide (HCN) is highly toxic; concentrations above about 280 ppm are immediately fatal. HCN is produced on burning all nitrogen- containing organic compounds, both natural and synthetic. The amount of HCN produced from a nitrogen-containing organic material tends to increase with increasing combustion temperature, especially above 600°C. The actual amount present in the smoke, however, varies widely because if sufficient air is available most of the HCN produced is itself burned in the fire.

Under similar fire conditions, polyurethane foams produce less HCN than the wool, nylon and polyacrylonitrile textiles commonly used in furnishings. Table 10-10 compares the toxic gases produced on burning polyurethane foams with those from burning wood and hardboard using the tubular furnace of DIN 53436.

The hazard from HCN is not thought to be as important as that from CO in actual fires but it may have a contributory effect.

Table 10-10 Toxic gases from polyurethane foam and wood. (By the tubular furnace, DIN 53436)

Material	Concentration of gas (ppm).			
	CO	CO_2	HCN	NO
Beech	6516	127300	29	75
Plywood	19090	52090	875	5
Hardboard	7400	104540	30	54
Polyisocyanurate rigid foam	3145	4990	380	0.6
Polyurethane rigid foam	2910	5700	300	0.55

(Air flow 3 litres/minute, temperature 500°C, furnace travelling counter to the air current at 10 mm/minute; samples were pieces of equal volume – each 200 x 10 x 10 mm.)

Acrolein is highly toxic. It is lethal in a short time at concentrations above about 10 ppm. It causes extreme irritation and lachrymatory action at levels below 1.0 ppm. It is formed in burning cellulosic materials, wood, cotton and paper, and also from oils and fats containing glycerol.

It is not formed in significant amounts on burning polyurethanes.

Large scale fire tests

Large scale fire tests of building constructions incorporating rigid polyurethane and polyisocyanurate foam insulation have been carried out in Europe and in the USA. The conclusions are that properly used rigid polyurethane and polyisocyanurate foam insulation does not add to the severity of fires and does not produce any serious toxic hazard in buildings. Tests of full-scale roofs made up with a range of insulating materials showed no correlation between the ratings of the insulants in conventional small-scale fire tests and their performance in an actual fire involving an insulated roof. The rate of spread of the fire across the roof was similar, for instance, with mineral wool insulation and with rigid polyisocyanurate foam insulation.

Figure 10-2   Fire test of a building construction.

In conclusion, although uncovered low density rigid polyurethane foam may burn very quickly and may increase the spread of a fire and the rate of burning, full-scale fire tests have shown that polyurethane foams, when used as an insulating infill with appropriate coverings, do not represent an increased fire hazard compared with many traditional building materials and most synthetic products. Polyisocyanurate foams and some flame-retardant rigid polyurethane foams can be used to make effective, flame-resistant, insulated building panels.

Further reading

In addition to the publications about handling isocyanates which are listed in table 10-11, there are a number of comprehensive booklets published by the International Isocyanate Institute Inc., (table 10-11). These cover a number of aspects of polyurethane production, chemical handling, waste disposal and fire testing. Additionally, many of the publications listed in the polyurethane bibliography (Appendix C), include useful comments on the safe handling of polyurethane materials.

Table 10-11 International Isocyanates Institute Inc., (III), publications

Brochure 1	Technical Information: **Recommendations for the Handling of Toluene Diisocyanate (TDI)** (November 1980 – Revised)
Brochure 2	The International Isocyanate Institute, Inc.: **A Report on the International Isocyanate Institute Inc.: its purpose, organisation and safety research activities** (September 1981 – Second Edition)
Brochure 3	Technical Information: **Recommendations for Waste Disposal from Polyurethane Foam Manufacture** (May 1979)
Brochure 4	Technical Information: **Recommendations for the Handling of 4,4'-Diphenylmethane Diisocyanate MDI Monomeric and Polymeric** (December 1982 – Reissued)
Brochure 5	Technical Information: **Analysis of Isocyanates in Air** (September 1982)
Bulletin 1	**Study on Microbial Degradation of Polyurethane Flexible Polyether Foams Under Waste Disposal Conditions**
Bulletin 2	**Model Fire Test on Corrugated Steel Roof Constructions Reflecting Real Fire Conditions – The Contribution of Combustible Insulating Material to Fire Risks**
Bulletin 3	**The Behaviour of Upholstered Furniture in Fire**
Bulletin 4	**The Toxicity of the Airborne Combustion Products of Polyurethane Foams**
Bulletin 5	**A Study of Rigid Foam Insulation for Steel Tanks (Guidelines and Precautions)**

11   The physical testing of polyurethanes

The last stage of polyurethane manufacture is to characterise the product by measuring its physical properties. Physical tests that characterise the polyurethane material are carried out using specimens of a standard size and shape. Other tests may be done on the complete manufactured article. The choice of test method depends on the objectives of the test. These are usually one of the following:

To characterise the material in a way that is easily understood and easily communicated to other people. This requires the use of standard test methods. Tests on new materials should define the type of material by quantifying the properties which give some guidance about its possible applications.

To control the quality of regular production. Effective quality control tests give a quick check on the reproducibility of production but they do not necessarily measure the standard properties of the material. Quality control tests often include non-standard tests specified by the purchaser of the product.

To check product durability, i.e. the durability of a material or the likely performance of an article made from it in use. Physical tests to evaluate the suitability of a product for its intended purpose include environmental exposure tests, abrasion tests, fatigue tests and creep tests. The simplest check on the likely performance of an article is to carry out tests that will simulate repeated use under extreme conditions of service. This often requires the performance testing of complete assemblies such as refrigerators, vehicle seats, building panels, as well as other articles in which the polyurethanes may play only a minor part such as polyurethane-coated articles, fabrics containing polyurethanes and vehicle suspension systems.

Characterising the properties

The characteristic properties of a material are obtained by using standard test methods on specimens of a standard size and shape under standard environmental conditions. The recommended tests for the basic properties are those defined by the standard test methods of the International Organisation for Standardisation (ISO). These tests are agreed modifications of national standard tests,

249

especially those of the American Society for Testing and Materials (ASTM), Deutsche Institute für Normung (DIN), Association Française de Normalisation (NF), and the British Standards Institution (BSI). Some important ISO tests for polyurethanes are listed in table 11-1 together with some references to similar national tests.

Table 11-1 Some standard test methods for polyurethanes

Title of standard test method	Product type	Internat. ISO no.	British BS no.	U.S.A. ASTM no.	German DIN no.	French NF no.
SI units and recommendations for their use	All	1000 (1973)	5555: 1981 (1989)	E 380-89 E 621-84	— —	— —
Conditioning of test specimens	All	291 (1977) 554 (1976)	2782: (1982)	D 618-61 (1981)	50014 50005	— — — —
Preferred test temperatures	All	3502 (1976)	2782: (1982)	D 618-61 (1981)	50014	— —
Constant humidity cabinets using aqueous solutions	All	R483 (1976)	3718: 1964 (1984)	E 104-85	50014	— —
Physical properties						
Determination of linear dimensions of test specimens	Rigid foams	1923 (1981)	4370: Part 1: 1988	D 1622-88	53420	— —
Determination of the apparent density of cellular materials	Rigid foams	845 (1984)	4370: Part 1: 1988	D 1622-88	53420	— —
Compression test of rigid cellular materials	Rigid foams	844 (1985)	4370: Part 1: 1988	D 1621-73 (1979)	53421	T56-101
Bending test for rigid cellular materials	Rigid foams	1209 (1976)	4370: Part 1: 1988	D 790-86 D 790M-86	53423	T56-102
Determination of tensile properties of rigid cellular materials	Rigid foams	1926 (1979)	4370: Part 2: 1973	C 203-85	53430	T56-103
Determination of shear strength of rigid cellular materials	Rigid foams	1922 (1981) R/1663	4370: Part 2: 1973	C 273-61 (1988)	53427	T56-118
Determination of apparent thermal conductivity by means of a heat flowmeter	Rigid foams	2581 (1975)	874: Part 2: 1988	C 518-85	52616	T56-124
Determination of apparent thermal conductivity by guarded hot-plate	Rigid foams	2582 (1980)	4370: Part 2: 1973	C 177-85	52612	X 10-21
Test for dimensional stability of rigid cellular materials	Rigid foams	2796 (1986)	4370: Part 1: 1988	D 2126-87	53431	T56-122
Determination of the temperature at which permanent deformation occurs under compressive load	Rigid foams	7616 (1986)	— —	D 621-64 (1988)	53424	— —
Determination of water absorption by immersion method	Rigid foams	2896 (1987)	— —	D 2842-69 (1975)	53433	— —
Determination of water vapour transmission of rigid cellular materials	Rigid foams	R/1663	4370: Part 2: 1973	E 96-80	53429	T56-105
Determination of the closed-cell content of cellular materials	Rigid foams	4590 (1986)	4370: Part 3: 1988	D 2856-87	— —	— —
Determination of friability of rigid cellular materials	Rigid foams	DP 6187 (1985)	4370: Part 3: 1988	C 421-88	— —	T56-109
Measurement of the dimensions of test specimens	Flexible foams	1923 (1981)	4443: Part 1: 1988	D 3574-86	53570	T56-119
Determination of tensile strength and elongation at break	Flexible foams	1798 (1976)	4443: Part 1: 1988	D 3574-86 D 3453-80 (1985)	53571	T56-108

Table 11-1

Determination of compression set	Flexible foams	1856 (1980)	4443: Part 1: 1988	D 3574-86	53572	T56-112
Determination of hardness (Indentation technique)	Flexible foams	2439 (1980)	4443: Part 2: 1988	D 3574-86	53576	T56-111
Accelerated ageing tests	Flexible foams	2440 (1972)	4443: Part 4: 1976	D 3574-86	53578	T56-117
Determination of compression stress/strain characteristic and value	Flexible foams	3386/1 (1979)	4443: Part 1: 1988	D 3574-86	53577	T56-110
Test for dynamic fatigue by constant load pounding	Flexible foam	3385 (1982)	4443: Part 5: 1980	D 3574-86	53574	T56-115 I56-114 T56-116
Determination of tear strength of flexible foam	Flexible foam	— —	4443: Part 6: 1980	D 3574-86	53575	T56-109
Determination of tear strength (trouser, angle and crescent test pieces)	Elastomers	34 (1979)	903: Part A3: 1982 2782: Part 3: 1986	D 624-86 D 1922-67 (1986)	53515 53507	— — — —
Determination of the indentation hardness of plastics by means of a durometer, (Shore° hardness)	Elastomers	868 (1985)	2782: Part 3: 1986	D 2240-86	53505	— —

Important aspects of reproducible physical testing are the preparation and conditioning of the test specimens and the environmental conditions during the actual test.

Conditioning specimens before testing
There are several reasons for conditioning samples before testing:

Materials for testing should have reached a stable state of cure.
The properties of many polyurethane materials change quite rapidly immediately after manufacture so that it may require several days before the rate of change becomes very slow and the measurement of physical properties can give reproducible results. The initial changes in properties or curing of the polymer are often accelerated by curing at an elevated temperature in the range from 30 to 120°C. As with all plastics (polymeric materials), polyurethanes exhibit changes in properties with changes in temperature, even when they are fully cured. These changes may be relatively large compared with those exhibited by some traditional materials such as metal, leather and wood. Polymers, including polyurethanes which contain hydrophilic groups such as amino groups and polyoxethylene segments, have reduced secondary intermolecular bonding when the amount of water vapour present in the atmosphere is high. The physical properties as measured change significantly with humidity and specimens for testing must, therefore, be allowed to reach equilibrium with a standard atmosphere. In the past, conditioning procedures used have varied widely depending on the climatic conditions at the place of test and the intended use of the product. The relevant ISO standard on the conditioning of specimens before testing, ISO 291 (1977), specifies two normal conditions:

Standard normal atmosphere: 23°C ± 2°C, 50% ± 5% relative humidity (R.H.) at 86 to 106 kPa atmospheric pressure. (Some ISO tests require 20°C ± 2°C, 65% ± 5% R.H.).
For tropical countries: 27°C ± 2°C, 65% ± 5% R.H. at 86 to 106 kPa atmospheric pressure.

A second 'close-tolerance' condition is also specified in ISO 291 (± 1°C and ± 2% R.H.) but the normal tolerances are quite adequate for use when conditioning most polyurethane materials. The conditioning period required by the standard test methods varies, with the type of test, from 12 hours to 88 hours. A conditioning period of 16 hours before test is convenient and is satisfactory for the routine testing of most polyurethanes.

Ambient conditions during testing. In general, the physical test measurements should be carried out in a standard atmosphere identical to that used to condition the test specimen. When this is difficult or costly it is common practice simply to maintain the test room temperature and test the specimen immediately after it is removed from the conditioning cabinet. This may give poor reproducibility. The procedure is satisfactory for some polyurethanes, but for flexible foams and elastomers it is best to maintain a humidity of 50% ± 10% in the testing laboratory. Special enclosures and procedures are required when physical properties are required at sub-normal or elevated temperatures. Preferred sub-normal and elevated temperatures are listed in ISO 3205 and the equivalent national standards (see table 11-1).

Mechanical conditioning before testing. Several standard test methods for flexible polyurethanes specify that the test specimen be subjected to a mechanical treatment before testing to ensure a test result which is more representative of the performance of the product in service.

Conditioning enclosures. Conditioning cabinets are essentially temperature-controlled ovens with a fan to circulate the air inside the oven and a means of controlling the relative humidity (R.H.) of the air. There are two methods of humidity control: by using trays of suitable saturated salt solutions which are exposed to the air circulating within the cabinet; or by the use of automatic moisture injection controlled by a humidity meter.

Properties of polyurethane materials

Polyurethane materials are characterised by their density or apparent density, and by their basic mechanical properties, i.e. by the effect on the polyurethane of mechanical stresses. The expression 'apparent density' is used for cellular materials because the density is calculated using the volume of the expanded material and not the smaller volume of the solid polymer. The physical tests take their name from the method of applying the stress to the specimen, e.g. tensile, compression, indentation, shear, flexural tests. In addition to the basic physical property tests, in which the material is stressed at a standard, slow rate, there are high-rate tests – such as impact tests and drop tests – and tests to measure the long-term behaviour under both static and/or varying loads.

Tensile properties. The tensile properties are important characteristics of the strength of any material. The specimen is stretched at a constant standard rate until it breaks.
The *tensile strength* is the maximum stress the material withstands before rupture.

Figure 11-1   Measuring tensile properties.

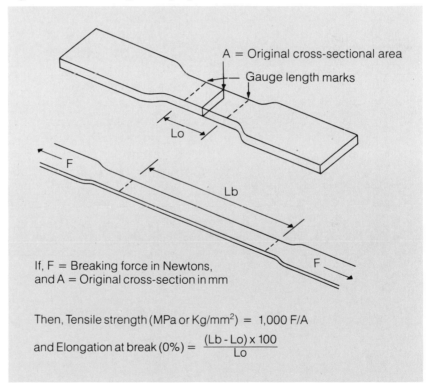

If, F = Breaking force in Newtons,
and A = Original cross-section in mm

Then, Tensile strength (MPa or Kg/mm^2) = 1,000 F/A

and Elongation at break (0%) = $\dfrac{(Lb - Lo) \times 100}{Lo}$

The *elongation at break* or the *ultimate elongation* is the maximum extension of the specimen at the point of rupture.
By plotting stress against strain (strain is usually expressed as the percentage deformation of the specimen) the same test yields the modulus of elasticity. This is the slope of the initial, straight-line portion of the stress/strain curve, i.e. it is the ratio of stress/strain in the region of ideal elasticity where Hooke's law holds. The stress/strain curve of some materials, including some rigid polyurethane foams and some high modulus elastomers, show a yield point – the

253

point at which non-elastic deformation starts. This indicates the start of irreversible changes such as the breaking of cell windows and struts in foams, the slipping of polymer chains and the tightening of looped chains in the soft segment domains of elastomeric polyurethanes. The *yield strength* is the stress at that point.

Figure 11-2   Stress/strain curves for various types of polymer.

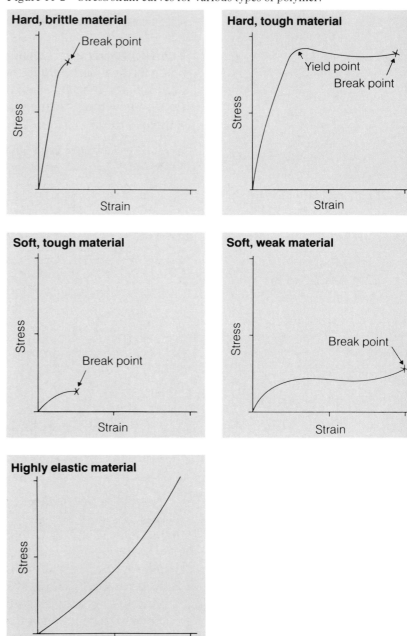

Stress/strain curves are also used to give some measure of the toughness of a material (figure 11-2). The integrated area under the curve is proportional to the energy-to-break per unit volume, which is a measure of the *toughness* of the specimen under the conditions of the test. Figure 11-1 illustrates diagrammatically a typical tensile test specimen and the derivation of the tensile strength and elongation at break. Dumb-bell shaped specimens are used for tensile tests on flexible foams, elastomers and plastics, and rigid foams, but the precise shape and size of the specimens and the rate of separation in the test differ with the type of material and sometimes with the particular standard test method. Some standards are not comparable. In particular, the tensile tests for rigid cellular materials given in ISO 1926 and BS 4370 : Part 2, give different results. This is not only because the test specimens differ in shape but also because different rates of strain are used (ISO, 5 ± 1 mm/min; BS, 10 ± 1 mm/min).

Table 11-2 Some standard test methods for composite articles containing polyurethanes

Title of standard test method	ISO number
Refrigerated cabinets, Technical specification:	
Part 1: General requirements	5160/1 (1979)
Refrigerated cabinets, Technical specification:	
Part 2: Particular requirements	5160/1 (1980)
Commercial refrigerated cabinets: Methods of test:	1992/1 (1979)
Part 1: Calculation of linear dimensions, areas, volumes	
Part 2: General test conditions	1992/2 (1973)
Part 3: Temperature test	1992/3 (1973)
Part 4: Defrosting test	1992/4 (1974)
Part 5: Water vapour condensation test	1992/5 (1974)
Part 6: Electrical energy consumption test	1992/6 (1974)
Part 8: Test for accidental mechanical contact	1992/8 (1978)
Specification and testing	
Part 1: General cargo containers	1496/1 (1978)
Part 2: Thermal containers	1496/2 (1979)
Particle board	
Definition and classification	820 (1975)
Determination of dimensions	821 (1975)
Determination of density	822 (1975)
Determination of moisture content	823 (1975)

Specimens of polyurethanes for tensile testing are usually produced by cutting or die-stamping from sheet material of standard thickness produced on a splitting machine. The specimens are mounted in holders – of the type and dimensions specified in the relevant standard method – on the tensile testing machine. These are power driven machines having one specimen holder which is driven at the required rate. The other holder is provided with a means of measuring the load or stress applied to the specimen. The strain is measured by following the gauge marks on the sample. This may be done by the operator, by photo-electronics, by a combination of manual following and automatic recording, or, for high strength materials, by mechanical methods. A typical tensile machine is illustrated in figure 11-3.

Figure 11-3   Tensile test machine.

Tensile modulus. The tensile modulus or the modulus of elasticity, is the slope of the initial, straight-line part of the stress/strain curve. It represents the region of approximately perfect elasticity where Hooke's law holds, (for the short-term stress/strain properties measured by standard procedures), i.e.

$$\frac{\text{Stress}}{\text{Strain}} = \frac{\delta}{\varepsilon} = E = \text{Young's modulus.}$$

The terms 100% modulus, 200% modulus, etc., (or "Conventional modulus" in Europe) are sometimes used to describe elastomers and microcellular polyurethane elastomers. The figures so described are not true moduli but are simply the stress required to produce strain values of 100%, 200%, etc.

The tensile strength and elongation at break can be qualitatively correlated with the polymer structure. Highly cross-linked materials such as those used for rigid foams, are strong and hard, but brittle. The stress at break is very high but fracture occurs before yield. Flexible polyurethanes, elastomers, foams and thermoplastics have much higher elongation at break. The initial, straight-line part of the stress/strain curve where the material exhibits approximately perfect elasticity, represents straightening and disentanglement of the long, flexible, molecular chains. Further extension of the polymer causes molecular slippage with the breaking of secondary bonds between adjacent polymer chains and may result in permanent set, i.e. a permanent change in the dimensions of the material.

Tear strength. The tear strength tests used for evaluating flexible polyurethane foams and elastomers are 'trouser' tear tests (BS 4443, ASTM D-3574 and DIN 53575) using a partially-slit test specimen (Figure 11-4), although the 90° angle test piece, die-cut from sheet material, (ISO 34, DIN 53515) is also used. The two legs of the specimen are gripped in the holders of the tensile test machine and separated at a standard rate. The test measures the energy required to tear the sample at the specified rate of separation. The energy required includes the energy needed to stretch the elastomer fully and therefore depends partly on the viscoelastic properties of the material. It is highly rate dependent. The tear strength of an elastomer is not directly related to the tensile strength of the material. Another tear strength test is sometimes used to measure both the resistance to puncture and the resistance to tear-initiation of high-strength sheet materials such as thermoplastic polyurethane foil (ASTM D-2582) but the results cannot be corrected to unit thickness because they are a property of the foil rather than of the polymeric material.

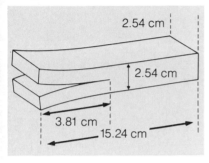

Figure 11-4   Test piece for tear strength determination.

Compression tests. In compression tests, a standard-sized specimen with parallel flat faces is held between two parallel plates of larger area than the specimen, and the force required to compress it at a constant rate is measured (figure 11-5). Stress/strain curves are plotted using the same terminology as that used for tensile tests. The size of flexible test specimens is not critical and the Standards allow considerable latitude. Moulded parts with parallel faces may be tested. In the compression-testing of rigid-foam materials it is most important to apply the force in a truly perpendicular fashion. The size and shape of the test sample are also very important. The faces of the rectangular parallelpiped test specimen must be parallel to ± 1%. Cubic samples are best in order to avoid buckling.

The compressive strength of rigid cellular materials is the value of the maximum compressive force divided by the initial surface area of the test specimen – but only if the maximum compressive force is reached before the strain or the relative deformation of the specimen reaches 10%. Otherwise the compressive stress is recorded at 10% strain. It is recommended that the test machine should simultaneously record both the applied force and the displacement of the specimen to give a stress/strain curve.

257

Figure 11-5   Compression hardness: method of measurement and the form of the stress/strain curve for flexible cellular polyurethanes.

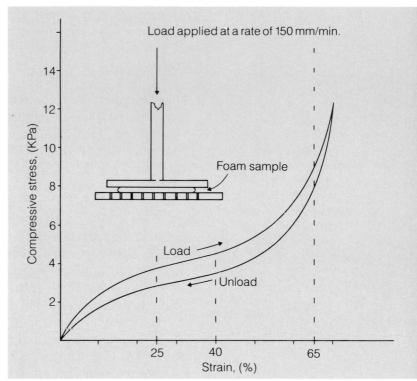

Indentation hardness tests. Indentation hardness is an important characteristic which is used to classify many flexible, cellular polyurethane materials. The principle standard tests are those using a large indentor of 200 mm diameter (ISO 2439 and equivalent tests, table 11-1) to classify low density, flexible polyurethane foams for use in upholstery, and indentation hardness tests on solid or microcellular elastomers and high density self-skinning foams, using the small conical, spring-loaded indentors of the Shore durometers (ISO 868, Type A, a truncated cone having an end 0.79 mm in diameter, and Type D, a sharp cone with a slightly-rounded end, 0.10 mm in radius). The depth of indentation under the force applied by a standard spring is used to measure the 'Shore hardness' of the material. There are many other identation hardness tests. Ball indentors, using a hardened steel ball 5 mm in diameter (ISO 2039, BS 2782, ASTM D785, DIN 53456) are sometimes used to characterise the surface hardness of structural-foam mouldings, surface coatings and RIM materials of high modulus. The Rockwell scale 'R' uses a similar 5 mm steel ball indentor and the Rockwell test procedure is also described in ISO 2039. Other indentation hardness tests are specified in product specifications such as those for automotive trim, crash padding and for packaging materials and parts. These are principally quality-control tests, designed to ensure reproducible production.

Figure 11-6   Indentation hardness: method of measurement and the form of the force/indentation curve for low density flexible foam.

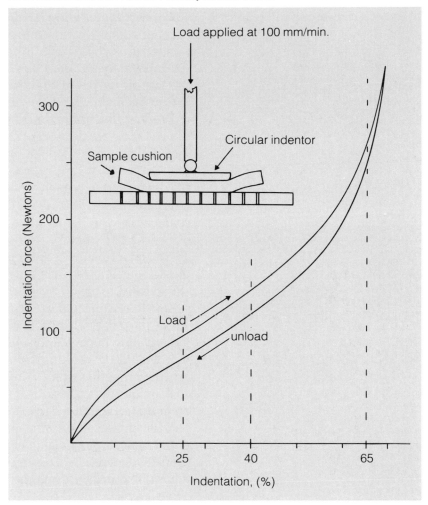

Indentation hardness/Indentation hardness index of flexible foam is a measure of the load-bearing properties of the foam for use in seating, bedding and other cushioning applications. In the test (figure 11-6) the foam specimen is indented at a standard, constant rate with an indentor which is smaller than the specimen. The results are often given as force/indentation curves (figure 11-6). These are sometimes confused with compression test stress/strain curves but, as discussed in Chapter 4, the tests are different in principle and there is no constant conversion factor. The standard definitions of the terms used to characterise foams by indentation are these (ISO 2439, BS 4443 : Part 2.):

– **Indentation hardness** is the total force, in newtons, required to produce a specified indentation of the sample under the standard conditions.

– **Material indentation hardness index** of a material is the result of

259

a test conducted by Standard Procedure 'A' on the standard size of test specimen without plying-up.

- **Product indentation hardness index** is the result of a test conducted by Standard Procedure 'A' on an article of non-standard size or shape or on plied-up material.
- **Material indentation hardness characteristics** are the results of a test conducted by Standard Procedure 'B' on the standard-size test specimen without plying-up.
- **Product indentation hardness characteristics** are the results of a test conducted by Standard Procedure 'B' on an article of non-standard size or shape or on plied-up material.

It is important to note that only the material indentation hardness index and the material indentation hardness characteristics are characteristic properties of the foam material rather than properties of a particular article.

The standard methods specify minimum conditioning regimes of the test specimens which may involve crushing by hand to remove closed cells, followed by mechanical conditioning in which the specimen is subjected to 3 indentations by 70% of the thickness before the indentation hardness is measured. The objective of taking indentation force readings only on the fourth cycle is to obtain reproducible test results that also give some indication of the load-bearing properties of the foam when used in upholstery or other cushioning applications. The force required to compress and indent the foam is greatest for the first cycle and falls thereafter, largely because of viscoelastic effects. Irreversible changes in cell structure may also occur, especially when the foam contains some closed or partially closed cells. It is important to distinguish between the figures obtained by using the two Standard Test Procedures. These are:

Procedure A: Determination of *Indentation hardness index*
Immediately after the third unloading the test specimen is indented at the standard rate to produce an indentation of $40 \pm 1\%$ which is maintained for a period of 30 ± 1 seconds when the indenting force is noted and the force released.

Procedure B: Determination of *Indentation hardness characteristics*
The indentation hardness characteristics are the loads, in newtons, required to indent a standard specimen to 25%, 40% and 65% of the original height. Because of viscoelastic relaxation, the 40% indentation hardness measured/obtained by this procedure is always lower than that determined by the simplified procedure A. Immediately after the third unloading, the test specimen is indented at the standard rate to $25 \pm 1\%$ which is held for 30 ± 1 seconds before noting the force. The indentation is then

immediately increased at the standard rate to $40 \pm 1\%$ where it is again held for 30 ± 1 seconds before noting the indentation force and increasing the indentation to $65 \pm 1\%$ and noting the force after 30 ± 1 seconds.

In general, materials should always be characterised by the indentation hardness characteristics. Procedure A gives only a limited indication of the load-bearing properties and is intended mainly for quality control purposes.
The relationship between the 25% and the 65% indentation hardness, the 'SAG Factor', is often used as a measure of the comfort characteristic of a foam seating cushion. Sag factors of about 3.0 or more are often specified in purchasing specifications for high-resilience seating.

Shear strength and stiffness. Because rigid foam is mostly used as the core material between facings having a relatively high tensile strength, the shear strength of rigid foam is an important characteristic of the material. Stiffness tests may be carried out on specimens of foam, and on foam-cored sheet with laminated or integral skins.

Shear strength is measured (ISO 1922) using a standard-sized specimen glued between rigid parallel plates which are mounted in a tensile test machine, (figure 11-7), and moved in opposite directions at a rate of 1 ± 0.5 mm/minute.
The maximum force applied before failure is recorded.

$$\text{Shear strength} \quad = \frac{\text{Maximum force}}{\text{Initial length} \times \text{width of test piece}}$$

The ***Bending test*** (ISO 1209) is a three-point loading test applied to a standard-sized bar of rigid cellular material, self-skinning foam or foam-cored laminate (figure 11-8). The loading edge is moved at 10 ± 2 mm/minute and the force corresponding to a deflection of 20 mm ± 0.2 mm is recorded. If the specimen breaks before deflecting 20 mm, the force and deflection at break are recorded. The apparatus is also used to determine the flexural strength of a material, although this is not included in the standard test of ISO 1209.

$$\text{Flexural strength} \quad = \frac{3 \times F \times l}{2 \times b \times h^2}$$

where F = maximum force recorded and l, b, h are the dimensions of the specimen as indicated in figure 11-8.

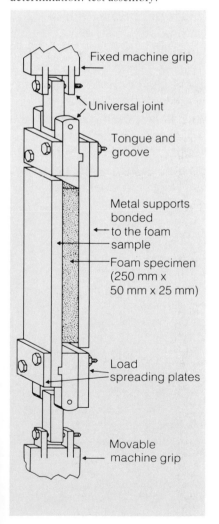

Figure 11-7   Shear strength determination: test assembly.

Fixed machine grip

Universal joint

Tongue and groove

Metal supports bonded to the foam sample

Foam specimen (250 mm x 50 mm x 25 mm)

Load spreading plates

Movable machine grip

Figure 11-8   Three-point bending test.

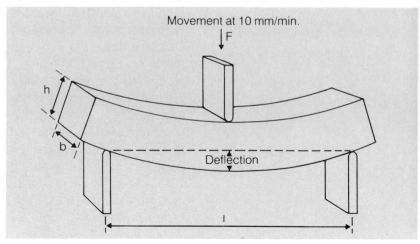

The stiffness of the specimen is recorded as the force, in newtons, required to give a deflection of 20 ± 0.2 mm

Thermal conductivity

The low thermal conductivity of rigid polyurethane foam is probably its most important physical property. Thermal conductivity changes with changes in temperature. The thermal conductivity of rigid polyurethane foam is usually measured at 0°C and 10°C although measurements are often also made at specially specified temperatures. The factors affecting the thermal conductivity of rigid polyurethane and polyisocyanurate foams and the effects of ageing are discussed in Chapter 7. There are two standard methods of measurement. The ISO method (ISO 2581) (figure 11-9) uses a calibrated heat-flow meter to measure the rate of heat flow through a specimen of foam subjected to a known temperature gradient.

Figure 11-9   Heat-flow meter.

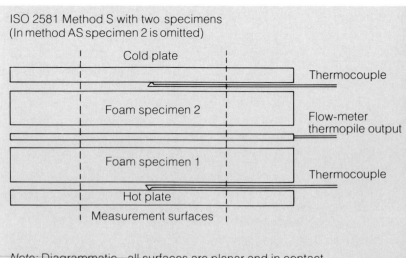

The specimen of foam is placed between isothermal hot and cold surfaces which differ in temperature by at least 15°C., and the rate of heat flow is measured by a heat-flow meter that has been recently calibrated using test specimens whose known thermal conductivity has been determined in a standard apparatus of the guarded hot-plate type. The steady-state output of the heat flow-meter is recorded. The thermal conductivity of the specimens is the density of the heat flow rate through the surface of the specimen divided by the temperature gradient. The ISO standard method describes the method of calibrating the apparatus and expressing the results in detail.

The guarded hot-plate apparatus and its use are described in ISO 2582, DIN 52612, BS4370: Part 2, and ASTM C-177. The methods are equivalent. Two specimens of the foam material are placed in contact with the opposite faces of an electrically-heated hot-plate (figure 11-10). The edges of the hot-plate and of the samples are guarded by electrically-heated collars held at temperatures that minimise heat losses from the edges. The electrical input to the hot-plate is adjusted until steady-state conditions are obtained when the power input equals the heat flow through the specimens. The thermal conductivity of the specimens is then calculated from the temperature gradient, the dimensions of the specimens and the power input.

Figure 11-10   Guarded hot-plate.

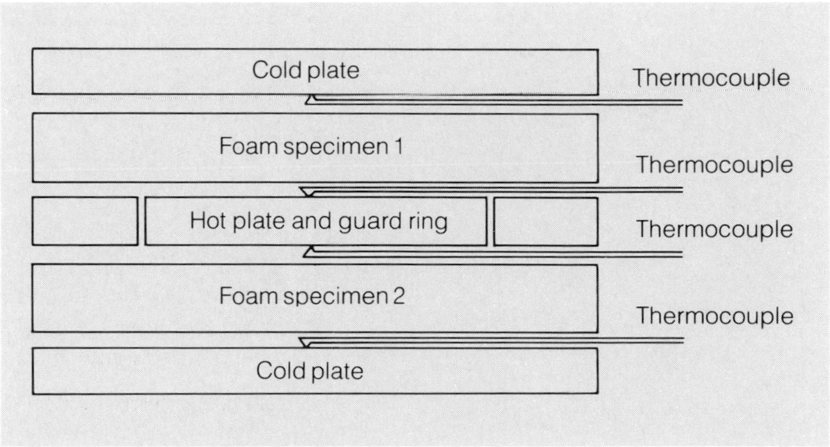

Cold plate	Thermocouple
Foam specimen 1	Thermocouple
Hot plate and guard ring	Thermocouple
Foam specimen 2	Thermocouple
Cold plate	

Water-vapour permeability

The water-vapour permeability of closed-cell rigid foams and of elastomeric polyurethane sheet or foil are both determined by similar gravimetric methods. A dessicant, calcium chloride, is placed in the bottom of a metal or glass beaker and the test specimen is sealed onto the top by sealing wax. The assembly is stored in a controlled standard atmosphere and weighed at regular intervals. When the rate of increase in weight becomes constant, the rate of water vapour permeation through unit area of the specimen in unit time is calculated.

Rigid cellular materials are tested by the method of ISO 1663. The foam specimen is sealed into the top of the beaker using a circular template to ensure that a reproducible area of the foam surface is exposed to the controlled humidity (figure 11-11) which may be either of the standard atmospheres, 38°C and 88.5% R.H. or 23°C and 85% R.H. The assembly is exposed to the controlled atmosphere for successive periods of 24 hours and weighed between each exposure until the rate of change becomes approximately constant.

Figure 11-11   Apparatus for determining the water vapour transmission rate of rigid foams.

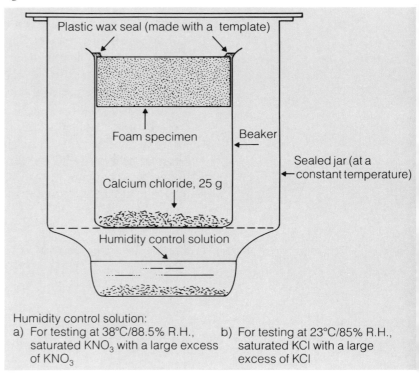

Plastic wax seal (made with a template)

Foam specimen

Beaker

Sealed jar (at a constant temperature)

Calcium chloride, 25 g

Humidity control solution

Humidity control solution:
a) For testing at 38°C/88.5% R.H.,
   saturated KNO_3 with a large excess
   of KNO_3
b) For testing at 23°C/85% R.H.,
   saturated KCl with a large
   excess of KCl

Closed cell content

The closed cell content of the foam affects several important physical properties, especially thermal conductivity, water vapour permeability and liquid water uptake. The closed cell content is easily and quickly checked and it is a useful quality control test. There is not yet an ISO method but the principles of the determination and the design of the apparatus are described in BS 4370 : Part 2 and in ISO DIS 4590. These tests measure the volume of air displaced by a foam specimen of known external dimensions. The foam specimen is placed in a closed chamber and the volume of the chamber is then increased by a known amount. The fall in pressure of the expanded air containing the test specimen is measured and compared with the pressure obtained by a similar volume expansion of the chamber without the foam specimen. The air displacement of the test specimen which is proportional to its closed cell content, is then calculated by the application of Boyle's Law.

Errors in the determination result from temperature variations during the test and from the cut, open, surface cells of the specimen. The temperature variations are minimised by expanding the air slowly over a standard time in order to standardise and minimise adiabatic cooling, and by enclosing the apparatus in a draught-proof case. For quality control purposes it is not necessary to correct the results for the error attributable to the cut surface layer of foam. However, when an accurate estimate of closed cell content is required, the correction factor is obtained by repeatedly slitting a foam specimen and determining the variation in the apparent closed cell content with the change in the ratio of surface area to volume of the specimen. Extrapolation of the relationship to zero surface/volume ratio gives the required correction factor.

Dimensional stability

The test for dimensional stability is important for polyurethane rigid foams having closed cells. Variations in the foam formulation or the manufacturing conditions that may affect the strength of the polymer or cause variations in the composition of the gas within the cells may result in foams that shrink or distort in service. The dimensional stability test is both an essential sorting test for new or modified foam formulations and a useful control of routine production quality. The standard test, ISO 2796, requires accurate measurement of the dimensions of the standard-sized foam specimens by the methods of ISO 1923, followed by storing the specimens for 20 ± 1 hours under one of eleven specified conditions. These conditions range from –55°C ± 3°C to 150°C ± 3°C. After 20 hours the specimens are conditioned at 23°C for 1 to 3 hours when the dimensions are remeasured and the specimens examined for distortion. The procedure is repeated to a total exposure to the test conditions of 48 ± 2 hours and/or 7 days and 28 days. For quality control purposes it is usual to simplify the procedure by using a fixed single exposure of the test specimens: typically specimens are stored at –15°C and others at 70°C and 90% to 100% R.H. for 24 hours. Foam for use as low temperature insulation is stored at a lower temperature, e.g. at –30°C for cold store insulation. The test results are reported as the percentage change in the width and the length under the specified conditions. In choosing foam specimens for dimensional stability tests it is important to include any parts known to be of below-average strength which are likely to have the lowest dimensional stability. These include, for instance, the sections of foam mouldings or foam blocks having below-average foam density or an excessively elongated ('stretched') foam cell structure.

Standard test specimens are flat slabs of foam, 100 mm x 100 mm x 25 mm thick.

Two sets of test samples are cut from the foam, one set at 90° to the other. The two sets of foam specimens thus represent two orientations in the foam. Where possible these two directions should be parallel and perpendicular to the direction of foam rise or to any known direction of cell structure anisotropy.

Resilience impact tests

All the procedures for mechanical tests discussed above are short-term property measurements carried out at convenient speeds derived from the tests developed for conventional engineering materials. As all polymeric materials are viscoelastic, these test methods represent one position only on a curve of mechanical properties as a function of the speed of testing. The work done in distorting a specimen, e.g. the area under the stress/strain curve – figure 11-2, is a measure of toughness. It may not, however, be a reliable indication of toughness in practical circumstances as when an article is dropped or is struck by a stone or other object. Impact tests are another, rather arbitrary, means of assessing the toughness or shock-resistance of a material. There are many kinds of impact tests. The principal types are pendulum impact tests, (Izod or Charpy test methods) and those using falling drop-hammers or falling darts. The Izod test (ISO R180) and the Charpy test (ISO R179) (BS 2782 and DIN 53453) measure the energy required to break a bar-shaped specimen of material. A pendulum of known mass is allowed to fall through a known height and, at its lowest point of swing, to strike the standard-sized specimen – the impact energy is then measured. Pendulum impact tests of this type, using both notched and unnotched samples are widely used to characterise solid plastics materials, but this type of test is basically unsuitable for use with most polyurethane materials with a cellular structure and an integral skin or other forms of composite structure. Falling-weight tests are preferred. There is no ISO standard test of this type and there are important differences between the various national standard tests (e.g. BS 2782, DIN 53443, ASTM D3029). There are also many standard tests for measuring the impact resistance of articles such as dust-bins, plastics pipes, crash helmets and for packing cases and other containers. Such tests are usually designed to simulate the impacts expected in service and are not specific to any particular material but to the function of the complete article. Impact testing of a polyurethane material on a laboratory scale is best carried out by comparison with results obtained with other materials of known performance in the applications envisaged. An example is the falling-dart impact tests of RRIM sheet material illustrated in Chapter 6 (figure 6-21) where RRIM is compared with rolled steel and aluminium sheet and with glass-fibre-reinforced polyester sheet moulding compound. Each sheet material is compared at the

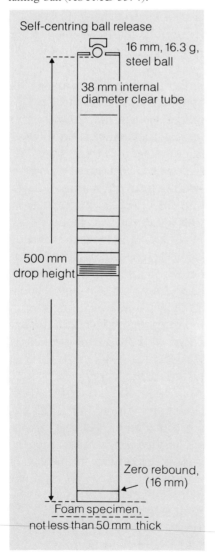

Figure 11-12   Impact resilience by falling ball (ASTM D 3574).

Self-centring ball release

16 mm, 16.3 g, steel ball

38 mm internal diameter clear tube

500 mm drop height

Zero rebound, (16 mm)

Foam specimen, not less than 50 mm thick

thickness normally used in the automotive industry.

The *resilience of flexible foams and cellular elastomers* is determined by impacting a specimen of the material with a known mass and measuring the rebound. Standard tests include those using a pendulum swinging through a pre-set arc and those using a guided falling weight. For low density flexible foams the preferred, simple method is the ball rebound test such as that of ASTM D 3574 – 1981, Test H. A steel ball of a standard size and weight is dropped onto the foam specimen from a standard height and the rebound height is measured (figure 11-12). The rebound height, expressed as a percentage of the drop height, is the ball rebound resilience value. The test is a useful for quality control and sorting. The impact resilience of elastomers, high-density and self-skinning foams may be determined by measuring the rebound of a pendulum (ISO 2650, BS 903 : Part A7) but for many cellular products, falling-weight impactors tend to give more consistent results (ASTM D 2632 - 1981).

The *dynamic cushioning performance* of cellular materials for use as shock absorbers in crash padding and packaging is best measured by an instrumented falling-weight apparatus. Flexible foams for use in packaging (BS 4443 : Part 3 : Method 9) are characterised by measuring the peak deceleration of a suitable mass when it is dropped onto a test specimen. The impactor has a flat base with an impacting area greater than that of the foam specimen. The Method 9 procedure requires a series of measurements using five drop hammers of different weights such that one weight gives approximately the minimum peak deceleration on impact and the other four drop weights are distributed above and below this value. All five drop weights are used at two impact velocities corresponding to free falls of 250 mm and 750 mm. Similar instrumented drop tests are used in the design of shock-absorbing padding. Many vehicle manufacturers test the dynamic cushioning performance of seat cushions by applying a sinusoidal vibration to a seat assembly supporting a typical load. The applied vibration may be over a frequency range from about 1 to 20 Hz or may simulate a real vehicle on an irregular road surface. The nature of the reponse of the load carried on the seating cushion to the applied load is a practical measure of the dynamic performance of the foam cushion.

The minimum peak deceleration is obtained when the impact of the drop hammer is sufficient to compress the foam to an extent that corresponds to the approximately linear portion of the stress/strain curve under the conditions of the test. The two heavier drop hammers will compress the foam until it begins to bottom, i.e. becomes fully compressed, when the rate of deceleration increases rapidly. The two drop hammers of lower weight are arrested in a shorter distance and therefore with rates of deceleration above the minimum peak deceleration. The BS 4443 test is intended as a quality control test and the actual weights of the five drop hammers are chosen by agreement between the supplier and the purchaser of the foam.

Long-term and accelerated ageing

Long-term tests include creep tests, compression set tests, the dynamic fatigue test, environmental resistance tests and accelerated ageing tests.

Compression set test
The compression set test is applied to flexible materials, especially cellular materials. The test (ISO 1856) is an accelerated creep test

which measures the residual deformation of a specimen which has been subjected to a given fixed deformation (50%, 75% or 90%) for a given time followed by a standard period of recovery. Preconditioned specimens, 25 mm thick, are compressed between steel plates to 50%, 75% or 90% of their original thickness using spacers (figure 11-13) and stored for 22 hours at 70°C. The specimens are then released and allowed to recover for a standard time at 23°C and their thickness remeasured. The compression set is expressed as the percentage loss in thickness resulting from the degree of compressive strain used. This simple compression set test is widely used for quality assurance and in purchasing specifications. It is an accelerated test and the results do not always correlate with those found in service at ordinary temperatures. An alternative standard procedure at normal temperature is also specified, but is rather slow for quality assurance purposes.

Figure 11-13 Compression-set jig.

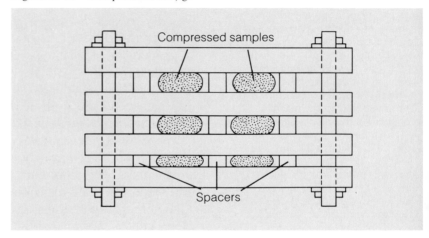

Simple, standard **creep tests** at room temperature, for low density flexible foams include Test I_1 of ASTM D 3574 and Method 8 of BS 4443. The ASTM test entitled *Static Force Loss at Constant Deflection* is a general test, useful for routine control. A standard foam specimen, 380 x 380 x the desired thickness – usually at least 50 mm – is compressed to 25% of the original thickness and held for 22 hours at 23°C ± 2°C and 50 ± 2% R.H. The specimen is allowed to recover for 30 minutes and the loss in thickness and in indentation force for 25% and 65% indentation is determined. The BS test is intended primarily for the quality assurance testing of foam for use in packaging applications. The test measures the creep of a foam specimen under a static load for 72 hours at normal temperature. The static load applied to the foam depends on the type of foam being tested. The stress applied to the foam should produce a creep strain of about 2.5% (Creep strain is defined as the change in strain of the test specimen after 72 hours as determined by measurements taken 15 minutes and 72 hours after loading.) The definition of

'Creep strain' is that usually used in plastics technology and is defined in BS 4618 as the total strain, which is time dependent, resulting from an applied stress. It is not the engineering or scientific definition of creep strain which is the degree of strain above the elastic limit of the material.

Rigid polyurethane and polyisocyanurate foams of low density, self-skinning and structural foams in general show much lower creep than flexible materials. There are no standard test methods for measuring the creep of these materials. For design purposes it is usually assumed that creep is negligible for stress levels less than about 20% of the yield stress. Creep tests for designs involving higher levels of stress may involve large numbers of specimens and tests conducted over at least one year. Creep may be measured in tension, compression, shear or flexure and creep varies with the applied stress and with temperature. Standards that describe creep tests suitable for obtaining design data are BS 4618, ASTM D 2290, ISO R899 and DIN 53444. BS 4618 is the most comprehensive.

Dynamic fatigue test

The dynamic fatigue test (ISO 3385, DIN 53574, BS 3379) for seating cushions is designed to evaluate the durability of flexible foams by applying an alternating load similar to that of a seated adult. A standard circular indentor giving a peak load of 750 Newtons is alternately applied and removed from a standard indentation hardness specimen at a rate of 75 ± 5 cycles/minute for 80,000 cycles. The loss in indentation hardness and in the thickness of the test specimen are determined. The test results show good correlation with the losses in hardness and thickness of seating cushions observed in actual service. The test is used to classify cushioning foams for use in very severe, average and light service conditions (BS 3379). Other dynamic fatigue tests include the roller-shear test of ASTM D3574 and the special tests which are included in some vehicle manufacturers' specifications for seating cushions.

Flexing tests

Flexing tests are carried out on polyurethane elastomers, microcellular and self-skinning flexible foams to determine and compare their resistance to flex cracking (ISO 131, BS 903 : Part A10) and to cut growth (ISO 133, BS 903 ; Part A11). Both tests are fatigue tests in which thin specimens are repeatedly bent using a De Mattia flex-test machine, developed for use in the rubber industry. The flexing test for polyurethane shoe-soling materials developed by the British Shoe and Allied Trades Research Association, is also a modified rubber test (Ross tester). A specimen, 4.8 mm thick, with a 2 mm chisel-cut in the centre is placed over a 9.5 mm diameter steel rod and flexed at a rate of 60 cycles/minute for 150,000 cycles. The test is usually carried out at a temperature of −5°C because flex-crack resistance falls with falling temperature.

The German Pirmasens flexing test is very similar but the chisel-cut is replaced by a needle puncture.

Wear and abrasion tests

There is no ISO standard test for abrasion or wear but several national standard tests are used to compare the abrasion-resistance of polyurethane materials. Abrasion-resistance tests determine the loss in weight, volume or thickness of a specimen which is rubbed with a specified abrasive material under a standard load. Relevant standard tests are:

– BS 903 : Part A9 : The determination of Abrasion Resistance of Vulcanised Rubber. Four methods are described, The Du Pont abrader, The Du Pont constant torque abrader and the Akron and Dunlop abraders. The Du Pont abraders are often used to compare the wear-resistance of polyurethane elastomers. The specimen is held in contact with a rotating abrasive disc.
– BS 3424 : Method 16 is for coated fabrics and is often applied to polyurethane-coated fabrics. The abrasion test machine or 'Rubfastness Tester' consists of a reciprocating table to carry the sample of coated fabric and a loaded knife to rub against the coating.
– ASTM D1242, Method B, uses an abrasive coated tape which is carried across specimens which are moving in the opposite direction. The volume losses of the specimens are compared.
– DIN 53516 describes an abrader, designed for use with vulcanised rubbers, which is also used with polyurethane elastomers. The test specimen is traversed over a rotating drum covered with a sheet of the selected abrading material. The test has been widely used for screening materials because very small test specimens are used.

Accelerated ageing tests

Independent of the effects of mechanical treatment are the chemical and physical changes in any organic material with time. The rate of such changes, the ageing of the material, varies with the environmental conditions of temperature and humidity and with the degree of exposure to light. The standard humidity and heat-ageing tests are accelerated tests which are used for screening and quality control. All accelerated tests may cause changes that would not occur in practice or may alter the relative rates of chemical ageing processes. Accelerated ageing may thus alter the relative ranking of materials compared with their real rate of ageing.

Humidity ageing tests

ISO 2440 refers to the ageing of flexible cellular polyurethanes. BS 4443 : Part 4 is based on the ISO methods but is applied to all flexible polymeric materials. ASTM D 3754, Test J Steam autoclave test

does not include the tests below 100°C and differs slightly in the limits of the higher autoclave temperature. The ISO and BS methods offer a choice of three accelerated ageing conditions:
(1) 85 ± 1°C and 100% R.H. for 20 hours,
(2) 105 ± 1°C (saturated steam) for 3 hours,
(3) 120 ± 1°C (saturated steam) for 5 hours (ASTM. 125 + 0 – 5°C).

Condition (1) is suggested for polyester-based polyurethanes; Condition (2) for polyether-based foams. The physical property examined is not specified because this will depend upon the material and its intended use. The most usual tests are those for load-bearing properties and/or tensile properties. Whichever the test used to monitor the ageing of the foam it is important to include reference specimens which are stored under standard conditions (23 ± 2°C/50 ± 5% R.H.) and tested at about the same time as those which have been aged. Test results are given as the percentage change in the property tested. It is usual to carry out ageing tests on at least three samples and to quote an average result.

Dry, heat-ageing tests
In ISO 2440, specimens are aged at 140°C ± 1°C for 16, 22, 72, 168, 240 hours or some multiple of 168 hours (tolerance ± 5% or 4 hours whichever is the lesser). The national standards differ only in the length of ageing: 22 hours is common to all. The percentage change in property is reported in the same manner as that for humidity ageing.

The effects of changes in temperature

Polyurethanes, in common with organic polymeric materials, particularly thermoplastics and elastomers, show significant changes in physical properties with changes in temperature. Elastomers become stiffer and may become brittle at low temperatures. High temperatures cause softening of both rigid and flexible materials and may also cause irreversible chemical changes or degradation of the polymer. There are many methods of measuring the upper and lower temperature limits to the useful performance of a material. The tensile and compression properties can be measured over a range of temperatures in temperature-controlled cabinets or in conditioning rooms. Similarly, tests for resistance to flexing and abrasion are carried out at both high and low temperatures (especially low) on polyurethane elastomers, microcellular and self-skinning foams in order to indicate the usable limits of their service temperature. The change in the dynamic stress/strain properties of elastomeric polyurethanes also gives a characteristic indication (Chapter 8, figure 8-3) of the useful operating temperature range. The torsion pendulum (ISO R537, DIN 53449, ASTM 2236) is a simple apparatus, widely used for the shear modulus and the energy absorbed on deforming a material. The test piece (ISO size, 60 mm x 10 mm x 1.0 mm) is clamped vertically by a fixed clamp at the upper end of the strip. The lower end is clamped to an oscillating pendulum whose angular movement can be measured, usually by means of a lightweight mirror rigidly attached to the pendulum to reflect a beam of light onto a scale. The shear modulus of the specimen material (at its natural frequency of oscillation), is obtained from the frequency and the amplitude of the oscillation of the pendulum. The rate of loss in the amplitude of the oscillation gives a measure of the damping factor, i.e. the work done in deforming the material at the rate of oscillation. This is known as the loss tangent (Tan δ). The test specimen is enclosed in a temperature controlled cylinder which is capable of being controlled to $\pm$ 1°C over the temperature range from $-$ 60°C to $+$ 250°C. The rate of change of the loss tangent with temperature change shows peak values corresponding to the glass transition temperatures (T_g) of the soft and hard segments of the polyurethane elastomer at the rate of heating or cooling used. An alternative method of measuring the change in loss factor with temperature is by measuring the ball-rebound or pendulum rebound resilience over a wide temperature range.

The change in stiffness, or the softening, of a polyurethane material is measured by one of several methods depending on the type of polymer and its applications. Compressive or bending loads are applied to a specimen and the temperature is increased at a standard rate until a given degree of distortion occurs.

Heat distortion temperature
The heat distortion temperature of rigid foam may be determined by the method of ISO 2799. This test method determines the temperature at which a rectangular parallelpiped test specimen carrying a static compressive load, shows a permanent deformation of more than 5%. The specimen, which is preferably a 50 mm cube, carrying a uniformly distributed load of 50N is placed in an oven at a temperature near to the expected heat distortion temperature (e.g. 100°C for PUR foam, 130°C for PIR foam) for 24 $\pm$ 1 hours. The specimen is then removed from the oven, conditioned for 30 minutes and the thickness measured. If the permanent deflection is less than 5%, the oven temperature is increased by 5°C and the test repeated.

This test does help to define the maximum permissible sevice temperature for intermittent, short duration exposure. Cantilever tests are used to determine the 'Heat sag' of RRIM and other materials at the temperatures used to stove car body paints. In that instance a cantilever specimen is exposed to the specified temperature for a fixed time, often 30 minutes, and the resulting deflection noted.

Fire testing

There are hundreds of methods of evaluating the effects of fire on materials and composite articles. The methods include national standard methods of test and tests specified by government departments, state, district and city authorities, by state and private companies. There are also standard tests recommended by associations and institutions such as those representing the electrical and vehicle industries. The basic objective of fire testing is to determine the likely contribution of a material or a composite article to the hazards present in a real fire. Fire tests are carried out on various scales, ranging from small-scale tests on the laboratory bench to large or full-scale tests involving whole rooms or buildings. Laboratory-scale tests are used to compare the ignitability and the rate of burning of materials under controlled conditions. They are useful as screening tests for new materials and to ascertain the effects of additives and of minor variations in structure on the combustibility of the polymer. *Laboratory-scale fire tests cannot be used to predict the effects in real fires.* There are many parameters affecting the course of a real fire and the hazards that result. The fire hazard also varies with the nature of the application area (e.g. in road vehicles, public buildings, ships, boats, private housing) and experience in one application area may be of little value elsewhere. This is the main reason for the multiplicity of fire tests. The ISO work on fire testing is mainly concerned with fire tests for building materials and building structures. The goal is a unified set of fire tests but it will be many years before this is attained. Meanwhile, national standards organisations have their own tests and in many

Figure 11-14   Stages in the development of a fire in a building.

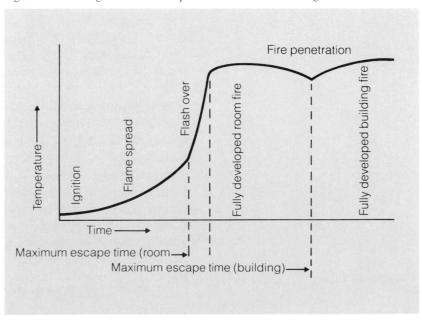

Hazards to life: depletion of oxygen and the presence of carbon monoxide: high air temperatures: smoke and lachrymators: heat: toxic gases other than CO: (Lack of vision, disorientation, panic, asphyxiation).

countries there are others developed by insurance companies to assess the fire risks associated with building materials, components and structures. The most widely used of these various tests are listed in table 11-4.

The objective of the ISO is to develop a series of fire tests related to the factors affecting the initiation and spread of a real fire. These include ignitability, flame spread and heat release. Figure 11-14 is a simplified temperature/time curve for one type of fire in a building, a conflagration – which is defined as a fire involving more than one room. The stages of such a fire, which are related to specific test methods, are:

Ignition: This is often caused by a relatively low-temperature source such as a smouldering cigarette or a match, or it may result from faulty electrical equipment such as an electrically-heated blanket or an over-heated television set.

Flame spread: Once ignited, combustible material burns, generating heat and a rapid rise in the temperature of the surroundings. If sufficient air is available, the fire spreads quickly and soon involves all the combustible material in the room. The decomposition of combustible material by heat may cause the evolution of flammable gases. These may ignite almost explosively, thus causing the fire to spread very rapidly. This phenomenon, termed 'flash-over', is often the important second stage of fire development.

Fully-developed room fire: The temperature developed in the third stage, when the room fire is fully developed, depends on the fire-load in the room and the availability of sufficient oxygen to sustain the fire. Room temperatures of over 1000 °C are possible, and the fire then penetrates ceilings, doors and walls to spread throughout the building. The rate of penetration through the walls, doors and ceilings will depend upon their degree of fire resistance as well as upon the temperature of the fire.

The terminology of fire testing

Earlier standard tests, such as the small-scale, horizontal burning test of ASTM 1692 now discontinued, used descriptive classifications, e.g. 'self-extinguishing', 'difficult to burn', *when tested by the method given in ASTM 1692.* Such classifications led to confusion and misunderstanding. To avoid ambiguity fire test results should always be expressed in a numeric fashion with a complete reference to the test method adopted, e.g. Class 2 by BS 476: Part 7; or Class M3 by NF P 92-503. In discussing the behaviour of a material in a fire or the results of any fire test it is important to use standard terminology. The recommended vocabulary is that of ISO 3261.

Some of the more important definitions are listed in table 11-3.

274

Table 11-3 **Some fire testing terminology**

Term	Definition
Afterflame/afterglow	Flaming/glowing of a material that persists after the removal of the ignition source
Calorific potential	The maximum calorific energy per unit mass which could be released by the complete combustion of a material
Char	The carbonaceous residue from the pyrolysis or incomplete combustion of a material. To form a char.
Combustible	Capable of burning.
Ease of ignition	The ease with which a material can be ignited under specified conditions.
Fire load	The calorific potential of all the combustible material in a space including facings of walls, floors and ceilings.
Fire resistance	The ability of a building element or construction to fulfil for a stated time the required function, integrity, thermal insulation, in a standard fire test.
Fire retardance	Property of a substance or a treatment applied to a material which significantly reduces its combustibility.
Flame spread rate	Rate of propagation of a flame front under specified test conditions.
Flammability	The ability of a material to burn with a flame under specified test conditions.
Mass burning rate	Mass lost per unit time under specified test conditions.
Optical density of smoke	The light absorbence of smoke expressed as the negative logarithm of the light transmittance.
Pyrolysis	The irreversible decomposition of a material by heat without oxidation.
Self ignition	Ignition of a material without the application of heat.
Smouldering	Slow combustion of a material without glowing or flaming but causing a rise in temperature often accompanied by smoke.

Small-scale laboratory tests

Small-scale tests used to compare the ease of ignition and the burning rate of small specimens of material are of varying degrees of severity depending on the orientation of the specimen, the size of the initiating flame and the degree of radiant energy supplied. The simplest tests use horizontal or vertical specimens of material held in a standard holder in a standard cabinet. ISO 3582-1978(E) is a test to determine the horizontal burning characteristics of small specimens subjected to a small flame. The test is based upon the withdrawn ASTM 1692. It is not intended for assessment of the potential fire hazards present in any actual use of materials. The test is carried out in a specified draught-free chamber using specimens with an area of 150 mm ± 1 mm x 50 mm ± 1 mm and thickness not less than 5 mm or greater than 13 mm. One specimen is supported horizontally in a specified holder and a standard wing flame is applied to one end of the specimen for 60 seconds. The burning characteristics of the material are observed and the burning rate, the extinction time and the extent of the specimen burnt are noted. Ten specimens are tested and the mean results reported. The test is intended mainly as a quality assurance test (figure 11-15).

Figure 11-15   ISO 3582: Horizontal burning test.

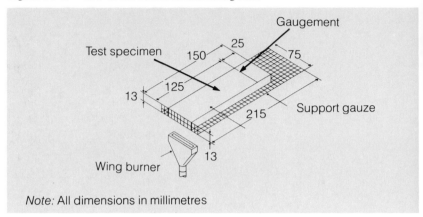

Note: All dimensions in millimetres

The arrangement for another widely used horizontal burning test is shown in figure 11-16. Originally devised to evaluate materials and components used in the interior of cars, trucks and buses, the test was introduced by the US National Highway Safety Administration and became a US Federal Motor Vehicle Safety Standard (FMVSS), number 302, in 1972. This test, or similar ones defined in ISO 3798 and DIN 75200, is now used in every country where cars are manufactured, although the performance requirements vary from country to country. FMVSS 302 requires five specimens, each 356 mm × 100 mm × the thickness of the component with a maximum of 1/2 inch (12.7 mm). A specimen is clamped in a support frame (figure 11-16), a flame from a bunsen burner is applied for 15 seconds and the rate of flame spread over the measured length is recorded. The maximum permitted rate of flame spread is 4 in/min (101.6 mm/min). The maximum permitted rate applies to each of the five samples, i.e. it is not the average rate.

Figure 11-16   US safety standard FMVSS 302: Clamp and burner.

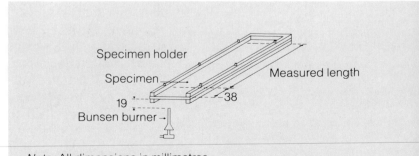

Note: All dimensions in millimetres

Figure 11-17 DIN 4102:
Brandschacht test.

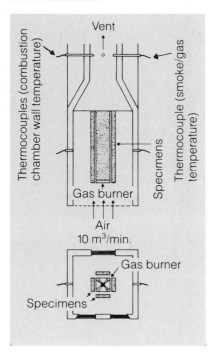

The small horizontal burning rate test of BS 2782 : Method 508A, is used to test thin plastic sheet (3 specimens, 150 mm × 13 mm × 1.5 mm) for use inside buildings and may be applied to polyurethane-coated materials and self-skinning foams. ISO R 1210 is a similar test.

Small-scale vertical burning tests. are more severe than horizontal tests because of the convection heating of the specimens. Small-scale vertical burning tests applicable to polyurethanes include the Underwriter's Laboratory Inc. 94 tests for specifications UL 94V-0, 94V-1 and 94V-2 which are applied for applications in the electrical field such as instrument housings in RIM and RRIM. Vertical tests are also applied to building materials. The best known of such tests are the intermediate scale Class A2 Brandschacht test of DIN 4102 – in which four 1-metre long specimens are suspended vertically to form a chimney over a small ring burner – and the Butler chimney test of ASTM D 3014.

The oxygen index test

The oxygen index test (ASTM D 2813, ISO 3216, BS 2782 : Methods 141A to 141D, NF T 51-071, etc.) employs a test specimen in the form of a square-sectioned bar which is clamped at the base and held vertically in a glass chimney of standard size. A known mixture of nitrogen and oxygen can be metered into the bottom of the chimney and the specimen may be ignited at the top. Tests are run to find the minimum concentration of oxygen required to support flaming combustion of the specimen under the conditions of the test. The sample must burn over at least 50 mm of its length and for at least 180 seconds. A fresh sample must be used for each test. Tests must be repeatable within 0.3% of the critical oxygen index (defined as the percentage of oxygen in the total oxygen/nitrogen mixture that just maintains flaming combustion). The oxygen index may also be measured as a function of ambient temperature, (ISO DIS 4589).

Figure 11-18 Butler Chimney test (ASTM D 3014)

Fire tests for building materials and building structures

In most countries, there are fire protection regulations covering the materials and the methods of construction used in the building industry with the fire testing of building materials being carried out by official testing laboratories. The regulations include fire protection requirements and fire performance test methods. The UK and the USA are exceptional in that testing may be done, and valid test certificates issued, by private test laboratories. Of the EEC countries, Belgium, France and Italy have fire regulations which apply over the whole country. Countries with a federal structure, Austria, West Germany and Switzerland, require the approval of each Länder or Canton before tests are accepted nationally. In the UK there are national regulations as well as local regulations

Table 11-4 Some standard fire-tests

Test	Test reference number
Small scale, laboratory tests Laboratory assessment of horizontal burning of small specimens subjected to a small flame	ISO 3582 (cellular materials); ISO 1210 (plastics); ASTM D 635; DIN 4102 Part 1; BS 4375 or BS 2782 (alcohol cup test); FMVSS 302 (vehicle interiors) = ISO 3795; DIN 75200 and JIS D1201; FAR Part 25 (Aircraft); UL Subject 94
Oxygen index test	ASTM D 2813; ISO 3216; BS 2782 : Part 1; NF T 51-071; NT Fire 013, (Note: No DIN test equivalent); GOST 21793
Laboratory assessment of vertical burning of small strips of material	ASTM D 3014 (Butler chimney); DIN 54332; DIN 4102 (B2 test); DIN 4102 (Brand-schacht); TGL 10685/11; FAR Part 25 (Aircraft); UL Subject 94
Smoke generation from small specimens of material	ASTM E 662 (NBS cabinet) = ISO Dev. test 5659; NT Fire 012; ASTM D 2843 (XP-2 chamber); BS 5111 : Part 1; GOST 12.1.017; DIN 53436 and 53437; NB Explt. X10-702; NF.T51-073
Fire tests for building materials and building structures Combustibility tests for building materials	ISO 1182; DIN 4102 Part 1; BS 476 : Part 4; ASTM E 136; NF P 92-501; GOST 17088 and ST SEV 382-76; NT Fire 001; NEN 3883
Ignitability of building materials	ISO/DP 5657 (ISO cone); BS 476 : Part 5; NT Fire 002
Surface flammability or spread of flame using a radiant heat source	ISO/DP 5658; BS 476 : Part 7; DIN 4102 Part 1; NF P 92-501 to 506; ASTM E 162; GOST 12.1.017; NEN 3883
Flame spread – tunnel tests	ASTM E 84 (Steiner tunnel) = UL Test 723; UK FOC Tests
Fire resistance tests on composite structures	ISO 834; ISO 3008/3009; BS 476 : Part 8 and Part 3 (roofs); ASTM E 119 and E 108; DIN 4102 Part 7 (roofs) and Parts 2 and 3; UL Corner tests; FMC corner test; UK FOC tests; NEN 3884
Calorific value or potential heat emission	ISO 1716; BS 476 : Part 11; DIN 51900 Part 2; NF M 03- GOST 17088
Fire tests for furniture upholstery Ignitability, cigarette/match and smouldering tests	BS 5852 : Part 1/Part 2:1990. ASTM 3453 State of California Bulletins 116-117
Mattress and cushion flammability	BS 6807:1990. US Federal Reg. 137; DOC FF 4-72 UK DOE/PSA Specification for upholstered bedding/institutional mattresses

Key to abbreviations

FMVSS	= Federal Motor Vehicle Safety Standard, (USA).
ISO	= International Organisation for Standardisation.
BS	= British Standards.
ASTM	= American Society for Testing and Materials.
DIN	= Deutsche Institut für Normung.
NT	= Nordtest, (Nordic Council for Denmark, Finland, Iceland, Norway and Sweden).
NF	= Association Française de Normalisation.
NEN	= Nederlandse Norm.
GOST	= USSR State Standards.
UL	= Underwriter's Laboratories Inc., USA.
FAR	= Federal Aviation Regulations, USA.
FOC	= UK Fire Officer's Committee.
TGL	= East German Standards.
DOC	= USA, Department of Commerce.
DOE/PSA	= UK Department of the Environment/Property Services Agency
JIS	= Japanese Institute for Standardisation.

Figure 11-19   Ignitability test (BS 476: Part 5)

applicable in Scotland, Wales, Northern Ireland and in the Greater
London Council area. The EEC Commission is working towards
unified EEC ratings for building materials but it is likely to be many
years before selected ISO tests are officially approved. Unified tests
are applied in Scandinavia (see Nordtest methods, table 11-4).
All countries have a standard test to determine the combustibility of
a material. This is usually a furnace test such as ISO 1182 (825°C) or
DIN 4102 (750°C) although France and Russia use calorimeter tests.
In the former, small cylindrical samples of the material under test
are dropped into the furnace. Depending on the amount of flame
and the temperature rise following the introduction of the sample,
and the weight of material surviving the test, the sample material is
classified combustible or non-combustible/difficult to burn.
Polyurethanes, in common with all other plastics materials and
organic substances in general, are combustible and the relevant tests
are those to determine ignitability, rate of burning and contribution
to the spread and development of fire. The amount of smoke and the
toxicity of the gases produced on burning may need to be assessed
by separate tests. The products of combustion are not a basic
property of any combustible material however, as their composition
will vary very widely depending on the conditions of combustion.

Ignitability tests
ISO test 5657 measures the ignitability of a specimen irradiated at
intensities of 5, 4, 3, 2, and 1 W/cm^2. The specimen is exposed to
the selected level of radiant energy for a maximum of 15 minutes. A
propane flame, 10 mm long, is applied for 1 second every 4 seconds.
The material is classified by the level of radiation required for
ignition. The ignitability test of BS 476 : Part 5 is a simple test in
which a vertical specimen is exposed to a gas flame, 10 mm long,
applied at an angle of 45°. It is used as a sorting test and is not
intended as a means of classification.

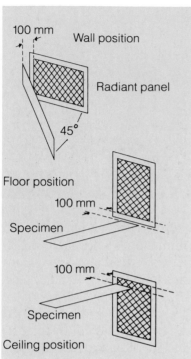

Figure 11-20   Wall, floor and ceiling panel tests to ISO/DP 5658: positioning of specimens for the spread of flame test.

100 mm   Wall position

Radiant panel

45°

Floor position

100 mm

Specimen

100 mm

Specimen

Ceiling position

Note: In each orientation the pilot flame impinges the centre line of the sample at a distance of 20 mm from the end of the specimen adjacent and facing the radiant panel

Ignitability/Surface spread of flame. These tests are an important part of the fire performance classification of materials for use in buildings. ***The ISO/DP 5658*** test uses a vertical, rectangular radiant panel, 300 mm × 450 mm fired by propane and adjustable to a surface temperature of 750°C and a radiation intensity of 6.2 W/cm^2 together with a propane pilot flame of length 80 mm which impinges on the specimen under test. It is intended to orientate the panel for use in testing wall, floor and ceiling panels (figure 11-20).

The specimen to be tested is loaded into the appropriate position and about 50 mm of the pilot flame is applied. The speed of the flame front is measured and the maximum travel recorded.

The BS 476: Part 7, Surface spread of flame test, uses a similar radiant panel and slightly larger test specimens than the ISO test (900 mm × 230 mm × 500 mm max., compared with 800 mm × 155 mm × 40 mm max.) and the specimen is mounted vertically at 90° to the panel compared with 45° in the ISO test. Depending upon the flame spread measured at 1.5 minutes and the final flame spread, materials are divided into four classes as follows:

Table 11-5   Surface spread of flame classification by BS 476

Classification	Flame spread at 1.5 minutes.		Final flame spread.	
	Limit (mm)	Tolerance (for 1 specimen)	Limit (mm)	Tolerance (for 1 specimen)
Class 1	165	25	165	25
Class 2	215	25	455	45
Class 3	265	25	710	75
Class 4	Exceeding Class 3 limits.			

The test radiator is illustrated in Chapter 7, figure 7-40.

In France, the ***NF P 92-501 Epiradiateur*** test is the basic method of classifying the fire performance of rigid materials of any thickness and is also used for flexible materials more than 5 mm thick. This test requires four specimens of 300 mm × 400 mm × thickness. Each sample is mounted in turn at 45° over and parallel to a 500W electric radiator. The specimen is 30 mm distant from the radiator and receives 3W/cm^2 of radiant energy over its lower part. Two butane gas pilot flames are applied, one to the lower and one to the upper surface of the specimen. Depending on the onset, height and duration of flaming combustion, the rise in temperature above the specimen, and on secondary effects such as the level of smoke generation, and the presence of burning droplets, after-glowing and after-flaming, the material is classified in five classes of increasing combustibility from M.0 to M.4.

The specimens to be tested should be in the form to be used in building, e.g. rigid polyurethane and polyisocyanurate foams should be tested as laminates with the rigid or flexible surface layers in place. When edge strips or other methods of enclosing the edges of laminates are used, these should also be in place. Materials that fail to reach classes M.1, M.2 or M.3 are subjected to a laboratory-scale spread of flame test, NF P 92-504. This is a simple horizontal burning test in which a bunsen flame is applied to the free end of a 400 mm × 35 mm specimen for 30 seconds and the rate of burning determined. The material is classified M.4 if it fails to meet the requirements of M.0 to M.3.

Fire resistance tests
Fire resistance tests are applied to elements of buildings such as floors, walls, ceilings, doors, or other components. The fire resistance of a building construction or component is defined as the ability of the structure to fulfil, for a stated period of time, its load-bearing, thermal insulation, or other function, in a standard fire test. Relevant tests include ISO 834 and the amendments ISO 3008 (Doors and shutters), and ISO 3009 (Glazed elements), DIN 834 and DIN 4102 Parts 2 to 7, ASTM E 119 and BS 476 : Part 8. The last mentioned test, which is based upon DIN 834 is typical. One face of the building element is exposed to furnace temperatures which increase with time at a standard rate, corresponding approximately to 556°C, 659°C, 821°C, 925°C, 1029°C, and 1190°C at 5, 10, 30, 60, 120 and 360 minutes. The component under test may fail structurally. Insulation failure is deemed to occur when the temperature of the outer, unexposed surface of the panel or wall increases by more than 140°C above its initial temperature. The use of polyisocyanurate foam insulation as in the core of fire-resistant panels can significantly improve the fire resistance of such panels, especially when the integrity of the char formed from the polyisocyanurate foam is improved by the incorporation of glass-fibre or other heat-resistant reinforcement.

Fire tests for upholstery and other furnishings
As discussed in Chapter 10, the commonest sources of ignition in all fires that result in fatalities are smoking materials, especially smouldering cigarettes or lighted matches. Resistance to ignition by matches and cigarettes is the subject of national regulations in the UK and in the USA. In France and in some Länder in Germany there are regulations governing the fire-performance requirements of furniture in some public buildings. In France there is also a regulation (Arrêt of 4/11/1975 and 1/12/1976) governing the total amount of nitrogen and chlorine contained in synthetic materials of all types that might be liberated as HCN or HC1, but the regulation is no longer enforced.
The main mandatory tests applicable to mattresses and upholstered furniture, which are the principal applications for low density flexible polyurethane foams, are:
US Department of Commerce, DOC FF 4-72, 'Flammability

Standard for Mattresses'. This test determines the resistance of a mattress to ignition by glowing cigarettes placed in a defined positions. The regulation requires that the charring of the mattress attributable to the cigarette does not extend more than 2 inches (5.1 cm) in any direction. The tests are repeated using two sheets. The smouldering cigarette is laid on a sheet stretched over and tucked under the mattress and then covered by a loose sheet (figure 11-21).

Figure 11-21   DOC FF 4-72: Flammability standard for mattresses: Cigarette test positions.

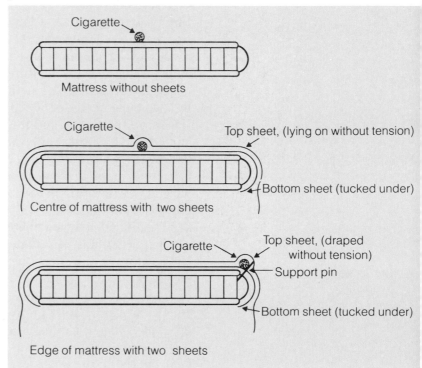

In addition to this Federal regulation, supplementary test procedures are specified in California, Maryland and other States. In the UK, the Upholstered Furniture Regulations, 1980 – introduced in two stages – require that all upholstered furniture meets the requirements of BS 5852 : Part 1. Otherwise, upholstered furniture must carry a label warning of the dangers of ignition caused by cigarettes and matches. The BS 5852 test provides two ignition sources, a freshly-lighted untipped cigarette of length and diameter 68 mm and 8 mm approximately and about 1 g in weight which smoulders at a defined standard rate, and a butane flame produced by a standard burner of 6.5 mm ± 0.1 mm internal diameter fed with 45 ± 2 ml/minute of butane. This ignition source is defined as source 1 and is designed to have a calorific output approximating to that of a burning (wooden) match (BS 5852 : Part 2, defines ignition sources 2 to 7. These are wooden cribs of increasing intensity). The smouldering cigarette test and the butane

282

flame test of BS 5852 : Part 1, are carried out using a test rig on which the cushion filling and cover(s) are assembled to simulate the vertical back and horizontal seat of an upholstered chair (figure 11-22). The smouldering cigarette is placed in position along the junction between the vertical and horizontal test pieces, allowing at least 50 mm from the nearest side edge or from any marks left by any previous test. The combustion is observed and any progressive smouldering or flaming recorded. The assembly fails the test if any progressive smouldering or flaming is observed at any time within one hour of the start of the test. A repeat test is required. The butane gas flame is applied for 20 ± 1 seconds axially along the junction of the vertical and horizontal test pieces. Flames, afterglow, smoking or smouldering that cease within 120 seconds of the removal of the butane flame are disregarded. Flaming or smouldering combustion after that time constitutes failure of the test. Again, a repeat test is specified. Finally, the test rig assembly is dismantled and the interior examined for signs of progressive smouldering. If any are present the assembly fails the test.

Figure 11-22   Vertical section of BS 5852 : Part 1 Test rig.

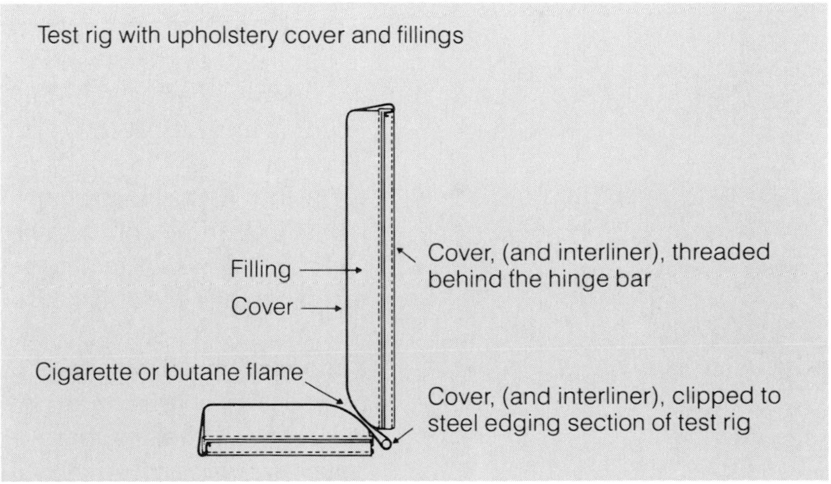

BS 5852 : Part 1 is intended to give guidance on the ignitability of upholstered furniture but, when more specific information is required, the principles of the tests should be applied to finished, actual furniture.

In the UK there are many other flammability tests for upholstered furniture and bedding. The most widely applied are the tests, for upholstery and bedding, specified by the Ministry of Defence (NES 714, 715) and those by the Department of the Environment – Property Services Agency (DOE/PSA), which are tests of varying severity depending on the intended use. The DOE/PSA test, FTS15: Ignitability of a Vandalised Mattress by a Flaming Ignition Source, is briefly described in Chapter 10. This test, in common with all the later flammability tests specified by departments of the

UK government, employs a standard ignition source selected from those of BS 5852 : Part 2. Some of the characteristics of these ignition sources are listed in table 11-6.

Table 11-6 The ignition sources of BS 5852 : Burning characteristics

Ignition source	Theoretical heat of combustion (kJ)	Flame height (mm)	Flame temp (°C)	Rate of burning (max. approx.)	Duration of flaming (seconds)
Cigarette	16	–	–	–	1200 smoulder
Gas flame 1	2	35–40	880	45 ml/min	20
Gas flame 2	12	140–150	890	160 ml/min	40
Gas flame 3	46	185–240	850	350 ml/min	70
Wood crib 4	142	150–245	710	6 g/min	195
Wood crib 5	285	250–335	725	10 g/min	180 approx.
Wood crib 6	1040	250–350	790	15 g/min	360
Wood crib 7	2110	345–490	795	32 g/min	390

The DOE/PSA FR and FTS specifications require fire tests on actual furniture in addition to the use of model test rigs. The control tests of the DOE/PSA require that upholstery and bedding for use in public areas must resist ignition by the cigarette and the wood crib number 5. For public areas of high risk, upholstery and bedding must resist test crib number 7.

There are no ISO flammability test methods for upholstered furniture. A working group (WG 4 of TC 136) is considering the subject. Work in conjunction with the ISO group is being done in Germany. It is expected to be several years before an ISO standard test procedure is published.

Secondary fire tests: smoke and toxic gases

Most deaths in fires are attributable to asphyxiation from the lack of oxygen in the hot air above a fire, or to inhalation of the smoke and toxic gases produced by burning organic materials. An assessment of the amount of smoke produced and its toxicity is an objective of many fire tests. Smoke is often assessed by the optical density of the smoke produced in small-scale tests, although this may not be a good guide to smoke production in a real fire. The measurement of the smoke density obtained on burning building materials is obligatory in most countries in Europe and similar regulations cover the testing of materials for use in road and rail vehicles and in aircraft.

The amount of smoke produced in a fire test may be measured by mechanical, electrical or optical methods. Mechanical methods necessitate filtration and weighing. Electrical methods use ionisation chambers. These are small devices in which the air between two electrodes is partially ionised by a low-energy radiation source. Smoke particles combine with ionised gas particles to reduce the conductivity of the air and, therefore, the current passing between electrodes. Ionisation chambers are most suitable for measuring low levels of smoke and are widely used as smoke detectors in automatic fire alarms. The most usual method of measuring the amount of smoke produced in a fire test is by an optical method using a photocell to measure the attenuation of a beam of light by its passage through the smoke. Some combustibility tests, such as the Steiner Tunnel test of ASTM E 84-81a and the Brandschacht test of DIN 4102, provide for continuous monitoring of the smoke passing through the exhaust from the apparatus. In Germany, such measurements are not part of the DIN 4102 classification of building materials but they are specified in other applications of the Brandschacht apparatus such as the combustibility tests required by the German Federal railways for the materials used in railway carriages. There is increasing interest in the measurement and comparison of both the amount and the rate of smoke production on burning plastics and other synthetic materials. The most important laboratory-scale methods for assessing materials for the amount of smoke produced in a fire are those based upon the use of the NBS and XP2 smoke chambers and of the experimental ISO smoke box. In these tests the smoke produced by burning a small sample is collected in the box enclosing the apparatus and the average optical density of the smoke is measured. The DIN 53436/53437 annular travelling furnace is used in Germany and elsewhere to measure the smoke produced from small samples of material by measuring the density of the smoke present in the continuous flow of gas over the sample in the furnace.

The XP2 Smoke Density Chamber is described in ASTM D 2843-77, 'Laboratory procedure for the Density of Smoke from the burning of decomposition of Plastics'. The test chamber consists of an aluminium box 300 mm × 300 mm × 790 mm and having a full-length glazed door. It is widely used in Europe, both in the form described in ASTM D 2843 and in various modified forms. The original ASTM method specifies the use of specimens 25 mm × 25 mm × 6 mm but, for testing foams, larger specimens such as those of 60 mm × 60 mm × 25 mm used in the Swiss modification of the method, are advantageous. The specimens are mounted horizontally on a wire-grid support and ignited by a single propane gas burner. The test is usually continued for 4 minutes and the smoke level is measured by the attenuation of a beam of light passing horizontally across the chamber. Three specimens are tested in turn and the mean light absorption is plotted against time from ignition of the specimen. The mean maximum smoke density is then obtained from the curve. The original design of the ASTM apparatus includes an 'EXIT' sign, 90 mm × 150 mm in size, fixed centrally to the back of the chamber, 480 mm above the floor, to assist in the visual observation and comparison of smoke levels. The original design of XP2 chamber is used in Germany, but only to assess the smoke level emanating from substantially-inorganic

Figure 11-23   Original XP2 smoke chamber.

building materials that have been rated Class A in the DIN 4102 furnace test. BS 5111 : Part 1 : 1974, 'Determination of Smoke generation characteristics of Cellular Plastics and Cellular Rubber Materials', employs the XP2 chamber to characterise the smoke generation of low density foams (densities up to 130 kg/m^3). The method uses five 25 mm cubes of material. It is useful for quality control purposes but it is of limited value in research and development work or for the comparison of different materials. An ISO version of the XP2 chamber is being developed.

The NBS Smoke Chamber is described in ASTM E 662-83. It is intended for use under laboratory conditions in research and development work but similar apparatus is specified in ISO DP 5659, NF EX 10-702 and in tests specified by French National Railways. The ASTM test method uses six specimens, each 76 mm × 76 mm × 25 mm (or the thickness of the material if less than 25mm). Three specimens are tested under smouldering conditions and three under flaming conditions. A specimen is mounted vertically, parallel to a radiant heater which radiates a flux of 2.2 Btu/s.ft^2 (2.5 W/cm^2) onto the central area of the specimen to obtain the smouldering condition. The source of ignition for flaming conditions is a six-tube, micro-burner positioned 6.4 mm away from, and 6.4 mm above, the bottom edge of the specimen. The micro-burner is fed by propane and is ignited for the duration of the test which is a maximum of 20 minutes, or 3 minutes after the minimum light transmission measured by the photometer. The photometer measures the attenuation of a beam of light transmitted from the bottom to the top of the chamber. The vertical position of the specimens gives poor reproducibility with specimens of materials that melt or drip, and a revised method is being evaluated using a horizontal specimen with a radiant heater which may be adjusted to give five levels of radiation flux.

The ISO smoke box consists of a decomposition chamber connected by an upper and lower duct to a measuring chamber. Twenty-five specimens, five for each level of irradiation in the decomposition chamber, are used. Each specimen, of size 165 mm × 165 × 70 mm or less, is mounted horizontally and partially covered with aluminium foil so that a circular area 140 mm in diameter is exposed to the radiation from an ISO cone radiator (ISO 5657 described earlier). Radiation intensities of 5, 4, 3, 2 and 1 W/cm^2 are used in separate successive tests. The smoke produced from the irradiated specimen (no pilot flame is used) passes into the measuring chamber where it is stirred by a fan and measured by the attenuation of a horizontal beam of light (figure 11-24). The test continues for 15 minutes. The smoke density is measured continuously and the time to flaming ignition is recorded for each level of radiation.

DIN 53436/53437 and NFT 51 073 use a travelling annular furnace (figure 11-25).

Figure 11-24   ISO smoke box: general arrangement.

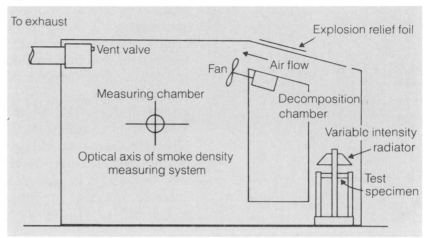

Figure 11-25   DIN 53436/53437: General arrangement of smoke density apparatus.

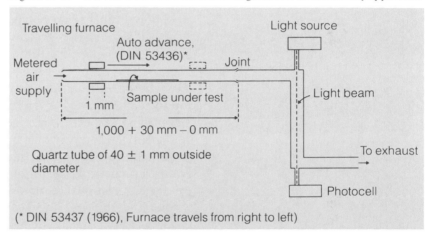

The test specimen of dimensions 270 mm × 5 mm × 2 mm
is placed in a quartz boat and placed horizontally in a quartz tube,
of 40 mm outside diameter and 1000 mm long.

The tube passes through the centre of the annular travelling furnace
which has been preheated to one of the specified test temperatures
(250°C, 300°C, 350°C, 400°C, 450°C, 550°C and, if necessary, 200°C
and 500°C). The air flow over the specimen is adjusted to 300 1/h
and the automatic furnace-advance mechanism actuated. After 15
minutes, the density of the smoke is recorded for 10 minutes. Unless
the first specimen gives a light attenuation of less than 5%, two or
three further specimens are tested at 250°C and the test temperature
is increased to 300°C and then, for subsequent tests, in steps of 50°C
to 550°C. The test is intended for the evaluation of inorganic, Class
A, building materials, but is used for some organic materials and
forms the basis of the DIN 53436 (draft) method for testing the
toxicity of the thermal decomposition products of Class A building
materials that has also been used for testing the decomposition

287

products of rigid polyurethane and polyisocyanurate foam-cored building materials. The NF T 51 073 uses a smaller version of the apparatus.

Larger scale fire tests are carried out on elements of buildings to test both the materials and the method of construction. Tests include surface spread of flame tests and fire resistance tests. The best known large spread of flame test for wall and ceiling panels is that described in ASTM E 84 -81a which was designated the Steiner Tunnel test by Underwriters Laboratories Inc., in honour of a pioneer of fire testing. This test (figure 11-26) requires specimens 24 ft ± 1/2 inch (7.32 m ± 13 mm) long, 20 1/4 ± 3/4 inches (514 ± 19 mm) wide and of the thickness and construction actually used in buildings. The twin gas burners, each 3/4 inch in diameter, are ignited and the rate of flame spread, the temperature rise and the amount of smoke produced are monitored. The distance travelled by the flame front is plotted against time from ignition of the burners. Depending on the area under the curve, materials such as rigid polyurethane and polyisocyanurate foam-cored panels are classified if the flame spread rating is less than 200 and the smoke density is less than 450, (Optical density by the NBS method).

Toxicity of the combustion products. The analysis of combustion products is discussed in ASTM E 800 – 81, 'Guide for the measurement of Gases present or generated during Fires'. Analysis of the combustion products from new materials is carried out in order to assess the likely hazard in a real fire. Laboratory-scale evaluations are of limited value for this purpose because the products of combustion may vary greatly with the scale and the temperature of a fire. The best guidance is obtained by analysing the products of large-scale fire tests. Analysis alone, however, does not give a satisfactory measure of the toxic hazard: this requires the study of the effects of the combustion products on a biological system. The problems of hazard assessment and the development of test methods are described in ISO technical report 6543 (1979). Laboratory-scale comparisons of the toxicity of the decomposition/ combustion products may be made using the methods of DIN 53436, Parts 1 – 3 (1981). This enables small specimens of material to be compared using the travelling annular furnace (figure 11-25). By varying the air flow (100 litres/hour or 300 litres/hour) and the furnace temperature, a comprehensive survey of likely combustion/ pyrolysis products is possible. The biological tests to DIN 53436 require the combustion products to be cooled and diluted with fresh air before exhausted into inhalation chambers containing rats that are exposed to the gases for 30 minutes. The concentrations of common combustion products, CO_2, CO, HCN, HCl from chlorinated polymers and the oxygen level should be monitored during the animal exposure.

Figure 11-26   Steiner tunnel test: general arrangement.

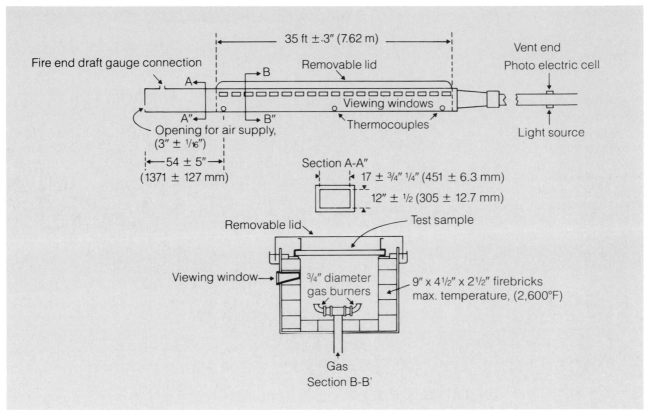

12 Some practical advice

No matter what the topic, once the necessary theory has been learned there follows a period during which it is necessary to gain practical experience in applying the theory to the manufacturing processes derived from it.

The aim of this chapter is to set down a miscellany of facts and advice which those newly engaged in the polyurethanes industry should find helpful.

The versatile world of polyurethanes is one in which new applications are being discovered every day. You can keep up with these developments by reading those journals devoted to the plastics industry in general and polyurethanes in particular. The bibliography at the end of this book will guide your further reading.

The reactive components

The amount of isocyanate required to react with the polyol and any other reactive additives, is calculated in the usual way to obtain the chemically stoichiometric equivalents. This theoretically stoichiometric amount of isocyanate may then be adjusted upwards or downwards, depending on the polyurethane system, the properties required of the polymeric product and known effects such as the scale of manufacture and the ambient conditions.

Isocyanate index (Index number)

The amount of isocyanate used relative to the theoretical equivalent amount, is known as the Isocyanate index or sometimes the Index number.

$$\text{Isocyanate index} = \frac{\text{Actual amount of isocyanate used}}{\text{Theoretical amount of isocyanate required}} \times 100$$

An isocyanate index between 103 and 108 for example, i.e. a 3% to 8% excess of isocyanate, is often used in flexible foam slabstock manufacture.

Calculating the ratio of the components required
The conventional way of calculating the ratio of the components

required for polyurethane manufacture is to calculate the number of parts by weight of the isocyanate that are required to react with 100 parts by weight (pbw) of the polyol and proportionate amounts of additives. The analytical data required for the calculation are the isocyanate value of the isocyanate and the hydroxyl value, residual acid value and water content of the polyol and other reactive additives. This information is given by the supplier with each batch of urethane chemicals. To avoid mistakes it is usual to print this information on each drum of material.

Isocyanate value. The isocyanate value is the weight percentage of reactive – NCO groups:

$$\text{Isocyanate value} = \% - \text{NCO groups} = \frac{42 \times \text{functionality}}{\text{Molecular weight}} \times 100$$

$$\text{or} \quad \frac{4200}{\text{Equivalent weight}}$$

Hydroxyl value (Hydroxyl number) The hydroxyl value (OHV) sometimes called the hydroxyl number of the polyol, is expressed in mg KOH/g of polyol. This convention arises from the method of determining hydroxyl values by acetylation with pyridine and acetic anhydride in which the result is obtained as the difference between two titrations with KOH solution. The OHV may thus be defined as the weight of KOH in milligrams that will neutralise the acetic capable of combining by acetylation with 1 g of the polyol. Polyols are sometimes characterised by quoting the weight percentage hydroxyl groups. This is related to OHV:

$$\% \text{ Hydroxyl groups} = \text{OHV} \times \frac{1.7}{56.1} = \frac{\text{OHV}}{33}$$

$$\text{or Equivalent weight} = \frac{56.1 \times 1000}{\text{OHV}}$$

Acid value The acid value is also expressed as mg KOH/g of polyol and is numerically equal to OHV in isocyanate usage. (see below).

Water content Water reacts with two – NCO groups and the

292

equivalent weight of water is thus:

$$\frac{\text{molecular weight}}{\text{functionality}} = \frac{18}{2} = 9$$

Worked examples
Example 1
How much TDI (48% NCO value), at an isocyanate index of 105, is required to make a foam blown with 3 parts of water per 100 parts of a polyester polyol having an OHV of 51 mg KOH/g and an acid value of 2.0 mg KOH/g?

$$\text{Equivalent weight of TDI} = \frac{42}{48} \times 100 = 87.5$$

$$\text{Equivalent weight of the polyester polyol} = \frac{56100}{(51+2)} = 1058$$

Therefore:
the amount of TDI required to react with 100 parts by weight of polyester polyol is:

$$\frac{100}{1058} \times 87.5 = 8.27 \text{ pbw of TDI}$$

The amount of TDI required to react with 3 pbw of water (the water content of the polyester polyol + the water added for the blowing reaction) is:

$$\frac{3}{9} \times 87.5 = 29.16 \text{ pbw of TDI}$$

The total TDI required to react with 100 pbw of polyol + 3 pbw of water is:

at 100 Index, $29.16 + 8.27 = 37.4$ pbw of TDI

at 105 Index, $37.4 \times \frac{105}{100} = 39.2$ pbw of TDI

Example 2
How much polymeric MDI (31.5% NCO value), at 98 Index, is required to react with 100 pbw of a polyether polyol with an OHV

293

of 28 mg KOH/g, an acid value of 0.01 mg KOH/g and a water content of 0.01% w/w, blended with 4.0 pbw of ethylene glycol chain extending agent and 2.0 pbw of m-phenylene diamine?

$$\text{Equivalent weight of the polymeric MDI} = \frac{42 \times 100}{31.5} = 133.3$$

$$\text{Equivalent weight of the polyether polyol} = \frac{56100}{28} = 2003$$

Equivalent weight of ethylene glycol =

$$\frac{\text{Mol.|Wt}}{\text{Functionality}} = \frac{63.08}{2} = 31.54$$

$$\text{Equivalent weight of } m\text{-phenylene diamine} = \frac{\text{Mol. Wt}}{\text{Functionality}} = \frac{108.14}{2} = 54.07$$

$$\text{Equivalent weight of water} = \frac{18}{2} = 9$$

Therefore:
the polymeric MDI required to react with 100 pbw of the polyol =

$$\frac{100 \times 133.3}{2003} = \underline{6.66 \text{ pbw}}$$

The polymeric MDI required to react with 4 pbw of ethylene glycol =

$$\frac{4 \times 133.3}{31.54} = \underline{16.91 \text{ pbw}}$$

The polymeric MDI required to react with 2 pbw of m-phenylene diamine =

$$\frac{2 \times 133.3}{54.07} = \underline{4.93 \text{ pbw}}$$

The polymeric MDI required to react with 0.1 pbw of water =

$$\frac{0.1 \times 133.3}{9} = \underline{1.48 \text{ pbw}}$$

Total MDI required at 100 Index = 29.98 pbw
at   98 Index = 29.38 pbw

The physical properties of some commercial isocyanates

Toluene diisocyanate (TDI)

Synonyms for toluene diisocyanate are:
– tolylene diisocyanate
– methyl phenylene diisocyanate

Table 12-1 Physical properties of toluene diisocyanate

Property	TDI isomer ratio, (2,4/2,6 ratio)		
	100/0	80/20	65/35
Physical state at room temperature	Liquid		
Colour	Colourless to pale yellow		
Odour	Characteristic, sharp, pungent		
Molecular weight	174.16		
Specific gravity at 15°C	1.222		
at 20°C	1.218		
at 25°C	1.214		
at 30°C	1.210		
Boiling point °C at 760 mm Hg	251	251	251
°C at 10 mm Hg	120	120	120
Freezing point °C	21.4 ± 1	14.0 ± 1	8.5 ± 1
Viscosity (mPa s at 10°C)	4.3	4.3	4.3
(mPa s at 20°C)	3.2	3.2	3.2
(mPa s at 30°C)	2.6	2.6	2.6
Flash point (°C Cleveland open cup)	135		
(°C Pensky-Martin open cup)	127		
Fire point (°C Cleveland open cup)	143		
Specific heat cal/g/°C at 26.5°C	0.374		
cal/g/°C at 25.0°C			0.379
Refractive index at 20°C	1.5684		
Explosion limits Lower (%, v/v at °C)	0.9 / 118°C		
Upper (%, v/v at °C)	9.5 / 150°C		
Vapour density (Air = 1)			
Saturated in air at 25°C	1.000025		
Saturated in air at 50°C	1.000152		
Saturated in air at 100°C	1.003		
Vapour pressure at 25°C (Table 12-2)	3.33 Pa		

Table 12-2 Vapour pressure / temperature for toluene diisocyanate

Temp.	Vapour pressure		Concentration of saturated vapour[1] in equilibrium with air	
°C	Pa	mm Hg	mg/m³	Parts per million (ppm)
−20	0.04	3.0×10^{-4}	3.0	0.4
0	0.33	2.5×10^{-3}	24	3.3
5	0.55	4.1×10^{-3}	38	5.3
10	0.88	6.6×10^{-3}	60	8.4
15	1.40	1.05×10^{-2}	95	13.0
20	2.10	1.6×10^{-2}	140	19.6
25	3.33	2.5×10^{-2}	215	30.1
30	5.05	3.8×10^{-2}	321	45.0
35	7.45	5.6×10^{-2}	466	65.2
40	11.10	8.3×10^{-2}	680	95.1
45	16.0	1.2×10^{-1}	967	135.4
50	22.6	1.7×10^{-1}	1349	188.8
60	45.2	3.4×10^{-1}	2616	366.2
70	86.5	6.5×10^{-1}	4856	679.8
80	157.0	1.18	8565	1199.1
90	293.0	2.2	15530	2174
100	492	3.7	25419	3559
110	798	6.0	40142	5625
120	1277	9.6	62594	8773
130	2022	15.2	96648	13531

[1] Maximum theoretically possible concentration in a closed system at uniform temperature for all ratios of 2,4-TDI : 2,6-TDI.

4,4'-Diisocyanato diphenylenemethane (MDI)

Synonyms for 4,4' diisocyanato diphenylmethane are:
- diphenylmethane diisocyanate
- 4,4' diphenylmethane diisocyanate
- *p,p*-diisocyanato diphenylmethane
- methylene diphenyl diisocyanate
- isocyanic acid, 4,4'-methylene diphenyl ester
- methane, bis(4-isocyanato phenyl)

Table 12-3 **Physical properties of 4,4'-diisocyanato diphenylmethane**

Property	MDI	Polymeric MDI composition
Physical state at room temperature	Solid	Liquid
Colour	White to pale yellow	Fawn to dark brown
Odour	None. Pungent at high temperatures	None to aromatic at room temperature
Molecular weight	250.26	Typically about 450
Specific gravity at 15°C	–	1.244
at 20°C	–	1.242
at 25°C	1.23 (solid)	1.239
at 40°C	–	1.224
at 50°C	1.19 (liquid)	–
Melting point (°)	38 to 43	0 (varies with composition)
Boiling point at 1 mm Hg, °C	170	–
at 760 mm Hg, °C	314	–
Heat of fusion (Cal/g)	24.3	–
Viscosity, (mPa s at 0°C)	–	6000 Typical values
(mPa s at 10°C)	–	1300 Typical values
(mPa s at 20°C)	–	400 Typical values
(mPa s at 25°C)	–	250 Typical values
(mPa s at 50°C)	4.7	–
Flash point (°C Cleveland open cup)	212 to 214	210 to 230
Fire point (°C Cleveland open cup)	–	220 to 250
Specific heat (cal/g/°C)	0.33 at 40°C	0.350 at 15°C 0.354 at 20°C 0.357 at 25°C
Vapour pressure, Pa at 25°C	6×10^{-4}	Less than 6×10^{-4}
Pa at 40°C	2.5×10^{-3}	Less than 2×10^{-3}
Pa at 70°C	1.3×10^{-1}	Less than 1×10^{-1}

1,5-Diisocyanato naphthalene (NDI)

Table 12-4 Physical properties of 1,5-diisocyanato naphthalene

Property	
Physical state at room temperature	Solid, (Flakes) [2]
Colour	White to yellow
Odour	Pungent
Molecular weight	210.19
Melting point ($^\circ$C)	129–131
Boiling point ($^\circ$C at 8 mm Hg) ($^\circ$C at 760 mm Hg)	190 263
Flash point ($^\circ$C Cleveland open cup)	155
Specific heat (cal/g/$^\circ$C)	0.287

Vapour pressure/temperature

Temp.	Vapour pressure		Concentration of saturated vapour in equilibrium with air [1]	
$^\circ$C	Pa	mm Hg	mg/m^3	Parts per million (ppm)
25	0.4	3×10^{-3}	31	3.6
50	1.7	1.3×10^{-2}	124	14.4
75	6.7	5×10^{-2}	4	52
100	22	1.7×10^{-1}	1400	160
125 [3]	**85**	**6.4×10^{-1}**	**4970**	**580**
150	170	1.3	9500	1100
175	670	5.0	34500	4000
200	2300	17	113000	13000

[1] Maximum theoretically possible concentration of vapour in a closed system in equilibrium at a uniform temperature
[2] Dust hazard
[3] Typically processed as a liquid, above this temperature

p-Phenylenediisocyanate (PPDI) and Trans-cyclohexane-1,4-diisocyanate (CHDI)

Table 12-5 **Physical properties of p-phenylenediisocyanate (PPDI) and trans-cyclohexane-1,4-diisocyanate (CHDI)**

Property	PPDI	CHDI
Physical state at room temperture	Solid, (flakes)	
Colour	Colourless	
Odour	Pungent	
Molecular weight	160.13	166.18
NCO content (% by wt., theory)	52.5	50.6
Melting point (°C)	94-95	59-62
Boiling point (°C at 760 mm Hg) (°C at 12 mm Hg) (°C at 25 mm Hg)	260 110-112 –	260 – 143
Density (g/ml)	1.170 at 100°C.	1.116 at 70°C
Vapour pressure/temperature	Similar to TDI	
Reactivity	PPDI is the most reactive aromatic diisocyanate available. [1] CHDI, in common with other aliphatic diisocyanates, is much less reactive than MDI and TDI. It is approximately twice as reactive as IPDI and HDI	

[1] Both PPDI and CHDI are presently available only in quantities up to a few tens of kilos

m-Tetramethylxylene diisocyanate (m-TMXDI) and p-Tetramethylxylene diisocyanate (p-TMXDI)

Table 12-6 Physical properties of m-tetramethylxylene diisocyanate (m-TMXDI) and p-tetramethylxylene diisocyanate (p-TMXDI)

Property	m-TMXDI	p-TMXDI
Physical state at room temperature	Liquid	Solid
Colour	Colourless	White
Molecular weight	244.3	244.3
NCO content (% by wt., theory)	34.4	34.4
Melting point (°C)	−10	72
Boiling point (°C at 3 mm)	150	150
Density (g/ml)	1.05	1.09
Vapour pressure (Pa at 100°C)	67	53
Viscosity (mPa s at 0°C)	25	–
(mPa s at 20°C)	9	–
(mPa s at 80°C)	–	8
Reactivity	Both m- and p-TMXDI are much less reactive than MDI and TDI. m-TMXDI is about as reactive as HDI and p-TMXDI is about as reactive as IPDI.	

Isophorone diisocyanate (IPDI)

Table 12-7 Physical properties of isophorone diisocyanate

Property	Value
Physical state at room temperature	Liquid
Colour	Colourless
Molecular weight	222.3
NCO content (% NCO by wt., theory)	37.8
Boiling point (°C at 10 mm)	153
Density (g/ml at 20°C)	1.062
Vapour pressure (Pa at 20°C)	0.04
(Pa at 50°C	0.93
Viscosity (mPa s at −20°C)	150
(mPa s at −10°C)	78
(mPa s at 0°C)	37
(mPa s at 20°C)	15
Reactivity	Much less reactive than MDI or TDI

A *Synonym* for isophorone diisocyanate is:
– 1-isocyanato-3-isocyanatomethyl-3,5,5-trimethylcyclohexane

Avoiding and correcting faults in polyurethane manufacture

Unlike the processes used for making many other plastics articles, the manufacture of articles from polyurethanes usually involves the simultaneous polymerisation and the shaping of the polymer. Although satisfactory control of this dual process is not difficult, it is important to remember that variations in the process conditions may affect the properties of the polymer as well as the appearance of the product. Faults in polyurethane manufacture are only very rarely due to variations in the basic raw materials, the diisocyanates, polyols and additives, but usually arise from uncontrolled changes in the manufacturing conditions.

The production of most polyurethanes requires the intimate mixing of one or more liquids with a liquid diisocyanate and the deposition of the resulting reaction mixture into a conditioned mould or cavity. The efficiency and reproducibility of the process depends on the effective mixing of precise ratios of the components and the maintenance of the processing temperatures and conditions within defined limits. Almost all faults in polyurethane manufacture result from uncontrolled changes in the principal stages of the production process:

– Excessive variation in the temperature of the component materials or the contamination of the materials with substances that affect the polymerisation such as water
– Variations in the metering pumps or other proportioning devices which give component ratios outside the required range
– Inadequate mixing of the components resulting from mechanical faults, maloperation, etc.
– Variations in the factory temperature or in the temperature of the moulds, cavities or other articles in contact with the reacting polyurethane mixture.

Changes in the temperature, ratio and mixing of the components affect the processing and properties of all polyurethanes but to different extents depending on the types of chemical system and process. For example, a given change in the ratio of the main chemical components, the diisocyanate and the polyol, has a much greater effect upon the properties of a high modulus RIM elastomer than the effect of the same change in ratio on the properties of a highly cross-linked rigid foam. Similarly, a physical change, such as slight air entrainment in the polyol stream may have an insignificant effect upon the structure of a low density flexible foam but will give visible faults on the surface of a self-skinning moulding. Some of the faults which may occur in any polyurethane reactive process are listed in table 12-8. The causes of common faults in some particular processes are listed in table 12-9 (flexible foams) and table 12-10 (rigid foams).

Reactive polyurethane processes

Table 12-8 **Possible faults in reactive polyurethane processes**

Fault	Possible cause	Remedial action(s)
Low hardness (soft rigid foam, abnormally soft flexible foam and elastomers).	Low ratio of isocyanate/polyol.	Check and adjust metering pumps. Check isocyanate filter(s). Check tank levels.
Low density/hardness	Excessive blowing, excess CFM-11 or CH_2Cl_2. Excess water, (possible water contamination). Too little overpacking (rigid foam moulding).	Compare the chemicals from the production machine with the retained samples of the batch and/or chemicals known to be satisfactory in production, by making small foams under standardised laboratory conditions. Adjust blowing agent or overpack as required.
High density/hardness Scorch	Insufficient CFM-11 or CH_2Cl_2 (evaporation loss?), and/or insufficient water. High ratio of isocyanate/polyol. Excessive overpacking, (cold-cure mouldings).	Compare the chemicals from the production machine with the retained samples of the batch and/or chemicals known to be satisfactory in production, by making small foams under standardised laboratory conditions. Adjust blowing agent. Check and adjust metering pumps. Check polyol filter(s) and tank levels. Reduce shot weight.
Heterogeneous foam structure (areas of coarse cells, streaks).	Insufficient mixing. Excessive air entrainment.	Check mixer [1]. Check pumps, valves, filters for leaks, especially on the feed side of the pumps.
Soft or sticky spots.	Surge at the beginning or the end of the dispensed shot.	Check material stream pressures, pressure and velocity balance (low pressure machines) [2]. Check piston operation, hydraulic valves, injection pressures (high pressure machines).

Table 12-8 Continued

High reaction rate (poor flow, premature gelation, high density).	High catalyst level.	Check the reactivity of the chemicals from the machine against standard materials by making foams in the laboratory. Adjust the catalyst level as required.
	High component temperatures, high mould temperatures. Overheating of mechanical mixer.	Check and adjust all temperatures to process standards. Check mixer bearings and coolant.
Low reaction rate (slow gelation, slow cure, loss of blowing gas, high density).	Low catalyst level.	Check the chemicals from the production machine against standard materials by making foams in the laboratory.
	Deactivation of catalyst. Low component temperatures, low mould temperatures. Insufficient mixing.	Check storage conditions. Adjust catalyst level. Check and adjust temperatures. Check mixer, check presence of rotor in mechanical mixer.

[1] A gradual deterioration in the foam structure during a production run may result from a change in mixing conditions caused by the build up of polymeric material in the mixer or the restriction of the mixer outlet by polymerising material. Clean the mixer, and adjust the rotor as necessary.

[2] Surge is a momentary deviation from the metered component ratio at the beginning or the end of the shot. On low pressure machines it is virtually eliminated by adjusting the restrictor valves in the recirculation return lines until the recirculation pressure is equal to the dispensing pressure, i.e. until there is little change in the pressure when the flow of the components is diverted from recirculation to dispense. On some machines it is also possible to vary the diameter of the entry ports into the mixing head to ensure that all chemical components enter the mixing chamber at similar velocities. (If pressure and velocity balancing does not satisfactorily eliminate surge effects it is possible that the mixer is unsuitable for the product or the shot weight in use.) On high pressure machines, surge is usually a symptom of mechanical malfunction such as slow piston operation due to polymer build up, low hydraulic pressure or faulty hydraulic valve operation, or to mis-matched injection pressures.

Flexible polyurethane foam processes

Table 12-9   Possible faults in flexible polyurethane foam processes

Fault	Possible cause	Remedial action(s)
General *Splits in the foam* (associated with a normal cell size and high permeability).	Tin catalyst – too low – deactivated	Check metering rate and adjust. Compare the catalyst from production machine against standard, freshly opened material by making foams in the laboratory.
	Conveyor speed too slow. Polyol and/or diisocyanate temperature too low. Silicone surfactant level too low.	Increase conveyor speed or reduce total output. Check and adjust the temperatures of the material streams. Compare the polyol component from the machine with one containing a standard amount of silicone surfactant, by making foams in the laboratory. Adjust the silicone level as indicated by the test foams.
	Incorrect amine catalyst blend ratio.	Compare the catalyst blend from the machine against standard material by making foams in the laboratory. Reduce the level of the blowing catalyst.
Splits in the foam (associated with an abnormally fine, broken structure).	Excessive air in the mix.	Check material streams for entrained air and remedy leaks.
	Stirrer speed too high.	Reduce stirrer speed in stages observing the effect on the foam structure.
	Exit nozzle from the mixer is too large in diameter.	Reduce mixer nozzle diameter in stages observing the effect on the foam.

Table 12-9 **Continued**

Closed cells or low permeability. Shrinkage (with normal foam structure).	Tin catalyst level too high.	Reduce metering rate of tin catalyst in 5% or 10% steps, observing the effect on the foam structure.
	Temperature of polyol and/or diisocyanate too high.	Adjust the temperatures of the chemical streams to the standard process temperatures.
	Incorrect amine catalyst blend ratio.	Check the catalyst blend from the machine tank against a standard catalyst blend, by making foams in the laboratory. Reduce the level of polymerisation catalyst or increase blowing catalyst as indicated by lab. test.
	Wrong diisocyanate blend on machines with mixed isomer feed systems.	Check metering rates of 80:20 and 65:35 TDI streams and adjust.
Closed cells or low permeability. Shrinkage (with abnormally coarse foam structure).	Stirrer speed too low.	Increase stirrer speed in steps and observe the effect on the foam structure.
	Mixer exit nozzle is of too small a diameter for the throughput.	Increase the mixer nozzle diameter in stages, observing the effect on foam structure.
Pinholes or shot-holes throughout the foam.	Air bubbles in the foam mix. Contamination of the material streams with suspended or insoluble isocyanate-reactive materials. Silicone oil or grease contamination.	Check source of air and correct. Check the chemicals from the machine (possibly after concentrating the contaminants by filtration, centrifugation, etc.), against standard materials, by making foams in the laboratory. Check machine filters. Clean and renew systems as required.
	Dispensing tube from the mixer is too long for the reactivity of the mix or for the output in use.	Clean dispense tube and check the effect on the foam structure. Adjust length or throughput as indicated.

305

Table 12-9 Continued

Slabstock flexible foam		
Scorch or high internal temperature during curing stage. (Block height approx. normal.)	Deficiency of polyol, (low pump speed or pump starvation).	Check pump speed and polyol flow rate(s). Check pump feed pressures throughout a production run.
Reduced block height of low density foam associated with high curing temperatures/ scorch.	Deficiency of CFM-11 or CH_2Cl_2.	Check metering rate of secondary blowing agent, feed temperature pressure and freedom from vapour bubbles and/or air locks.
Reduced block height accompanied by increased TDI vapour release at block cut-off.	Deficiency of water for primary blowing.	Check metering pump setting and delivery under operating pressure. Check feed tank level and the filters and valves in the pump feed line.
Excessive recession of foam after cell opening.	Insufficient tin catalyst. Abnormally fine/open structure. Stirrer speed too high or excess air nucleation. Incorrect amount of tertiary amine catalyst.	Check metering rate. Check catalyst activity by making foams in the laboratory. Check under *Splits* above. Check the catalyst blend by making foams in the laboratory.
Moulded flexible foam cushions *Splitting*	Tin catalyst too low or deactivated. Excessive overpacking of the mould. Material stream temperatures too low.	Check the activity of the catalyst from the machine by making foams in the laboratory and replace or correct as indicated. Check the weight of foam dispensed into the mould and adjust to standard. Check and adjust.
Shrinkage	Tin catalyst level too high. Material stream temperatures too high. Insufficient mixing.	Check flow rate and adjust. Check and adjust. Check and clean mixer.

Table 12-9 Continued

Surface voids	Vent holes blocked. Poor distribution of foam mix in the mould.	Clear. Adjust to standard distribution pattern.
Loose skin and coarse surface texture.	Mould temperature too high at filling station.	Increase cooling air/water flow at the mould conditioning station.
Thick, hard skin	Mould temperature too low at filling station.	Reduce cooling air/water flow at the mould conditioning station.
Cold-cure mouldings *Surface and subsurface voids*	Leaking mould-split line.	Renew mould sealing strip.
High closed cell content difficult to remove by crushing.	High overpack. Low diisocyanate flow level (abnormally low isocyanate index)	Check dispense weight and adjust. Check diisocyanate pump and flow rate. Adjust to standard formulation.
Coarse surface cell structure	Mould temperature too high.	Check mould temperature at filling station and adjust to standard.
Cushion hardness outside specification	Isocyanate index incorrect. Overpack too high or too low.	Check metering ratios and correct to within ± 1% of standard. Check cushion weights and total shot weights. Adjust as required, assuming cushion hardness varies approx. with shot weight.

Rigid polyurethane foam processes

Table 12-10   Possible faults in rigid polyurethane foam processes (using polymeric MDI)
Foam-cored sandwich construction by foam injection
(panels, refrigerators, freezers, boats, etc.)

Fault	Possible cause	Remedial action(s)
Soft foam, shrinkage at low temperatures. a) Normal colour of foam	"Stretched" foam at end of rise resulting from low overpack i.e. low shot weight or low CFM-11 content.	Compare material from production machine with standard material by making foams in the laboratory. Compare rise height/density with standard. Adjust CFM-11 if low. Alternatively increase the weight dispensed into the mould by 5% and check effect on foam appearance.
b) Light coloured foam	Low isocyanate/polyol ratio.	Check isocyanate filter. Check both isocyanate and polyol metering rates under operating pressure. Adjust to recommended ratio.
Brittle dark coloured foam.	Excess isocyanate (incorrect metering).	Check and clean the polyol filter. Check metering of both the isocyanate and the polyol pump under operating pressures. Adjust metering ratio. (NB: replace polyol pump if metering rate falls significantly under pressure).

Table 12-10 **Continued**

Heterogeneous foam structure with streaks, coarse foam structure.	Insufficient mixing or nucleating of the foam mix.	Increase stirrer speed. Clean mixer. Increase the level of mixing/ nucleating air to the mixer.
	Contamination of foam materials with oil, grease, silicone oils, or incompatible foam systems.	Compare the materials from the production machine against materials by making foams in the laboratory from newly opened containers. Check effect of likely contaminants on lab. scale foam making. Clean, flush out machine and refill avoiding contamination.
	"Surge" or gross variation in the ratio of the foam components.	Pressure balance the material flows (see foot-note to table 12-8) or adjust the line pressures on high pressure machines to yield uniform cell structure.
	Contamination with solvent from the cleaning solvent feed fitted to some mixers.	Check and clean the non-return valve on the cleaning solvent feed.
	Contamination with water in the mixing/ nucleating air supplied to the mixer.	Ensure adequate separation of condensed water from the air supply.
Voids or craters in the foam.	Air trapped by the rising foam.	Clean or reposition the vent holes. Change angle of jig or mould to alter the flow pattern in the cavity. Change the position of the injection point.
	CFM-11 condensation on the facing under foaming pressure.	Increase the surface temperature of the facing to a minimum of 30-35°C.
	Insufficient blowing of the foam. Low CFM-11 content (loss due to storage at elevated temperatures in open vessel).	Compare materials from the machine against fresh standard materials by making foams in the laboratory. Add CFM-11 to obtain standard height/ density.

Table 12-10 **Continued**

Voids or craters (continued)	Contaminated facings.	Clean facings. Avoid contamination with oil, grease, silicones, fatty acids, etc.
	Excessively high level of mixing/nucleating air.	Restrict the air flow to the mixer to the minimum required to give an acceptable cell structure. Clean mixer and replace air valves if necessary to stabilise the air flow rate.
Poor adhesion of foam/facing.	Facing temperature too low.	Increase jig or platen temperature or increase the dwell time to reach equilibrium temperature before injecting foam mix.
	Insufficient overpack to maintain good contact with the facing.	Increase fill weight and/or check and increase the CFM-11 level.
	Excess isocyanate in foam mix.	Check metering ratios and adjust to standard ratio.
	Facing material contaminated with grease, oil, or other substance that prevents contact between the rising foam and the facing material.	Clean, and if necessary degrease, the facing surface before foaming.
	Excessive moisture in porous or absorbent facings.	Dry facings before use.
Distortion of cabinet, panel, etc., on removal from the jig or press.	Too short a jig dwell time.	Increase jig dwell time in steps until foam sandwich becomes stable.
	Too high an overpack of the foam.	Reduce shot weight in steps of not more than 5% and check effect on the stability of the foam-filled unit. Alternatively, reduce the CFM-11 content of the foam mix.

Continuous laminate manufacture

Foam faults caused by metering an incorrect ratio of isocyanate/
polyol, by contamination of the foam chemicals or the facing
materials, by inadequate mixing or the wrong level of blowing agent,
are common to all foam processes. Table 12-11 lists the faults which
may occur in the continuous manufacture of rigid foam sandwich
panels using ICI/Viking continuous lamination machines. Many of
the faults described, however, are characteristic of continuous
lamination processes and are not peculiar to a particular type of
machine.

Table 12-11 **Possible faults in rigid polyurethane foam processes
continuous lamination**

Fault	Possible cause	Remedial action(s)
Irregular foam structure with characteristic layering effect.	Distortion of the foam at the nip by the extrusion and rolling of foam which is gelling too slowly.	Increase the catalyst level to increase the rate of gelation. Increase the lay-down gap to allow a longer interval before the nip. Reduce the conveyor speed to a longer interval before the nip. Increase the platen temperature to increase the rate of gelation.
A general reduction in the thickness of the board towards the edges.	Poor foam mix distribution, underfilled edges	Reduce the rate of traverse reversal to increase the amount of foam mix deposited towards the edges.
Foam entering the nip is thicker near the edge, i.e. up to about 10 cm from the edge.	Poor foam distribution, overfilled edges.	Increase the rate of traverse reversal to reduce the amount of foam mix deposited towards the edges.
(Note: Secondary effects of overfilled edges depend on the state of gelation of the foam entering the nip and upon the type of system, polyurethane or polyisocyanurate, for example):		
Smeared, layered, irregular foam structure near the edges of the board.	Overfilled edges	Increase the rate of traverse reversal to reduce the amount of foam mix deposited towards the edges.
Intermittent, longitudinal voids adjacent to the lower facing.		
Inclusions or voids beneath the top surface across the laminate.	Overfilled edges which support the floating platens and prevent good contact of the upper flexible facing/ foam.	

Table 12-11 **Continued**

Lateral waviness of the top surface.	Excess top paper facing concertinas to form ridges. The result of the top belt moving through the conveyor section faster than the bottom belt (uneven belt stretch).	Reduce the pressure on the top belt nip roller or take up the slack in the bottom belt with the adjustable roller.
Random general unevenness of the board.	Bottom facing not lying flat because of uneven tensioning, or insufficient vacuum on the laydown platen. Occasionally caused by a distorted paper facing which has been stored in wet conditions and dried.	Check tracking and tensioning of the bottom facing. If paper distorted, discard and replace.
Loose areas between the top facing and the foam where the foam appears 'hairy' and dark in colour.	The top paper facing is pulled away, after contacting the foam, by shear forces from an incorrectly aligned nip roller.	Check that the nip roller is 0.4 mm below the level of the first top platen, is rotating during manufacture and is precisely parallel to the platens.
Higher density, dark-coloured layer of foam adjacent to the top facing.	Rolling of soft foam at the nip causes liquid foam mix to override the foam layer.	Increase the foam cure at the nip by: – increasing the laydown gap – increasing the catalyst level – increasing the platen temperatures. Reduce the conveyor speed and/or reduce the nip pressure.
Persistent longitudinal concavities in the lower facing.	Foam or other material stuck to the surface of the bottom platen or a badly distorted heated platen.	Clean the laydown platform and heated platen. Replace distorted plates.

Table 12-11 **Continued**

Wavy edges (paper-faced board)	Paper with 'slack' edges.	Store paper under controlled humidity conditions before use to avoid edge expansion which causes 'slack' edges.
	Inconsistent deposition of foam mix at the edges due to variable traverse reversal.	Clean or repair faulty traverse reversal mechanism, as necessary.
Soft, undercured foam edges.	Lower temperature at the edges of the laminate inhibiting the rate of cure.	Depending on the type of machine, improve the insulation to reduce the heat losses from the edge of the platens, or increase the heat input to the edge heaters.
Uneven top surface usually accompanied by loose top facing.	Low tack of foam surface at the nip.	Shorten the laydown gap or increase the conveyor speed. Reduce the catalyst level. Reduce the platen temperature.
	Front floating platens are not floating correctly. Air pockets trapped between the top facing and the foam surface by uneven laydown of the foam.	Check and overhaul platen linkages and operating mechanism. Improve foam laydown by: Cleaning nozzles and adjusting mixing/nucleating air flow. Check and correct uniformity of traverse operation. Check that the nozzle diameter is correct for the foam output in use.
Longitudinal ridges in the laminated boards.	Tension ridges in the lower paper facing.	Reduce the tension on the lower facing and/or feed rolls to obtain flat continuity with the lower platen.

Table 12-11 **Continued**

Low-strength woolly foam layer adjacent to the lower facing.	Excessive moisture driven out of the lower facing on heating reacts with the foam mix to give a blown, soft foam layer.	Remove the water by preheating the lower facing or, preferably, use a facing with a low water vapour permeability such as a polyethylene-coated paper.
Blowholes, visible as small blisters on the surface of the rising foam when using polyethylene coated paper.	Moisture in the paper being expelled through pin holes in the polyethylene coating.	Use better quality, thicker, polyethylene coating, or run with minimum possible platen temperatures.
Paper-faced board becomes curved ex conveyor press.	Greater shrinkage of one facing resulting from more pre-drying of the paper. Difference in the top and bottom belt speeds.	Apply more pre-heat to the facing which becomes concave – or less heat to the opposite facing. Adjust or repair the nip roller pressure regulators. Adjust for equal top and bottom belt tension.
Board becomes curved ex machine or on stacking.	The density gradient across the foam thickness is too high i.e. the density at the bottom face is too low and the density at the top face is too high. Uneven thickness (thick edges, thin centre).	Reduce the platen temperature and/or increase the catalyst level. Incorrectly adjusted or faulty traverse mechanism. Correct setting or repair.

Moulded microcellular elastomers (shoe soling, etc.)

Most of the problems which arise in both the manufacture of microcellular polyurethanes and those which arise later – from variations in their properties – result from using an incorrect ratio of the isocyanate and polyol blend components. Satisfactory product quality and uniformity requires that the blend component ratios are maintained within ± 1% of the standard process ratios. Metering the component flows to this accuracy requires accurate temperature control of the materials. Wide variations in the ratio of the isocyanate/polyol blend will yield wide variations in the abrasion and flexing resistance of the product and may also affect hydrolysis resistance.

The second most common cause of production problems is variation in mixing due to the use of dirty, maladjusted or unsuitable mixers. Routine cleaning and adjustment of the mixing head is essential to satisfactory production on most machines. The frequency of the cleaning and checking operation will depend upon the type of machine and the chemical system being used but four times each eight-hour shift is not excessive.

Table 12-12 **Possible faults in moulded microcellular elastomers**

Fault	Possible cause	Remedial action(s)
Streaks of lighter-coloured material, areas of surface pinholes. Colour variations.	Dirty, maladjusted or unsuitable mixer. Surge at the beginning or the end of the shot.	Ream out the barrel (Desma and VTE machines, check and adjust rotor clearance), clean rotor. Check stream pressures, pressure, balance on recirculation machines. Check for air pockets in the material lines.
	Air bubbles in the materials metered to the mixing head.	Check for bubbles in the feed tanks and in the feed lines to the metering pumps. Check possible air ingress via the rotor-shaft bearing on some machines.
	Mixer exit nozzle too large for the output in use.	Fit smaller diameter nozzle.
	Contamination from previous coloured batch, or leakage from dispensing valve on variable colour machines.	Flush out or change affected lines or clean or renew non-return valves on the pigment lines.

Table 12-12 Continued

Shrinkage of parts ex mould, Reduced hardness, tendency to underfill – less "spew".	Low ratio of isocyanate/polyol blend stream.	Check flow rates under operating pressures. Check isocyanate filter(s). Check the temperatures of the materials in the metering pump feed lines.
Tendency to overfill – excessive "spew", tendency to swell on demould – oversize mouldings. *Increased hardness,* lower flex performance.	High ratio of isocyanate/polyol blend stream	Check flow rates under operating pressures. Check polyol stream filter(s). Check the temperatures of the materials in the metering pump feed lines.
Slow reaction rate, low green strength.	Overheating of polyol blend components, excessive time at melt-out temperature (polyesters). Polyol blend time-expired in storage or stored at high ambient temperatures.	Control heating cycle to avoid overheating. Avoid excessive delay (more than 48 hours) between melting and use of solid polyester polyol systems. If high ambient storage cannot be avoided, use three component systems rather than two component polyester systems.
Tendency to overfill – excess flash with high reaction rate.	Contamination of the polyol blend component with water.	Foam test the materials from the machine in the laboratory against new standard materials. Water contamination gives faster rise and greater height. Check air dryers on the machine tank. Check compressed air water separator and dryer.
Variation in fill, reaction rate, green strength and shrinkage during production	Separation of polyol blend.	Homogenise each drum before use. Ensure constant agitation of machine tanks by suitable slow speed stirrer to prevent separation of blend components.
Slow reaction rate, soft mouldings with poor green strength.	Overheating or prolonged heating of the isocyanate component yielding insoluble MDI dimer.	Avoid prolonged heating of the iso-cyanate component. See supplier's recom-mendations for melting out and storage above the melting point.

316

Self-skinning parts

In addition to the above faults, the main problems with self-skinning part manufacture are the occurrence of surface defects. On most machines the reject rate is sensitive to the shot weight and the moulded part size because the relationship between the mixer dimensions and the volume of material dispensed is an important factor in avoiding air entrainment and surface defects.

Table 12-13 **Possible faults in self-skinning mouldings**

Fault	Possible cause	Remedial action(s)
Streaks of lighter coloured skin, sometimes with surface pin-holes.	Air bubbles in the material(s) metered to the mixing head.	Check materials in the feed lines to the metering pumps for presence of entrained air. Find air leak and eliminate.
	Mixer and/or mixer outlet too large for the shot size.	Try smaller nozzle and/or smaller mixer. Alternatively increase throughput and decrease dispense time to a minimum of 1 second.
	Air entrained from the mixer at the beginning of each dispense cycle.	Alter the mould orientation to shift the fault or 'tide mark' to a hidden face of the moulded part. Use reverse-scroll mixing rotor to expel the air. Use in-mould coating.
Thick, porous skin.	Excess water in the polyol blend, in the third component or in the pigment system.	Compare the chemical components from the machine against standard materials in a laboratory foam test. Check free rise height and reactivity. Correct material having excess water. Trace the source of contamination and eliminate it.

317

Table 12-13 **Continued**

Thick, solid skin.	Mould temperature too low. Excessive overpacking of the mould. Low level of CFM-11 and/or CH_2Cl_2.	Check and correct. Reduce shot weight in steps of 5% and observe the effect. Make foams in the laboratory to compare the chemical components from the machine against standard materials. Adjust for similar free rise height by adding CFM-11 only to the sample from the machine.
Thin skin, easily peeled from the part.	Mould too hot at the filling position.	Reduce the mould temperature – increase the dwell at the cooling station or increase the cooling air flow.
Thin skin, low reactivity, low green strength.	Chain extender level too low. Catalyst level too low. Separation of polyol blend components.	Make laboratory foams to check the chemicals from the machine against standard materials. Low rate of reaction, low rate of gelation and long tack free time confirms formulation error. Unless analytical facilities are available, drain machine tank and refill with fresh delivery after homogenising. Ensure tank agitator is operating.
Surface voids. a) At the top of the mould.	Air trapped above the rising foam. Wrongly positioned or too small a vent. Mould under-fillled.	Add or enlarge the uppermost vent hole or alter the mould filling or rise angle. Increase the shot weight in 5% steps and observe effect on the void.
b) Distributed over the upper face of the part.	Material temperature(s) too low. Air trapped above the rising foam because the mould is not tilted sufficiently.	Check and correct material temperature.

13   New developments

Part 1 The elimination of chlorofluorocarbons (CFCs) from polyurethane foams[a]

Introduction

The chlorofluorocarbon CFC-11 has been the most important physical blowing agent for polyurethane foams. Other non-reactive blowing agents are methylene chloride – widely used in flexible foams – and CFC-12 which has been used in much smaller quantities for the nucleation and frothing of rigid foams. Most rigid polyurethane foam, however – as described in Chapter 7 – has been blown by the vaporisation of CFC-11. This is because CFC-11 has the ideal physical and chemical properties for the application, (table 13-1). The low reactivity and high stability of CFCs, however, ensures that they remain unchanged in the atmosphere for many years. In 1974, two American scientists, Roland and Molina, included CFCs in their model atmosphere – along with some 60 trace chemical compounds. They assumed that CFCs reach the stratosphere where they are broken down by high energy UV radiation to yield chlorine atoms which catalyse the conversion of ozone to oxygen, thus adding to the depletion of the ozone layer by polluting gases. The depletion or thinning of the ozone layer reduces its effectiveness as a shield against harmful UV radiation. In the spring of 1985, Farman of the British Antarctic Survey reported a hole in the ozone layer over the South Pole. This direct measurement, together with the accumulating evidence of the role of chlorine in ozone depletion resulted in an international agreement, *The Montreal Protocol on Substances that Deplete the Ozone Layer* which became operative on 1 January, 1989. This agreement required that the production and consumption of a range of fully halogenated CFCs, including CFC-11 and CFC-12, should not exceed the amounts used in 1986, reducing to 50% of these levels by 1 July, 1998. The protocol included provision for revision, and the revision process is expected to be completed during 1990. The revised protocol will call for faster reductions in the production and use of CFCs, probably aiming at a complete ban by the year 2000. ICI supports these objectives with major investments in the research, product development and the manufacturing plant required for the reduction and elimination of CFCs from polyurethane products. The many references to the value of CFC (CFM) blowing in Chapters 1 to 11 show the magnitude of the problem.

(a) In the main part of the book CFCs are referred to as CFMs. For Chapter 13 the abbreviation CFC is used as that is the current convention.

The use of CFCs may be avoided by substituting alternative blowing agents together with modifying the foam-making chemicals and/or the foam reaction, depending upon the type of foam and its application. These options are often severely restricted by the process and application requirements. The replacement of CFC-11 in rigid polyurethane foam presents the greatest challenge. This is because – unlike flexible foam where any CFC-11 used is released into the atmosphere during the foam production process – 90 to 95% of the CFC-11 used in rigid polyurethane foam manufacture is retained within the closed cells of the foam which is virtually impermeable to CFC-11 vapour. It is because the cells contain CFC-11 vapour that rigid polyurethane foam has the lowest thermal conductivity of any widely available insulant (see table 7-10, p.165). The very low thermal conductivity of rigid polyurethane foam makes an important contribution to energy conservation and to reducing the production of carbon dioxide from the burning of fossil fuels, which contributes to global warming via the greenhouse effect.

Table 13-1 **Why CFC-11 is the ideal blowing agent for rigid polyurethane foam**

1 – CFC-11 is an easily handled liquid at room temperature, (b.p.23.8°C)

2 – CFC-11 has a high blowing efficiency – up to 95% in rigid foam

3 – CFC-11 vapour has a very low thermal conductivity of 0.0064 W/mK (0°C), resulting in rigid polyurethane foam being the most effective insulant

4 – CFC-11 is soluble in most polyols and commercial isocyanates – it is very easy to handle

5 – CFC-11 has a low solubility in the polyurethane polymer and has no significant plasticisation effects

6 – CFC-11 has a negligible rate of diffusion out of the foam

7 – CFC-11 is non-flammable

8 – CFC-11 has a low order of toxicity

9 – CFC-11 is chemically stable and has no corrosive effect upon polyurethane processing machinery

10 – CFC-11 is a cost effective material – manufactured by a single-step reaction

Alternative non-reactive blowing agents

(Chapter 3, pp.46-48)

The most promising alternatives to CFC-11 are the hydrochloro-fluorocarbons HFA-123 and HFA-141b, but these HFAs are unlikely to be available for commercial use until 1993 at the earliest because production must await the completion of long-term toxicity testing. Compared with CFC-11, both HFA-123 and HFA-141b have a greatly reduced potential for ozone depletion. HFA-123 is non-flammable, but the use of HFA-141b alone may require flame-proof foam-making plant. Alternatively, a non-flammable mixture of HFA-141b and HFA-123 may provide a satisfactory replacement for CFC-11. HFA-22 has been used to replace CFC-12 in froth-foam processes

Table 13-2 (Table 3-8 p. 48 continued) Non-reactive blowing agents for polyurethanes

Blowing agent	Trifluoro-dichloro-ethane (HFA-123)	Dichloro-monofluoro-ethane (HFA-141b)	Difluoro-monochloro-methane (HFA-22)	Trichloro-fluoro-methane (CFC-11)
Molecular weight	153	117	85.5	137.5
Boiling point at 1 atmos., °C	28.7	32.0	−40.8	23.8
Occupational exposure limit (existing in many countries)	Not available – under test		500-1,000 ppm	1,000 ppm
Density of liquid at 25°C, g/ml	1.46	1.23	1.21	1.48
Lifetime in the atmosphere, years	1.9	8.7	19	75
Relative ozone depletion potenial, (R.O.D.P.)	0.02	0.09	0.05	1.0
Relative halocarbon global warming depletion, (H.G.W.P.)	0.02	0.09	0.27	1.0
Thermal conductivity, (vapour at 60°C, W/mK)	0.0127	0.0125	0.0130	0.0093
Commercial supply, (earliest likely year)	1993	1993	Now	Now

and has been recommended at low levels to replace CFC-11 in conventional foam, but having a low boiling point (-41°C) it is more difficult to handle. The replacement of CFC-12 with HFA-134a (CH_2FCF_3) has also been suggested but, unlike HFA-22, it is not yet commercially available.

Rigid polyurethane foams
(Chapter 7, pp. 127-130)

The very low thermal conductivity of CFC-11 vapour together with its very low diffusion coefficient through the cell walls results in the extremely low thermal conductivity of both freshly made and aged rigid polyurethane foam, (table 7-1, first two entries, p. 128). CFC-11 also has additional, useful effects in the highly cross-linked and economical foam systems based upon high functionality polyols. The CFC reduces the viscosity of the polyol blend – simplifying machine metering and mixing – and during the foam reaction, the vaporisation of the CFC absorbs excess exothermic heat. The technical options to reduce or replace the use of CFCs in rigid polyurethane foams include increased water-blowing – combined with formulation changes – and substitution by HFA-123 or HFA-141b. Each of these options will tend to result in foams of higher thermal conductivity compared with similar foams blown with CFC-11.

In the long-term, the availability of HFA-123 or HFA-141b is believed to be essential for the manufacture of rigid polyurethane foam having a very low thermal conductivity throughout a long service life. Until HFAs become commercially available, the most readily available method of reducing the use of CFC-11 in rigid polyurethane foam manufacture is via dual blowing with carbon dioxide, formed by the reaction of additional water with an equivalent, additional amount of isocyanate.

Reducing CFC-11 by additional blowing with carbon dioxide

Rigid polyurethane foam systems have been developed over some thirty years for use in specific applications. Simply reducing the CFC-11 content and increasing the water content to increase the level of carbon dioxide in a foam system has major effects upon that system including a higher viscosity of the polyol component, higher foam reaction temperatures giving longer demould times, and higher initial thermal conductivity.

The development of rigid polyurethane foam systems blown by a mixture of carbon dioxide and a minimum amount of CFC-11 has therefore required much chemical and systems application development to eliminate these defects. For example, increasing the ratio of carbon dioxide/CFC blowing requires a reduction in the hydroxyl value of the polyol to compensate for the changed polyol percentage and to maintain a constant isocyanate usage.

As the proportion of carbon dioxide to CFC-11 is increased the resulting increase in thermal conductivity is not linear (table 13-3). Foam systems have been developed in which up to 50% of the CFC-11 is replaced by an equimolar amount of carbon dioxide[b] with minimum loss of insulation efficiency. In these systems, the increase in the thermal conductivity as a result of heat transfer through the gas phase has been almost entirely compensated by improvements in the cellular structure of the foam, thereby reducing heat transfer by radiation. (The effects of cell structure on heat transmission are discussed in Chapter 7, pp. 163-167). These new foam systems, containing about 5% to 6% rather than about 12% to 13% of CFC-11, were introduced to domestic appliance makers in Europe during 1989. They are now widely used in refrigerator and freezer manufacture where they have replaced wholly CFC-blown systems without significant modification of the production machinery, the rate of production or the performance of the product. Similar foam technology has been developed to allow a 50% reduction in the CFC-11 used to make foam-cored panels and laminates for the construction industry.

It is more difficult to use high levels of carbon dioxide in polyisocyanurate rigid (PIR) foams (polyurethane-modified polyisocyanurate foams, p. 131) because the high reaction temperatures, together with the polyureas resulting from the increased water/isocyanate reaction, tend to give unacceptably brittle isocyanurate foams with poor adhesion to substrates. In PIR foam systems substitution of up to 25% of the CFC-11 is possible before adverse effects are observed. The further replacement of CFC-11 in PIR foam requires the substitution of an alternative, non-reactive blowing agent such as HFA-141b.

Wholly 'water-blown' foams

As the amount of CFC-11 is further reduced and the proportion of 'water blowing' is increased so that carbon dioxide becomes the major gas within the cells of the foam, there is a marked increase in thermal conductivity on ageing (table 13-3) – unless the carbon dioxide-blown foam is enclosed within an impermeable vapour barrier. This is because, in the unprotected foam, the carbon dioxide will diffuse out of the cells of the foam quite rapidly, (table 13-4) to be replaced by the inward diffusion of air. Foams

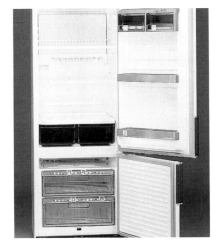

Figure 13-1 An HFA blowing agent was used for the rigid foam insulation in this trial refrigerator / freezer.

(b) Although the gas within the cells of the freshly made foam contains approximately equal numbers of carbon dioxide and CFC-11 molecules, the relatively large molecules of CFC restrict the thermal conductivity of the mixture. The change in thermal conductivity as the foam ages depends largely upon the permeability of the polyurethane polymer of the cell walls. The precise rate at which the constituents of a gas mixture will pass through the cell walls depends mainly upon the size of the gas molecule, its solubility in the polymer and its partial pressure within the cells of the foam. CFC-11 and HFAs such as HFA-123 and HFA-141b have extremely low rates of permeation whereas carbon dioxide and air have a relatively high rate. The rate at which carbon dioxide is lost from uncovered foam is greater than the rate at which air can permeate into the foam. This is the cause of the shrinkage observed in unsuitably formulated, carbon dioxide blown, foams.

Table 13-3 Typical effects of reducing CFC-11 and increasing carbon dioxide on the thermal conductivity and ageing of rigid polyurethane foam

Reduction in CFC-11 (%)	Conductivity of the foam,	
	Initial value (W/mK at 10°C)	Aged value[c] (W/mK at 10°C)
0	0.017	0.022
20	0.018	0.023
50	0.0185	0.0245
75	0.021	0.027
100	0.022	0.032

(c) 12-18 months

Table 13-4 Approximate effective diffusion coefficients of gases from within the cells of rigid polyurethane foam at 70°C

Gas	Effective diffusion coefficient D_{eff} ($10^{-13}m^2s^{-1}$)
CFC-11	3.2
HFA-123	3.5
HFA-141b	5.0
CO_2	greater than 7000
Air	greater than 700

Figure 13-2   Thermal conductivity of rigid polyurethane foam at 0°C versus time (long term ageing) showing effect of foam cell size.

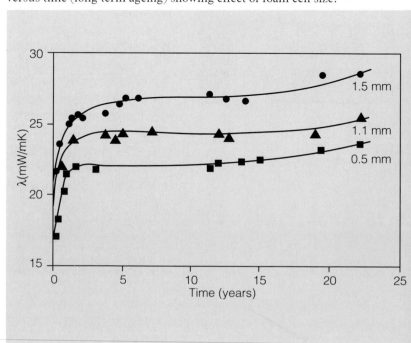

blown only by carbon dioxide have been developed and systems are available which are suitable for both domestic appliance and building construction applications. By optimising the polyols systems to obtain an efficient cell structure, 100% carbon dioxide blown foams have been obtained with initial conductivities around 0.023 W/mK at 10°C, some 20% to 25% higher than the thermal conductivities of foams blown mainly with CFC-11 or with equimolar mixtures of CFC-11 and carbon dioxide. In unprotected foams, a very rapid increase in thermal conductivity occurs as the carbon dioxide – which diffuses rapidly out of the cells – is replaced by the slower inward diffusion of air. There are, however, many applications (Chapter 7, pp. 136-150) where these ageing effects are irrelevant because the foam is the core of a composite structure having impervious outer skins. The change from CFC-11 blown foam to a wholly carbon dioxide blown system required an increase in the thickness of the insulating foam to maintain the minimum heat loss. Barrier layers and sandwich constructions intended to prevent the loss of carbon dioxide from insulating foam should – because of the high permeability of rigid polyurethane foam to carbon dioxide – be subjected to long-term ageing tests, preferably under service conditions. However, for most applications, the complete replacement of CFC-11 by carbon dioxide is not an attractive option.

Replacing CFC-11 with HFA-123 or HFA-141b

Intense development work to eliminate completely the use of CFC-11 in rigid polyurethane foams is continuing. It has been shown that HFA-123 or HFA-141b can completely replace the 5% to 6% of CFC-11 in dual blown foam systems based on CFC-11 and carbon dioxide. After reformulation to optimise the foam properties, foam made using HFA-123 or HFA-141b has a thermal conductivity similar to that obtained using CFC-11. Similar results were obtained after accelerated ageing at 70°C.

Both HFA-123 and 141b have a greater solvent action than CFC-11 and fridge and freezer liners made from ABS and high impact polystyrene may require improved protective coatings. As poly-olefins are resistant to these HFAs, one option is the co-extrusion of polyethylene and polystyrene to yield a thermoformable sheet suitable for refrigerator liners.

The assessment of foams containing HFAs 123 and 141b is continuing in order to characterise their performance fully before they become commercially available. All the work done so far suggests that the performance of foams blown with HFA-123 or 141b will not differ significantly from those blown with CFC-11.

Low density flexible foams
(Chapter 4, pp. 56-57; p. 62)

Slabstock foams

As described in Chapter 4, the primary blowing agent for low density flexible foams is the carbon dioxide produced by the reaction of isocyanates with water (water blowing). The range of flexible foam density and hardness obtained solely by water blowing is limited by the amount of heat and by the polyureas produced by the reaction of water with isocyanates, (pp. 55-57). The use of

Figure 13-3   In the long term, flexible slabstock foam will probably be mainly water blown.

secondary, non-reactive blowing agents – CFC-11 and/or methylene chloride – extends the range of flexible foams to lower levels of hardness and density, (figure 4-25). In the absence of non-reactive blowing agents, which absorb some of the exothermic heat of the foaming reaction, the maximum amount of water that may be used to blow unfilled slabstock foams is about 4.6 to 4.8 parts of water to 100 parts of polyether polyol – depending on the block size and the formulation – corresponding to a minimum foam density of 21 to 22 kg/m^3. Lower density foams and a wide range of soft foams normally cannot be made from conventional flexible foam polyols by water-blowing alone (Chapter 4, pp. 55-57, pp. 80-81 and figure 4-25). Methods of reducing CFC usage in the short term include the recovery and recycling of the CFC-11 released from the foam and the substitution of methylene chloride for CFC-11. Long term options are listed below although it is expected that in the future most flexible slabstock foam will be water blown.

Methods of extending the range of water-blown slabstock foams[c]

1) Modify the water-blown foam process to yield softer foams
2) Non-reactive blowing using methylene chloride
3) Increase chemical blowing by using the isocyanate/formic acid reaction, ('AB' Process) to replace water-blowing partially
4) Auxiliary blowing with CFC-11 but recycling the maximum amount of the CFC-11 used
5) Use alternative blowing agents, (HFA-123 or HFA-141b)

(c) All are used to some extent, except 5

Long-term options to avoid using CFCs

i) Process modifications to produce softer, water-blown foams

As the density of conventional flexible foam slabstock is reduced by increasing the amount of water blowing, the increase in the polyurea content of the hard block increases the stiffness of the foam polymer. The foam becomes less resilient with a lower SAG or support factor, i.e. the relationship between the 25% and the 65% indentation hardness (Chapter 11, p. 261). Several formulation modifications are used, alone or in combination, to make softer, low density, water-blown foams. These modifications include the use of hydrophilic polyoxypropylene/polyoxyethylene polyols containing 15 to 20% of polyoxyethylene groups. High resilience (HR) foams with satisfactory physical properties at densities down to 24 kg/m^3 are based upon improved modified polyols having a partially grafted, stable, dispersed phase consisting of a vinyl copolymer with a high ratio of polystyrene/polyacrylonitrile, together with improved surfactants.

ii) Chemical blowing using the isocyanate/formic acid reaction, ('AB' Process)

The 'AB' process uses the reaction between formic acid and isocyanate:

Figure 13-4

$$2(HCOOH) + R.(NCO)_2 \rightarrow H_2N.R.NH_2 + CO_2 + CO$$
$$+$$
$$R.(NCO)_2 \rightarrow \text{Substituted urea}$$

Compared with the reaction with water, the reaction of diisocyanates with formic acid produces twice the volume of gas, an equal weight of polyureas and a similar amount of heat. The 'AB' process, therefore, yields lower density, softer foams than similar wholly water-blown formulations. To obtain foam densities below about 22 kg/m^3, partial blowing by the 'AB' process requires a proportionate reduction in water blowing to avoid an increase in the total exothermic heat produced. The minimum foam density obtainable by the 'AB' process appears to be in the range from 17 to about 20 kg/m depending on the scale of the production process. Using the reaction of formic acid with isocyanate to replace secondary blowing with CFC-11 is environmentally beneficial. Replacing 100 parts by weight of CFC-11 by the 'AB' process yields about 25 parts of an equimolecular mixture of carbon dioxide and carbon monoxide. The disadvantages of the 'AB' process arise from the hazards involved in handling formic acid and from the presence of carbon monoxide in the foam-making plant. Formic acid (TLV 5ppm) is corrosive and a respiratory irritant. Carbon monoxide is highly toxic (Chapter 10, p. 245) with a TLV of 50 ppm. The release of carbon monoxide from the foam is slow and a high proportion is emitted in the block cut-off and storage areas. Additional ventilation is therefore required in these areas compared with conventional production, and carbon monoxide monitoring equipment is essential. Formic acid requires acid-resistant storage tanks, pumps, lines and metering equipment.

iii) Auxiliary blowing with HFAs

Although HFA-123 and HFA-141b are unlikely to be available in commercial quantities earlier than about 1993, both HFAs have been compared with CFC-11 as blowing agents for low density (14 kg/m^3) soft slabstock foam. Both HFA-123 and HFA-141b gave satisfactory foams. However, even when these HFAs become commercially available, economic and environmental restraints are likely to prevent their widespread use as blowing agents for flexible foam.

Short-term options to reduce the release of CFC-11

i) Recycling CFC-11

All the CFC-11 used as an auxiliary blowing agent for flexible foam is released to the atmosphere after foaming. The CFC-11 may be recovered from the ventilation exhaust by absorption on activated carbon from which it is desorbed and recovered by steam distillation. However, only a part of the total CFC-11 used to blow the foam can be easily recovered because a substantial proportion is released slowly in the block storage and curing area. The peak emissions of CFC-11 occur when the foam cells open just after maximum rise, when the side paper or polyethylene sheet is stripped, and when the blocks are cut off at the end of the foam machine. It has been estimated that about 40% of the CFC-11 used is emitted at these three points on the foam production machine, a further 20% during the transfer of the blocks from the machine to the curing area and the remaining 40% during the 10 to 12 hours in the curing zone. A 40% recovery of the CFC-11 used to blow flexible foam is obtained by the use of activated carbon filters in the exhaust vents from a conventional, slabstock foam, production machine. Recovery of the major part released during curing is difficult because of the high dilution with venting air. Over 80% recovery of CFC-11 is possible by the use of a fully enclosed machine such as the 'Vertifoam' machine (figure 4-12), followed by an enclosed curing chamber in which the release of the remaining CFC-11 is accelerated by a stream of air at a controlled temperature.

Recycling processes are unlikely to prevent the release of significant amounts of CFC-11 into the atmosphere and are therefore most unlikely to affect agreements to phase out the use of CFCs. Recycling may become important to the permitted use of methylene chloride and the economic use of the alternative HFAs.

ii) Non-reactive, auxiliary blowing using methylene chloride

Methylene chloride is a technically feasible substitute for CFC-11 in the production of most grades of conventional flexible slabstock foam with the exception of some low density supersoft grades. It is widely used in the USA because of its low cost compared with CFC-11. Methylene chloride is more toxic than CFC-11 and is listed in many countries as "An industrial substance suspect of carcinogenic potential in man". Having greater solubility in the foam polymer and a higher boiling point than CFC-11, methylene chloride is retained longer in the hot, freshly made foam and a high proportion is emitted in the block cut-off and foam curing areas. Additional ventilation and monitoring equipment is required compared with that needed when using CFC-11. Methylene chloride has a short lifetime in the atmosphere and its ozone-depletion potential is extremely low, nevertheless venting to atmosphere may well be increasingly restricted in the future by legislation controlling solvent emissions and limiting the use of chlorinated solvents.

Moulded foams *(Chapter 4, pp. 71-76)*
Auxiliary blowing with CFC-11 reduces density and hardness with the minimum cost and therefore has been widely used in hot-cure, cold-cure and HR foam moulding, especially to make soft foams for the squabs (backs) of vehicle seats. The possible options for slabstock foam manufacture – CFC recovery, formic acid/isocyanate blowing and the use of methylene chloride – are of little practical value in moulded cushion manufacture. Relatively higher levels of water-blowing are possible however, because of the greater rate of heat loss in the moulding process compared with large blocks of foam. It is expected that the use of CFC-11 in hot-cure moulding will be replaced by increased water-blowing, and reformulation using

lower functionality, more hydrophilic polyols. The net result will be some increase in the density and cost of soft, hot-moulded cushions.

Cold-cure and high resilience (HR) foam moulding

The volume of hot-cure moulding production is diminishing as an increasing proportion of automotive and high quality furniture seating cushions are moulded from HR foam and, especially, from cold-cure moulding systems based on MDI variants. The several advantages of MDI-based, cold-cure, flexible foam moulding are described in Chapter 4, p. 76. All-MDI-based foam systems allow the adjustment of the load-bearing properties of the foam over a wide range by simply changing the isocyanate index. This property of MDI-based foam systems has been developed to permit the manufacture of very soft moulded cushions solely by water-blowing at very low isocyanate indices, (Chapter 12, p. 291).
The moulded foams produced by this technology have a very soft, latex-like, handle, and good recovery from compression.
When using high water levels to produce blowing, some adjustments to the polyol structure are necessary to retain the desired properties in the foam.

Self-skinning, flexible and semi-rigid foams
(Chapter 5, pp. 88-93)

In integral skin foam the traditional technology relies on the blowing agent condensing at the colder mould surface to produce a solid elastomeric skin. A variety of options currently being evaluated includes substitution with methylene chloride, HFA-123 or HFA-141b. These would not require any change in technology. Other possibilities, such as HFA-22, which have been demonstrated to be potential solutions, would require machinery modifications to accommodate the very low boiling point of the blowing agent. For some applications, increasing the water level to give a fully carbon dioxide-blown foam also appears to be a viable option.

Part 2 Developments in polyurethane processes and materials

Flexible slabstock or blockfoam
(Chapter 4, pp. 61-71)

The average density of flexible slabstock foams, (table 4-1, p. 56), is tending to increase. This trend towards higher densities follows the reduction in CFC usage, together with a demand for higher quality upholstery, but the main reason, especially in the USA and UK, is the production of foams to meet tougher flammability regulations (Chapter 11, p. 281-284).

Combustion modified high resilience (CMHR) foams
The greatest change is in the UK, where the effect of the *Furniture and Furnishing (Fire) (Safety) Regulations 1988* is to prohibit the sale of domestic furniture, beds, mattresses, cushions and pillows containing any flexible polyurethane except "Combustion Modified" (CM) foams. The permitted CM foams must pass the test specified in BS 5852: Part 2: 1990, with not more than 60 g loss in weight – in addition to the cigarette test of Part 1. The test method of BS 6807: 1990 is used for mattresses. The government regulations state that the foam test specimen of specified size is covered with the specified flame retardant, polyester-fibre cover and subjected to ignition source 5, (table 11-6, p. 284). As a result of this legislation, most of the slabstock foam now produced in the UK is CM foam in the density range from about 24 to 40 kg/m^3, with a mean average of about 32 kg/m^3.

The most popular slabstock foam for furniture cushioning is CM high resilience (CMHR) foam. CMHR foam is defined in BS 3379: 1990 as a high resilience foam that meets the flammability requirements of the 1988 regulations. A typical CMHR foam includes melamine or intumescent graphite, which, on exposure to a flame, forms a heat resistant char.

MDI-based cold-cure moulding systems *(Chapter 4, pp. 73-78)*
Most – about two-thirds – of moulded flexible foam is now cold-cure foam, with an increasing proportion based on MDI variants rather than TDI. The biggest market for moulded flexible foam is in the vehicle industry, mostly as seat cushions for automobiles and public transport; only a little over 10% is used in furniture and bedding. Worldwide competition in car manufacture has created a demand for higher quality interior trim and especially for high quality seating at the minimum cost. This demand has increased the use of

'dual hardness' or 'multi-hardness' HR foams, particularly all-MDI-based foam systems for moulded seat cushions, headrests and sound-absorbing trim. It has also led to the adoption of 'foam-in-fabric' moulding for high volume production in both Europe and America.

Foam-in-fabric moulding *(Chapter 4, pp. 78-79)*

Automotive seating was traditionally produced by the cut-and-sew method, in which flexible polyurethane foam parts were wrapped in pre-cut and pre-sewn fabric covers. Deterioration could be rapid as pockets formed after frequent use and it was difficult to achieve a consistently perfect shape from seat to seat. Foam-in-fabric (FIF) technology has been evaluated for some years. Known as pour-in-place foaming in the USA, it was already applied in the 1970s for the production of high-class office furniture and other simply-shaped items such as tractor seats. This involved injection of foam onto PVC sheets.

At that time, various manufacturers tried FIF technology in the production of car seats but there were severe problems, largely owing to the inappropriateness of the materials then available. Since then, close cooperation between car companies, mould makers, equipment manufacturers and chemical suppliers has resulted in solutions which have made FIF seating an increasingly attractive alternative.

All-MDI-based flexible polyurethane foam was the key to success and the advantages of these systems led to a rapid penetration of the industry. There are five clear benefits which have led to this success.

– Environmental
MDI has a lower vapour pressure than TDI – as can be seen in table 13-5 – and all-MDI-based foams allow the use of water-based release agents.

– Increased productivity
Because of the fast cycle times which can be achieved, more units can be produced from the same number of moulds. All-MDI-based foams' fast curing is especially valuable in thin-layer sections, where problems are often encountered with TDI foam – particularly if immediate trimming is needed. Just-in-time requirements are more easily met as all-MDI-based systems can be used for seat assembly within six hours.

Figure 13-5   Foam-in-fabric production techniques open new perspectives for styling.

Table 13-5   **Vapour pressure of MDI and TDI, mm Hg x 10^{-3}**

	Temperature (°C)		
	25	40	70
MDI, mm Hg × 10^{-3}	0.005	0.03	1.4
TDI, mm Hg × 10^{-3}	25.0	83.0	650.0

– Dual hardness

Automotive seating requires the bolster, or side of the seat, to be harder than the mid section. With all-MDI-based systems, index latitudes are much wider. Hardness is easily adjusted by varying the index and does not require the use of expensive reinforced polyols. Table 13-6 shows a typical example.

– More durable foams

Durability parameters of all-MDI-based systems are generally superior to those made with TDI-based foam. Typical MDI properties can be seen in table 13-7. Increased durability is especially noticeable in humid ageing tests – shown in figure 13-8 – making MDI foams very suitable for use in hot, humid climates.

Figure 13-6 Foam-in-fabric technology simplifies manufacture and increases design flexibility.

Table 13-6 **Hardness variation (65% CFD) (Chrysler mini-van front seat)**

| | Cushion | | Backrest | |
	Bolster	Centre	Bolster	Centre
Density (kg/m³)	50	50	50	50
Compression force deflection 65% (N)	28	15	23	18

Table 13-7 **Durability and comfort parameters for all-MDI-based foam**

Typical properties	
50% compression set (%)	6.5
50% humid ageing compression set (%)	
Hardness loss	13.5
Thickness loss	0.5
Flex fatigue (%)	
Hardness loss	24.0
Thickness loss	2.0

– Consistent open-cell structure

In FIF production using barrier technology (see the next section), vacuum crushing is applied to the high proportion of seats which include metal inserts. Dual-hardness foam manufacture using index change only to achieve hardness variation, requires consistent open-cell foam throughout, especially where impermeable film barriers are used. Systems are available which produce, consistently, open-cell foams that can be processed through vacuum crushers. This allows dual-hardness foam manufacture without foam shrinkage or surface-deformation problems.

Current FIF technology

FIF production was initially developed using barrier technology in which textile composites were laminated with a non-penetrable film backing. Increasing interest in FIF led to the development of a

Figure 13-7   The Ford Fiesta uses foam-in-fabric seating produced by the barrier method.

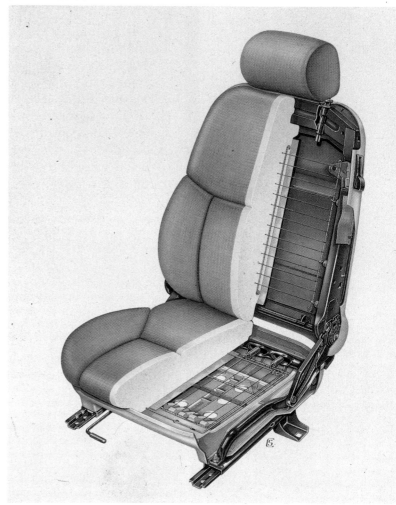

Figure 13-8   Influence of humidity on hardness.

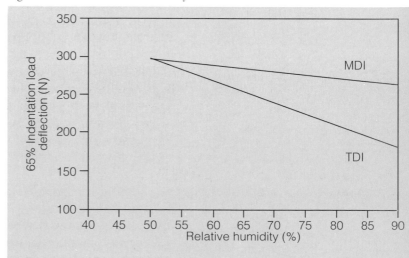

Figure 13-9 Ford selected foam-in-fabric seating after long-term investigations of alternatives.

second technology, known as non-barrier, where no impermeable barrier layer is used.

The barrier method. A blank of laminated (fabric plus film) seat cover is placed in the mould and drawn by vacuum to the bottom of the cavity. The polyurethane system is injected into the open mould, allowing precise distribution and dual hardness. The mould is then closed. The finished part is demoulded and vacuum crushed.

The non-barrier method. Textile composites with no film barrier are used – typically involving a 2- to 5-mm thick polyester or polyether slabstock layer backing. The absence of a barrier means the chemical system must not penetrate the unprotected layer, so fast-reacting all-MDI-based systems are preferred. The most appropriate are cream-foam systems. In the production of a head restraint for example, these are injected into the pre-sewn textile composite pocket and the fast reaction time means they can be demoulded in 60 to 90 seconds. No crushing is needed.
An alternative non-barrier technology uses textile composites with special slabstock layer backing and no protective film laminate, the slabstock layer acting as a barrier. Foam is injected into the pre-sewn fabric cover – kept in place using specially developed techniques.

Reaction injection moulding (RIM) developments
(Chapter 6, pp. 114-126)

Developments in the chemical systems, and in the control of the RIM process machinery, have made RIM polyurethane products – from soft, energy-absorbing elastomers to rigid structural materials

– increasingly competitive with both injection-moulded thermoplastics and fibre-reinforced thermosets for the manufacture of large parts.

Mat-moulding RIM (MMRIM) or structural RIM (SRIM)

provides an economical alternative to glass-fibre reinforced polyester (GRP) construction, yielding a strong, light-weight moulded part. Large parts – such as a pick-up truck cargo lid – may be moulded using epoxy-lined GRP moulds.

The MMRIM process consists of injecting a low viscosity polyurethane or polyisocyanurate reaction mixture into a closed mould containing pre-cut, glass-mat reinforcement. Both fast-reacting SRIM and slower systems – to allow the production of large parts with low output machines – are used. The faster SRIM systems are injected by high output machines in order to allow the free-flowing reaction mixture to displace the air from the glass mat before the onset of polymerisation causes the viscosity to rise. Demoulding in less than 1 minute after injection is possible and rigid reinforced products are made having densities in the range from about 700 to 1,000 kg/m^3.

Polyurea RIM systems

Polyurea RIM systems have been developed which can satisfy the automotive market need for damage-resistant front and rear ends, and the longer-term potential market for the replacement of metal body panels.

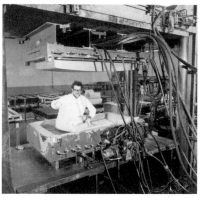

Figure 13-10 Glass-mat-reinforced rigid foam was used for the cargo lid of this pick up truck.

Polyurea RIM materials have become established for the production of large and complex automotive parts by fulfilling necessary technical and thermal-resistance requirements – and by enabling dependable and consistent production of parts to be achieved at the rates demanded by the industry.

The benefits of these materials include high-temperature stability, low moisture absorption, good chemical and solvent resistance, dimensional stability and excellent paintability. Collaborative efforts by chemical suppliers, automotive manufacturers and moulders have demonstrated that RIM polyureas exhibit short mould residence times and excellent release properties. Thus they allow large-scale automated production with minimal cycle times – with the addition of part demoulding using robots, the overall cycle time can be reduced to less than one minute.

When developing polyurea RIM formulations for the replacement of metal parts, the essential properties include:

- short cycle times
- excellent processability
- high flexural moduli (*ca.* 2000 MPa)
- no sag or distortion at elevated temperatures (up to 200°C)
- coefficients of thermal expansion as low as possible (*ca.* 50 mm/mm x 10^{-6}/°C)
- acceptable impact strengths, especially at sub-ambient temperatures (*ca.* −30°C)
- paintable surfaces having a Class A finish

Conventional polyurea elastomers are block copolymers characterised by a dual-phase behaviour, derived from a combination of a rubbery softblock phase and an amorphous hardblock phase. The softblock imparts toughness, and is largely responsible for the impact resistance and tensile elongation performance. These soft phases have sub-ambient glass transition temperatures (Tg_s), and consist primarily of relatively high molecular weight polyether chains.

Hardblocks control such properties as flexural modulus, thermal resistance and dimensional stability. The hard phases have high glass transition temperatures (Tg_h), and are formed by the reaction of aromatic diamines such as diethyl toluene diamine (DETDA) with MDI. Polyurea hardblocks contain high concentrations of aryl urea linkages having high levels of hydrogen bonding.

Polyurea RIM materials may be formulated to cover a very broad stiffness range, from low modulus elastomers for front- and rear-end applications to high modulus plastics for body panels. Variations are obtained by changing the relative softblock-to-hardblock concentrations, depending upon the intended end use.

The presence of primary aliphatic amino groups makes conventional

RIM polyureas extremely reactive and fast curing, without the need for catalysts. Isocyanate-amine reaction profiles will differ as a result of factors such as system hardblock content, softblock functionality, softblock equivalent weight, choice of isocyanate variant, component temperatures and mould temperatures.

As a class of materials, they may exhibit poor flowability – resulting in mould filling problems. These inherent system characteristics must be offset by the use of large, high output RIM dispensing machines, capable of very high injection rates. High reactivities and high injection rates in turn create a need for higher mould-clamping pressures.

The problems are accentuated when moulding large and/or complex automotive components. Even relatively small RIM parts may also present difficulties – since mixing quality and ratio accuracy (hence, part-to-part reproducibility) needs to be ensured, while using very short injection times. Thus, conventional polyurea RIM materials must be processed using specific dispensing rate, mould size and mould clamping conditions. If the optimum combination of these parameters is not used, unacceptably-high scrap or reject rates will result.

Previously- and currently-available systems based on conventional polyurea technology have, to varying degrees, compromised physical and mechanical performance, for example, stiffness and high-temperature performance, in order to obtain better processing characteristics.

A new generation of novel polyurea RIM intermediates, which can be formulated into systems offering excellent processability, without sacrificing physical properties and thermal stability has recently been developed. Systems that exhibit all the technical benefits of conventional polyurea RIM materials, and at the same time are easily processed into front- and rear-end components and body panels, are thus available.

These use MDI-based isocyanate prepolymer technology, in combination with imino-functional isocyanate-reactive components.

Polyurea prepolymer technology

The novel MDI prepolymer variants overcome the brittleness and cracking problems found with conventional high-stiffness polyureas using normal low mould-temperature conditions.

They have been demonstrated to permit successful processing of polyurea RIM materials – specifically including high modulus body panel systems with hardblock contents up to as much as 70% – to produce components with excellent cure and green strength over a very broad stiffness range. This improved processing is achieved while maintaining the physical properties and high temperature stability of polyureas.

Figure 13-12   Comparative reaction profiles.

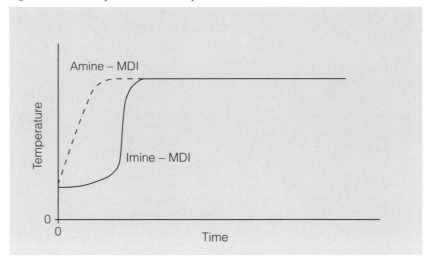

Figure 13-13   Flexural modulus versus hardblock content, unfilled polyureas.

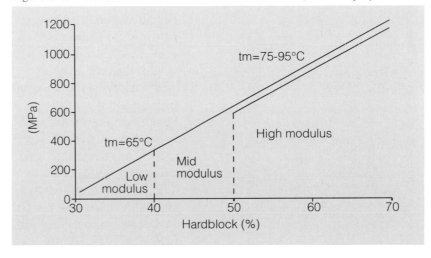

Isocyanate-reactive components: imines

The high reactivity and flow problems of conventional polyureas
have been overcome through the use of imino-functional isocyanate-
reactive components.

By using imines, mould filling may be accomplished more simply,
owing to the reaction profiles of these unique polyurea systems.
Typically, these exhibit an initial induction period, during which
time the reaction between the isocyanate and the imino-functional
species is delayed or retarded. The reaction then accelerates
noticeably, showing the same type of "snap cure" as conventional
polyurea RIM systems. Thus the use of imino-functional reactants
has imparted improved system flowability without increasing
system cycle time. The different reaction profiles are illustrated in
figure 13-12.

341

Figure 13-14   Flexural modulus versus hardblock content, filled polyureas.

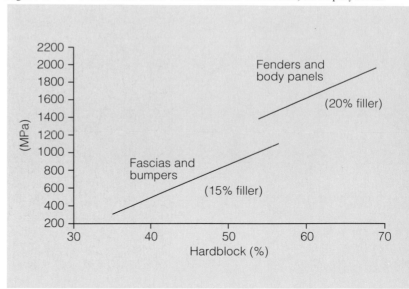

Broad application range

In figure 13-13, the flexural modulus range for unfilled polyurea RIM systems is shown as a function of hardblock content. The modulus range presented extends from approximately 50 MPa – for a 30% hardblock system – to 1200 MPa for a 70% hardblock system. Low and mid-modulus systems up to 600 MPa, or around 50% hardblock, can be moulded with excellent green strength at a mould temperature of 65°C. High modulus systems from 600 MPa (50 to 70% hardblock) require higher mould temperatures, but not exceeding 95°C.

The flexural modulus range for filled polyurea RIM systems is shown in figure 13-14, which indicates that the new ICI polyurea isocyanate prepolymer and imine technologies offer a wide scope of formulation flexibility.

Figure 13-15   Polyurea RIM bumper for the US version of the Peugeot 405.

Figure 13-16   Polyurea RIM systems have been developed for complete soft front or rear ends.

Reinforced polyurea RIM systems may be formulated for automotive applications ranging from soft front- and rear-ends to mid-modulus self-supported bumpers and high modulus body panels. All of the materials exhibit the physical and mechanical properties and technical benefits of polyureas, but their processability is greatly improved over conventional polyurea RIM systems.

Bumpers

Table 13-8 shows the physical property range for polyurea RIM bumpers filled with 15% of hammer-milled fibre. Polyurea RIM bumpers are mid-modulus formulations processable at 65 to 70°C mould temperature – with excellent flow characteristics, dimensional stability and physical and mechanical properties.

Table 13-8   **Polyurea RIM bumpers – physical property range** [d]

Filler type	Milled fibre
Filler level (%)	15
Thickness (mm)	3
Density (kg/m³)	1100-1150
Flexural modulus (MPa)	500-1000
Tensile strength (MPa)	20-25
Elongation (%)	125-75
Heat sag (mm) (160°C, 60 min, 150 mm O/H)	3-5
CLTE (x 10⁻⁶/°C) (−30°C to +70°C)	55-70
Falling dart impact (J)	
+20°C	60-70
−20°C	40-50
Heat distortion temp. (°C at 1.82 MPa)	130-160

[d] Postcured at 160°C for 30 min.

Figure 13-17   The PSA group has carried out manufacturing trials for polyurea RIM body panels.

Initial practical experience with polyureas under production conditions has shown the advantages of these materials for moulding a variety of external automotive parts such as bumpers, spoilers, and soft front and rear ends. The high temperature resistance of these systems – up to 160°C – is again substantially better than that of earlier glycol- and amine-extended systems. Partial on-line finishing of bumpers in conventional automotive paint lines therefore becomes feasible, allowing accurate colour matching with the car body.

Polyureas for body panels

Body panel systems, filled with 20% hammer-milled glass fibre perform, as shown in table 13-9. Moduli up to 2000 MPa combine with excellent mechanical properties, thermal stability and processability at mould temperatures below 95°C. Their high stiffness and thermal stability are important attributes, making these materials on-line paintable.

Table 13-9   Polyurea RIM body panels – physical property range [e]

Filler type	Milled fibre
Filler level (%)	20
Thickness (mm)	3
Density (kg/m^3)	1100-1150
Flexural modulus (MPa)	1400-2000
Tensile strength (MPa)	25-35
Elongation (%)	75-50
Heat sag (mm) (160°C, 60 min, 150 mm O/H)	2-4
CLTE (x 10^{-6}/°C) parallel (−30 to +70°C)	40-60
CLTE (x 10^{-6}/°C) perpendicular (−30 to +70°C)	90-110
Falling dart impact (J)	
+20°C	35-45
−20°C	10-20
Demould shrinkage (%)	0.5-0.6
Properties after exposure (160°C, 60 min)	
Expansion (%)	0.50-1.25
Shrinkage (%)	0.15-0.30

[e] Postcured at 160°C for 30 min.

Organic fillers are generally incorporated – to increase stiffness and improve dimensional integrity through lower CLTE, water absorption and heat sag. Filled polyurea body panel systems exhibit the same stiffness as the major competitive thermoplastics in an unfilled state – while the presence of filler leads to lower shrinkage and better dimensional stability.

Water absorption. Plastic materials exhibiting low moisture absorption allow the automotive engineer freedom to design narrow-tolerance body panels. Imine-based polyurea vertical panel materials – unfilled, 20% flake glass-filled, and 21% mica-filled – were subjected to 1- and 10-day water immersion, in accordance with ASTM D570-81. An amine-extended polyurethane-urea system with equivalent modulus was measured as a control sample. Since moisture absorption has been found to decrease with increasing density, sample densities were held constant (at production specifications). The results of the water absorption test are shown in figure 13-18. Growth of the polyurea was reduced by approximately 40% compared with the control system. Flake glass and mica fillers further reduced water absorption.

Figure 13-18   Percent change in length for vertical panel (materials subjected to 1- and 10-day room temperature water immersion).

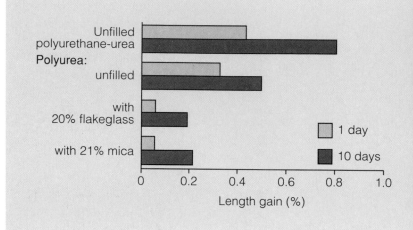

Substrate stability at 200°C. Paint-curing cycles for wings or door panels placed on a vehicle made primarily of steel, which will be subjected to electrodeposited coating systems for corrosion protection, include limited material exposure at 200°C. Laboratory thermal stability testing and visual observation offer alternative means of evaluating plastic part performance in a paint-cure oven. The chemical bonds formed by the reaction of imines with MDI are stable at elevated temperatures. The dynamic mechanical spectra (DMTA) of polyurea RIM body panel systems show a high and flat rubbery plateau region in the storage modulus (E') versus temperature curves. These plateaux generally extend beyond +200°C before substrate softening causes the profiles to decrease.
Polyurea body panels also typically exhibit low temperature glass transitions at around −45 to −50°C in the tangent delta versus temperature damping curves. The mechanical spectrum of a polyurea body panel system is shown in figure 13-19. Again, it can be seen that high stiffness and thermal stability make these materials eminently suitable for on-line painting in tandem with steel components.
Thermogravimetric analysis (TGA) of polyurea RIM body panels based on imines showed no significant weight losses, even at temperatures up to +250°C. In contrast, amine-extended polyurethane-urea samples of equivalent modulus lost eight percent by weight over the same temperature range (+25 to +250°C).
A typical body panel TGA profile is given in figure 13-20.
Differential scanning calorimetry (DSC) thermogram profiles of imine-based polyurea body panels usually reveal high temperature glass transitions at around +220 to +225°C, corresponding to the

346

Figure 13-19   Dynamic mechanical spectrum of a polyurea body panel system.

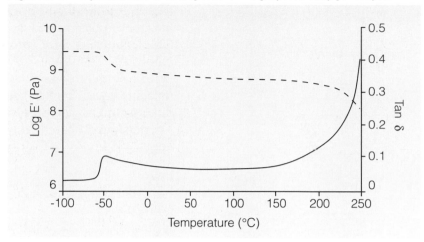

amorphous aromatic hardblocks. DSC thermogram analyses have also shown that the onset of polymer decomposition occurs at around +250°C, exactly where rapid weight loss occurs in the TGA profiles. A typical body panel DSC thermogram is presented in figure 13-21.

Figure 13-20   Thermogravimetric analysis profile of a polyurea body panel system.

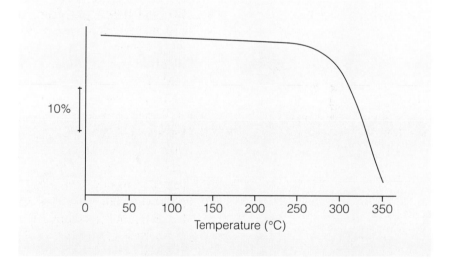

On-line painting. Polyurea RIM elastomer mouldings can be painted with approved automotive primers and top coats. Either solvent vapour cleaning or water (acid or alkaline) washing procedures can be used for degreasing prior to painting. The superior high temperature properties of polyurea low- and mid-modulus bumper systems make it possible to paint partially on-line, ensuring good colour matching with surrounding metal components.

Polyurea RIM body panel materials are a viable plastic alternative to steel for the automotive industry. They not only exhibit a good balance of physical and mechancial properties, but also show excellent potential for automated production and total on-line painting.

With continuous co-operation between chemical suppliers, machine manufacturers, paint suppliers, car producers and part manufacturers, polyurea RIM body panel materials will surely play a major role in shaping the future of exterior automotive components.

Figure 13-21   Differential scanning calorimetric thermogram of a polyurea body panel system.

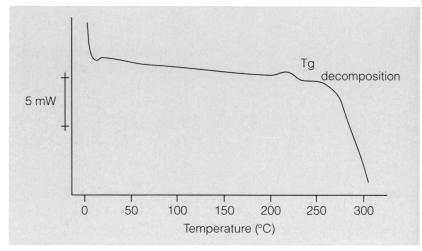

Bibliography

1. Saunders, J.H. and Frisch, K.C., *Polyurethanes: Chemistry and Technology,* Vols. I and II, Interscience, (John Wiley & Sons), New York, (1962, 1964).
2. Buist, J.M., and Gudgeon, H., (Eds.) *Advances in Polyurethane Technology,* Maclaren and Sons, London, (1968).
3. Twitchett, H.J., *Chem. Soc. Rev.,* 3, (2), 209, (1974).
4. Goodman, I. and Rhys, J.A., *Polyesters,* Vol. 1, The Plastics Institute, London, (1965).
4. Frisch, K.C. and Reegen, S.L., (Eds.) *Advances in Urethane Science and Technology,* Vols. 1 to 7, Technomic Pub. Co., Westport, Conn., USA. (1971-1979).
5. Wright, P., and Cummings, A.P.C., *Solid Polyurethane Elastomers,* Maclaren and Sons, London, (1969).
6. David, D.J., and Staley, H.B., *Analytical Chemistry of the Polyurethanes,* Interscience Publ., (John Wiley & Sons), New York, (1969).
7. Ferrigno, T.H., *Rigid Plastic Foams,* Reinhold Publ. Co., New York, (1963).
8. Allport, D.C., and Janes, W.H., (Eds.) *Block Copolymers,* Applied Science Publishers, London, (1973).
9. Buist, J.M. (Ed.) *Developments in Polyurethane – 1,* Applied Science Publishers, London, (1978).
10. Becker, W.E., (Ed.) *Reaction Injection Moulding,* Van Nostrand Reinhold Co., New York, (1979).
11. Woods, G., *Flexible Polyurethane Foams: Chemistry and Technology,* Applied Science Publishers, London, (1982).
12. Goodman, I., (Ed.) *Developments in Block Copolymers,* Applied Science Publishers, London, (1982).
13. Hilyard, N.C., (Ed.) *Mechanics of Cellular Plastics,* Applied Science Publishers, London (1982).
14. Brown, R.P., (Ed.) *Handbook of Plastics Test Methods,* Second Edition, George Godwin Ltd., in association with the Plastics and Rubber Institute, London, (1981).
15. Troitzsch. J., (Ed.) *International Plastics Flammability Handbook,* Carl Hanser Verlag, Munich; Macmillan, London. (1983), (German & English versions).

16. Oertel, G., (Ed.) *Polyurethane Handbook*, Hanser Verlag, Munich, (1985).
17. Sweeney, F.M., *Reaction Injection Moulding Machinery and Processes*, Marcel Dekker, New York, (1987).
18. Sparrow, D.J., and Thorpe D., "Polyols for Polyurethane Production", Chapter 9 in *Telechelic Polymers, Vol. 2, Synthesis and Applications*, Edited by Goethals, E. J., CRC Press, (1988)
19. Macosko, C.W., *RIM Fundamentals of Reaction Injection Moulding*, Carl Hanser Verlag, Munich; Oxford University Press, New York, (1989).

Glossary

Acid value
A measure of the residual acidity of a substance, e.g. a polyester. Measured in mg KOH/g needed to neutralise the acidity.

AGS acids
An abbreviation for a mixture of adipic, glutaric and succinic acids, which is obtained from nylon waste.

Addition
Polymerisation: a polymerisation reaction where no by-products are produced. (cf **condensation** polymerisation).

Alicyclic
An aliphatic (substance) containing non-aromatic ring structures, of carbon atoms.

Aliphatic
An organic (substance) containing straight – or branched – chain arrangements of carbon atoms.

Aromatic
An organic (substance) usually containing one or more benzene ring structures. In some substances, typical "aromatic" chemical properties are also conferred by other ring structures.

Block copolymer
A polymer that contains linear polymer sequences of two or more different polymer molecules in one polymer chain. The polymer blocks are sometimes called *segments*, particularly when they are small in size (cf **graft** copolymer).

Blowing agent
A substance incorporated in a mixture for the purpose of producing a foam. For polyurethanes, this is usually either carbon dioxide generated from the isocyanate / water reaction, or is a low boiling organic liquid volatilised by the heat of the polyurethane-forming reactions.

Branching
Lateral extension points in a polymer chain.

Cast, casting
The filling of essentially open moulds with liquid mixtures of polyurethane reactants or liquid monomers and allowing them to polymerise in situ. A **CPU** is a cast polyurethane.

Catalyst
A substance which accelerates the reactions of chemicals without itself being consumed.

Cell structure
Open cells   – interconnecting cells in a foam.
Closed cells   – cells enclosed by continuous membranes and struts.
Cell opening – the breaking of intercellular membranes. This can occur naturally during foam rise, or later by foam crushing.
Cell count   – the number of cells per linear centimetre.

Chain extenders
Substances which lengthen the main chain of a polymer molecule by

causing end-to-end attachments. Chain extenders usually have a *functionality* of 2.

Component	When applied to polyurethane manufacture, a component is one of at least two reactive materials which are mixed together to form a polymer.
Condensation polymerisation	A polymerisation reaction in which two or more molecules combine with the production of by-products, such as water.
Covalent	A chemical bond between two separate atoms in a molecule involving the sharing of electrons by two atomic nuclei.
Critical velocity	The rate of flow of a liquid at which smooth laminar flow changes to turbulent flow.
Crosslinking	Formation of bridges between different polymer chains.
Crystallinity	State of repeated, and ordered, molecular organisation.
Cure	Refers to the completeness of chemical reaction processes.
Domain	A region in a polymer matrix which contains molecular segments of similar types.
Elastomer	A synthetic rubber-like material capable of rapid, reversible extension.
Filler	An inert material added to a polyurethane reaction mixture. Fillers are usually solid, particulate materials such as glass, silica or barytes.
Flame retardant	An added substance which inhibits the initiation and/or spread of flame.
Fluorocarbon	A substance containing fluorine and carbon with other chemical elements. The most common fluorocarbons used in polyurethanes are the volatile liquids CFM-11 (CCl_3F) and CFM-12 (CCl_2F_2). CFM is an abbreviation of chlorofluoromethane. CFC is an abbreviation for chlorofluorocarbon.
Functionality	The number of reactive groups in a chemical molecule.
GRP	Glass-reinforced polyester resin.
Graft	A polymer arrangement in which polymeric side chains of one chemical composition are attached to a polymer backbone molecule of a different chemical composition.
Green strength	The strength of a moulded part at the time of demoulding – a subjective measure of the ease of demoulding and handling a part.
HDI	An abbreviation for hexamethylene diisocyanate and 1,6-diisocyanato-hexane.
HMDI	An abbreviation for hydrogenated MDI and 4,4′-diisocyanato-dicyclohexylmethane.
Hydrogen bonding	Points of attraction between molecules, caused by polarised links involving hydrogen atoms.

Hydroxyl	An alcoholic group (–O–H). The reactive group in polyols. The hydroxyl value or number is a quantitative measure of the concentration of hydroxyl groups, usually stated as mg KOH/g, i.e. the number of mg. of potassium hydroxide equivalent to the hydroxyl groups in lg. of substance. A *primary hydroxyl group* has the structure –CH_2–O–H, and is more reactive to isocyanates than a *secondary hydroxyl group*, which has the structure –CH–O–H.
IPDI	An abbreviation for isophorone diisocyanate and 1-isocyanato-3-isocyanatomethyl-3,5,5-trimethylcyclohexane.
Impact strength	Measure of resistance to fracture when a sudden force is applied.
Initiator (starter)	A substance with which ethylene oxide or propylene oxide reacts to form a polyether polyol.
Injection point	The point at which a polyurethane reaction mixture is inserted into a mould cavity.
Isocyanate	A substance containing an isocyanate (–N=C=O) group. A polyisocyanate contains more than one isocyanate group.
Isocyanurate	A trimeric reaction product of an isocyanate having the ring structure.

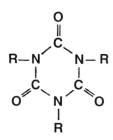

Isocyanurates are characterised by their good thermal stability.

Isomers	Molecules of the same atomic composition and molecular weight but of different geometric configuration.
MDI	An abbreviation for 4,4'-diphenylmethane diisocyanate and 4,4'-diisocyanato diphenylmethane.
Microcellular	An elastomer of cellular structure having a density between about 0.3 and 1.0 g/ml.
Mixing head	A device, usually attached to liquid metering arrangements, which intimately mixes two or more components of a polyurethane reaction formulation.
Molecular weight	The sum of the atomic weights of all the atoms in a molecule. Each atom has an *atomic weight*, which is related to its mass compared with that of one atom of carbon-12 (mass = 12).
Monomer	A substance which is capable of conversion into a polymer.
Mould	Enclosure, often metal, in which a polyurethane mixture reacts to give a shaped article.

NDI

An abbreviation of 1,5-naphthalene diisocyanate and 1,5-diisocyanatonaphthalene.

Organic

A chemical substance based on carbon. Organic substances exlude simple binary substances such as carbon dioxide, some ternary compounds such as metallic cyanides and metallic carbonates.

PET

An abbreviation for poly(ethyleneterephthalate).

PPDI

An abbreviation for *p*-phenylenediisocyanate and 1,4-diisocyanato-benzene.

Plastics

Synthetic polymeric materials.

Polar

A molecule having positive or negative electrical charges permanently separated.

Polyester

A polymer containing ester linkages ($-O-\overset{\overset{O}{\|}}{C}-$).
The polyesters used in polyurethane technology contain reactive hydroxyl end groups.

Polyether

A polymer containing ether linkages ($-\overset{|}{\underset{|}{C}}-O-\overset{|}{\underset{|}{C}}-$).

The polyethers used in polyurethane technology contain reactive hydroxyl end groups.

Polymer (Copolymer)

A high molecular weight substance, natural or synthetic, which can be represented as a repeated small unit (monomer). A copolymer contains more than one type of monomeric unit.

Polyol

A substance containing several hydroxyl groups. A diol, triol, and tetrol contain 2, 3 and 4 hydroxyl groups respectively.

Polyurea RIM

Conventional polyurea elastomers are block copolymers characterised by a dual-phase behaviour, derived from a combination of a rubbery softblock phase and an amorphous hardblock phase.

Polyurethane

Polymeric substance containing many urethane linkages ($-\overset{\overset{H}{|}}{N}-\overset{\overset{O}{\|}}{C}-O-$). Abbreviated as PU.

Prepolymer

A polyurethane reaction intermediate made by reacting isocyanate with a polyester or polyether, in which one component is in considerable excess of the other.

Reaction

A chemical change in which two or more atoms or molecules give a new substance.

RIM/RRIM/SRIM

Reaction Injection Moulding and Reinforced and Structural RIM. Processes which involve the rapid metering, and mixing of polyurethane reaction ingredients, followed by their injection into a mould. (See also Polyurea RIM).

Saturated

An organic substance in which all the carbon – carbon bonds have their full complement of hydrogen atoms.

Segment	A short part of a polymer chain composed of identical units. A hard segment has a high softening point or melting point and is rigid at room temperature; a soft segment has a low softening point or melting point and is flexible or rubbery at room temperature.
Self skinning	A foam reaction mixture which forms a skinned surface on being moulded at a specified temperature and pressure.
Slabstock	Rigid or flexible polyurethane foam made in the form of a continuous block, usually of approximately rectangular cross section.
Stabiliser	An additive which tends to keep a substance from changing its form or chemical nature. Stabilisers may protect against hydrolysis, oxidation, light etc.
Synergistic	The property of mutual reinforcement of an effect by two (or more) additives, e.g. catalysts.
System	Often used to describe the supply of all the chemical components needed to produce a polyurethane product. A *systems house* is a commercial organisation which specialises in selling formulated polyurethane ingredients to end users.
TDI	An abbreviation for toluene diisocyanate.
TMXDI	An abbreviation for *m*- and *p*-tetramethylxylene diisocyanate and 1,4- and 1,3-di(isocyanato-dimethyl-methyl)-benzene.
Thermoplastic	A polymeric material which may be formed and reformed by the application of heat and pressure.
Thermoset	A material which may be formed and polymerised by the application of heat and pressure. Once reacted by the application of heat, thermosets cannot be reformed.
Urethane	The chemical group $(-\overset{\displaystyle H}{\underset{\displaystyle |}{N}}-\overset{\displaystyle O}{\overset{\displaystyle \|}{C}}-O-)$. Also a corruption of *polyurethane*.

Index*

* All trademarks are printed
  between inverted commas

Photographs used in this publication
have been prepared with assistance from
the following companies:

Adidas
Afros Cannon
Alfa Romeo
Allibert
Ashland Chemical
Avalon
Bains Harding
BL
Britax
Cagiva
Chantiers Naval Jullien
Chevrolet
Chrysler
Cocon Kunststof Chemie
Compak Systems
Crompton Plastics
Desma
Fairfax Fibreglass
Faram
Finiplast
Ford
GEC-Henley
General Motors
Giroflex
Grace, W.R., Teroson
Greben
Gusbi
Hairlok
Harrison & Jones
Holden Hydroman
Hotpoint
Ilford
IPS Packaging Systems
Isotermica
Kartell
Koflach
Krauss Maffei
Kunz
Landini
LEC Refrigeration
Lotus Cars
Macchi Arturo
Mann & Hummel
Marley Foam
Marmor Specialties
Peugeot
PFI (Pirmasens)
Plastics and Machinery
Polyair
Polyuretani Clemente
PTI (Viking)
Reliant Motor
Renault
S.R. Industrie
Saab
Saitec
Salford University
Shaw Pipe
Systime Computers
Teppich-Werk Neumünster
Thomson-Brandt
Tyfoam
United Foam
Vapotherm
Vertifoam
Vetrotex
Zanotta

Printed in the Netherlands by Steens Schiedam b.v.